State-of-the-art Laser Gas Sensing Technologies

State-of-the-art Laser Gas Sensing Technologies

Special Issue Editors

Yufei Ma
Aurore Vicet
Karol Krzempek

MDPI • Basel • Beijing • Wuhan • Barcelona • Belgrade • Manchester • Tokyo • Cluj • Tianjin

Special Issue Editors

Yufei Ma
Harbin Institute of Technology
China

Aurore Vicet
University of Montpellier
France

Karol Krzempek
Wroclaw University of Science
and Technology
Poland

Editorial Office
MDPI
St. Alban-Anlage 66
4052 Basel, Switzerland

This is a reprint of articles from the Special Issue published online in the open access journal *Applied Sciences* (ISSN 2076-3417) (available at: https://www.mdpi.com/journal/applsci/special_issues/Laser_Gas_Sensing).

For citation purposes, cite each article independently as indicated on the article page online and as indicated below:

LastName, A.A.; LastName, B.B.; LastName, C.C. Article Title. *Journal Name* **Year**, *Article Number*, Page Range.

ISBN 978-3-03928-398-9 (Pbk)
ISBN 978-3-03928-399-6 (PDF)

© 2020 by the authors. Articles in this book are Open Access and distributed under the Creative Commons Attribution (CC BY) license, which allows users to download, copy and build upon published articles, as long as the author and publisher are properly credited, which ensures maximum dissemination and a wider impact of our publications.

The book as a whole is distributed by MDPI under the terms and conditions of the Creative Commons license CC BY-NC-ND.

Contents

About the Special Issue Editors vii

Yufei Ma, Aurore Vicet and Karol Krzempek
State-of-the-Art Laser Gas Sensing Technologies
Reprinted from: *Appl. Sci.* **2020**, *10*, 433, doi:10.3390/app10020433 1

Jiaqun Zhao, Ping Cheng, Feng Xu, Xiaofeng Zhou, Jun Tang, Yong Liu and Guodong Wang
Watt-Level Continuous-Wave Single-Frequency Mid-Infrared Optical Parametric Oscillator Based on MgO:PPLN at 3.68 μm
Reprinted from: *Appl. Sci.* **2018**, *8*, 1345, doi:10.3390/app8081345 5

Wei Wang, Linjun Li, Hongtian Zhang, Jinping Qin, Yuang Lu, Chong Xu, Shasha Li, Yingjie Shen, Wenlong Yang, Yuqiang Yang and Xiaoyang Yu
Passively Q-Switched Operation of a Tm,Ho:LuVO$_4$ Laser with a Graphene Saturable Absorber
Reprinted from: *Appl. Sci.* **2018**, *8*, 954, doi:10.3390/app8060954 13

Deyang Yu, Yang He, Kuo Zhang, Qikun Pan, Fei Chen and Lihong Guo
A Tunable Mid-Infrared Solid-State Laser with a Compact Thermal Control System
Reprinted from: *Appl. Sci.* **2018**, *8*, 878, doi:10.3390/app8060878 21

Yufei Ma
Review of Recent Advances in QEPAS-Based Trace Gas Sensing
Reprinted from: *Appl. Sci.* **2018**, *8*, 1822, doi:10.3390/app8101822 35

Karol Krzempek
A Review of Photothermal Detection Techniques for Gas Sensing Applications
Reprinted from: *Appl. Sci.* **2019**, *9*, 2826, doi:10.3390/app9142826 51

Bo Li, Dayuan Zhang, Jixu Liu, Yifu Tian, Qiang Gao and Zhongshan Li
A Review of Femtosecond Laser-Induced Emission Techniques for Combustion and Flow Field Diagnostics
Reprinted from: *Appl. Sci.* **2019**, *9*, 1906, doi:10.3390/app9091906 71

Zhenhui Du, Shuai Zhang, Jinyi Li, Nan Gao and Kebin Tong
Mid-Infrared Tunable Laser-Based Broadband Fingerprint Absorption Spectroscopy for Trace Gas Sensing: A Review
Reprinted from: *Appl. Sci.* **2019**, *9*, 338, doi:10.3390/app9020338 97

Zhenhai Wang, Pengfei Fu and Xing Chao
Laser Absorption Sensing Systems: Challenges, Modeling, and Design Optimization
Reprinted from: *Appl. Sci.* **2019**, *9*, 2723, doi:10.3390/app9132723 131

Fei Wang, Shuhai Jia, Yonglin Wang and Zhenhua Tang
Recent Developments in Modulation Spectroscopy for Methane Detection Based on Tunable Diode Laser
Reprinted from: *Appl. Sci.* **2019**, *9*, 2816, doi:10.3390/app9142816 159

Biao Wang, Hongfei Lu, Chen Chen, Lei Chen, Houquan Lian, Tongxin Dai and Yue Chen
Near-Infrared C_2H_2 Detection System Based on Single Optical Path Time Division Multiplexing Differential Modulation Technique and Multi-Reflection Chamber
Reprinted from: *Appl. Sci.* **2019**, *9*, 2637, doi:10.3390/app9132637 175

Zhifang Wang, Shutao Wang, Deming Kong and Shiyu Liu
Methane Detection Based on Improved Chicken Algorithm Optimization Support Vector Machine
Reprinted from: *Appl. Sci.* **2019**, *9*, 1761, doi:10.3390/app9091761 187

Hanquan Zhang, Mingming Wen, Yonghang Li, Peng Wan and Chen Chen
High-Precision $^{13}CO_2/^{12}CO_2$ Isotopic Ratio Measurement Using Tunable Diode Laser Absorption Spectroscopy at 4.3 µm for Deep-Sea Natural Gas Hydrate Exploration
Reprinted from: *Appl. Sci.* **2019**, *9*, 3444, doi:10.3390/app9173444 203

Xiaorui Zhu, Shunchun Yao, Wei Ren, Zhimin Lu and Zhenghui Li
TDLAS Monitoring of Carbon Dioxide with Temperature Compensation in Power Plant Exhausts
Reprinted from: *Appl. Sci.* **2019**, *9*, 442, doi:10.3390/app9030442 217

Xue Zhou, Jia Yu, Lin Wang and Zhiguo Zhang
Investigating the Relation between Absorption and Gas Concentration in Gas Detection Using a Diffuse Integrating Cavity
Reprinted from: *Appl. Sci.* **2018**, *8*, 1630, doi:10.3390/app8091630 233

Chi Wang, Yue Zhang, Jianmei Sun, Jinhui Li, Xinqun Luan and Anand Asundi
High-Efficiency Coupling Method of the Gradient-Index Fiber Probe and Hollow-Core Photonic Crystal Fiber
Reprinted from: *Appl. Sci.* **2019**, *9*, 2073, doi:10.3390/app9102073 245

Xinhua Wang, Jihong Ouyang, Yi Wei, Fei Liu and Guang Zhang
Real-Time Vision through Haze Based on Polarization Imaging
Reprinted from: *Appl. Sci.* **2019**, *9*, 142, doi:10.3390/app9010142 255

About the Special Issue Editors

Yufei Ma (Prof. Dr.) received his PhD degree in physical electronics from Harbin Institute of Technology, China, in 2013. He was a Visiting Scholar at Rice University, USA, from September 2010 to September 2011. He is currently Professor at the Harbin Institute of Technology, China. His research interests include optical sensors, trace gas detection, laser spectroscopy, and optoelectronics. He has published more than 100 publications and given more than 10 invited presentations at international conferences.

Aurore Vicet (Associate Prof. Dr.) is Associate Professor at Montpellier University, France, in charge of spectroscopic developments on tunable lasers. She is involved in the study, simulation, and characterization of single-frequency semiconductor lasers based on distributed feedback for spectroscopic applications. She is also involved in the study and development of tunable laser spectroscopic systems based on resonant photoacoustic techniques, using both lasers or LEDs, and relying on quartz-enhanced photoacoustic technique and Si-based oscillators.

Karol Krzempek (Dr.) obtained his PhD on Nonlinear Frequency Conversion-based Mid-infrared Laser Sources at the Faculty of Electronics at Wroclaw University of Science and Technology (WUST), in 2016. At present, he is continuing his research in this area at the WUST. His main research contributions include the design and optimization of CW and pulsed fiber-based laser sources working in the 1, 1.5, and 2 μm wavelength regions with subsequent nonlinear mixing of the emission of such coherent sources. Other research interests include sensors relying on photothermal gas detection techniques as well as efficient use of hollow-core fibers as low-volume gas cells in laser spectroscopy applications. He has participated in 11 research grants and is co-author of 13 national patents. Throughout his academic career, Karol Krzempek has received numerous national scholarships and awards. Recently, Karol Krzempek contributed as a co-editor for MDPI and *Frontiers in Physics*. He has contributed to more than 70 scientific works and is actively involved in development of the Laser Spectroscopy Group in WUST.

Editorial

State-of-the-Art Laser Gas Sensing Technologies

Yufei Ma [1],*, Aurore Vicet [2] and Karol Krzempek [3]

[1] National Key Laboratory of Science and Technology on Tunable Laser, Harbin Institute of Technology, Harbin 150001, China
[2] IES, Univ. Montpellier, CNRS, 34000 Montpellier, France; aurore.vicet@umontpellier.fr
[3] Laser and Fiber Electronics Group, Wroclaw University of Science and Technology, 50-730 Wroclaw, Poland; karol.krzempek@pwr.edu.pl
* Correspondence: mayufei@hit.edu.cn; Tel.: +86-451-8641-3161

Received: 21 November 2019; Accepted: 6 January 2020; Published: 7 January 2020

1. Introduction

The increasing desire to detect and monitor in different fields [1–4] such as in environmental air, life sciences, medical diagnostics, and planetary exploration demand the development of innovative sensing systems. Laser spectroscopy-based techniques have the advantages of high sensitivity, non-invasiveness and in situ, real-time observation [5–7]. Because of these merits, we introduced state-of-the-art laser gas sensing technologies in this Special Issue. A total of 30 papers was received for consideration of publication. Among them, six manuscripts were rejected by the editor in the initial check process without peer review. The remaining manuscripts were all reviewed by at least two reputed reviewers in related fields from the USA, France, Italy, Germany, Russia, and so on. Finally, 16 manuscripts were accepted for publication in *Applied Sciences-Basel*. We would like to thank all of these numerous reviewers for their effort.

2. Main Content of the Special Issue

The recent advance in laser sources and detectors has opened up new opportunities for laser spectroscopy-based sensing and detecting techniques. Furthermore, the new technique has helped to promote its applications. Therefore, in this Special Issue, papers focus on novel laser sources and advanced sensing methods and their applications.

With respect to the laser sources aspect, three papers are concerned. All of them are related to mid-infrared lasers, which are beneficial to laser spectroscopy methods due to the strongest fundamental absorption bands of gas molecules located in this wavelength region. The first paper, authored by J. Zhao, P. Cheng, F. Xu, X. Zhou, J. Tang, Y. Liu, and G. Wang presents a continuous-wave single-frequency singly-resonant mid-infrared optical parametric oscillator (OPO) with emission wavelength at 3.68 µm [8]. The output power of more than 1 W indicated the high output level. Therefore, such a source is especially beneficial to power related laser-based gas detection techniques, such as photoacoustic and photothermal spectroscopy [9,10]. The second paper submitted by W. Wang, L. Li, H. Zhang, J. Qin, Y. Lu, C. Xu, S. Li, Y. Shen, W. Yang, Y. Yang, and X. Yu reports a pulsed Tm,Ho:LuVO$_4$ solid-state laser with a repetition rate of 54.5 kHz and an output power of 1034 mW. The emission wavelength shifted from 2075.02 nm to 2057.03 nm when the operation mode was switched from continuous wave to Q-switched [11]. The last paper in this section, authored by D. Yu, Y. He, K. Zhang, Q. Pan, F. Chen, and L. Guo, is about a compact thermal control system for a tunable mid-infrared solid-state laser, which could be used to improve environmental temperature adaptability and solve heat dissipation problems for mid-infrared lasers [12].

In the gas sensing aspect of this Special Issue, Y. F. Ma presents a review paper about recent advances in the quartz tuning fork based on photoacoustic detection [13], while K. Krzempek summarizes the research progress in gas sensing by photothermal spectroscopy [14]. Both techniques are based on the photoacoustic effect. Another review paper concerned with femtosecond laser-induced emission spectroscopy and its application in combustion and flow field diagnostics was presented by B. Li, D. Zhang, J. Liu, Y. Tian, Q. Gao, and Z. Li [15]. The last three review papers, authored by Z. Du, F. Wang, and X. Chao, respectively, mainly focus on direct laser absorption spectroscopy, especially in the mid-infrared region [16–18]. All the above review papers presented a full discussion with regard to the related technical field of gas sensing. The remaining papers report on the technical research of gas detection based on direct laser absorption spectroscopy [19–25]. The target analytes were acetylene (C_2H_2) [19], methane (CH_4) [20], oxygen (O_2) [21], and $^{13}CO_2/^{12}CO_2$ isotopic ratio [22]. The corresponding sensors were used for the monitoring of power plant exhausts [23] and vision imaging [24].

Author Contributions: Y.M.: writing original draft; A.V. and K.K.: reviewing and editing. All authors have read and agreed to the published version of the manuscript.

Funding: National Natural Science Foundation of China (No. 61875047 and 61505041), Natural Science Foundation of Heilongjiang Province of China (No. YQ2019F006), Fundamental Research Funds for the Central Universities, Financial Grant from the Heilongjiang Province Postdoctoral Foundation (No. LBH-Q18052).

Acknowledgments: We would like to sincerely thank our Section Managing Editor, Marin Ma (marin.ma@mdpi.com), for all the efforts she has made for this Special Issue and Xiaoyan Chen, Senior Editor over the past few months, both of them from the MDPI Branch Office, Beijing.

Conflicts of Interest: The authors declare no conflict of interest.

References

1. Ravishankara, A.R.; Daniel, J.S.; Portmann, R.W. Nitrous oxide (N_2O): The dominant ozone-depleting substance emitted in the 21st century. *Science* **2009**, *326*, 123–125. [CrossRef] [PubMed]
2. Milde, T.; Hoppe, M.; Tatenguem, H.; Mordmüller, M.; O'Gorman, J.; Willer, U.; Schade, W.; Sacher, J. QEPAS sensor for breath analysis: A behavior of pressure. *Appl. Opt.* **2018**, *57*, C120–C127. [CrossRef] [PubMed]
3. Ma, Y.F.; Lewicki, R.; Razeghi, M.; Tittel, F.K. QEPAS based ppb-level detection of CO and N_2O using a high power CW DFB-QCL. *Opt. Express* **2013**, *21*, 1008–1019. [CrossRef] [PubMed]
4. Bradshaw, J.L.; Bruno, J.D.; Lascola, K.M.; Leavitt, R.P.; Pham, J.T.; Towner, F.J.; Sonnenfroh, D.M.; Parameswaran, K.R. Small low-power consumption CO-sensor for post-fire cleanup aboard spacecraft. In Proceedings of the Next-Generation Spectroscopic Technologies IV, Orlando, FL, USA, 12 May 2011; Society of Photo-Optical Instrumentation Engineers: Bellingham, WA, USA, 2011; Volume 8032, p. 80320D.
5. He, Y.; Ma, Y.F.; Tong, Y.; Yu, X.; Peng, Z.F.; Gao, J.; Tittel, F.K. Long distance, distributed gas sensing based on micro-nano fiber evanescent wave quartz-enhanced photoacoustic spectroscopy. *Appl. Phys. Lett.* **2017**, *111*, 241102. [CrossRef]
6. He, Y.; Ma, Y.F.; Tong, Y.; Yu, X.; Tittel, F.K. Ultra-high sensitive light-induced thermoelastic spectroscopy sensor with a high Q-factor quartz tuning fork and a multipass cell. *Opt. Lett.* **2019**, *44*, 1904–1907. [CrossRef]
7. Ma, Y.F.; He, Y.; Tong, Y.; Yu, X.; Tittel, F.K. Quartz-tuning-fork enhanced photothermal spectroscopy for ultra-high sensitive trace gas detection. *Opt. Express* **2018**, *26*, 32103–32110. [CrossRef]
8. Zhao, J.; Cheng, P.; Xu, F.; Zhou, X.; Tang, J.; Liu, Y.; Wang, G. Watt-Level Continuous-wave single-frequency mid-infrared optical parametric oscillator based on MgO:PPLN at 3.68 µm. *Appl. Sci.* **2018**, *8*, 1345. [CrossRef]
9. Krzempek, K.; Dudzik, G.; Abramski, K. Photothermal spectroscopy of CO_2 in an intracavity mode-locked fiber laser configuration. *Opt. Express* **2018**, *26*, 28861–28871. [CrossRef]
10. Rousseau, R.; Loghmari, Z.; Bahriz, M.; Chamassi, K.; Teissier, R.; Baranov, A.N.; Vicet, A. Off-beam QEPAS sensor using an 11-µm DFB-QCL with an optimized acoustic resonator. *Opt. Express* **2018**, *27*, 7435–7446. [CrossRef]
11. Wang, W.; Li, L.; Zhang, H.; Qin, J.; Lu, Y.; Xu, C.; Li, S.; Shen, Y.; Yang, W.; Yang, Y.; et al. Passively Q-switched operation of a Tm,Ho:LuVO4 laser with a graphene saturable absorber. *Appl. Sci.* **2018**, *8*, 954. [CrossRef]
12. Yu, D.; He, Y.; Zhang, K.; Pan, Q.; Chen, F.; Guo, L. A tunable mid-infrared solid-state laser with a compact thermal control system. *Appl. Sci.* **2018**, *8*, 878. [CrossRef]

13. Ma, Y. Review of recent advances in QEPAS-based trace gas sensing. *Appl. Sci.* **2018**, *8*, 1822. [CrossRef]
14. Krzempek, K. A review of photothermal detection techniques for gas sensing applications. *Appl. Sci.* **2019**, *9*, 2826. [CrossRef]
15. Li, B.; Zhang, D.; Liu, J.; Tian, Y.; Gao, Q.; Li, Z. A review of femtosecond laser-induced emission techniques for combustion and flow field diagnostics. *Appl. Sci.* **2019**, *9*, 1906. [CrossRef]
16. Du, Z.; Zhang, S.; Li, J.; Gao, N.; Tong, K. Mid-infrared tunable laser-based broadband fingerprint absorption spectroscopy for trace gas sensing: A Review. *Appl. Sci.* **2019**, *9*, 338. [CrossRef]
17. Wang, F.; Jia, S.; Wang, Y.; Tang, Z. Recent developments in modulation spectroscopy for methane detection based on tunable diode laser. *Appl. Sci.* **2019**, *9*, 2816. [CrossRef]
18. Wang, Z.; Fu, P.; Chao, X. Laser Absorption sensing systems: Challenges, modeling, and design optimization. *Appl. Sci.* **2019**, *9*, 2723. [CrossRef]
19. Wang, B.; Lu, H.; Chen, C.; Chen, L.; Lian, H.; Dai, T.; Chen, Y. Near-infrared C_2H_2 detection system based on single optical path time division multiplexing differential modulation technique and multi-reflection chamber. *Appl. Sci.* **2019**, *9*, 2637. [CrossRef]
20. Wang, Z.; Wang, S.; Kong, D.; Liu, S. Methane detection based on improved chicken algorithm optimization support vector machine. *Appl. Sci.* **2019**, *9*, 1761. [CrossRef]
21. Wang, C.; Zhang, Y.; Sun, J.; Li, J.; Luan, X.; Asundi, A. High-efficiency coupling method of the cradient-index fiber probe and hollow-core photonic crystal fiber. *Appl. Sci.* **2019**, *9*, 2073. [CrossRef]
22. Zhang, H.; Wen, M.; Li, Y.; Wan, P.; Chen, C. High-precision $^{13}CO_2/^{12}CO_2$ isotopic ratio measurement using tunable diode laser absorption spectroscopy at 4.3 µm for deep-sea natural gas hydrate exploration. *Appl. Sci.* **2019**, *9*, 3444. [CrossRef]
23. Zhu, X.; Yao, S.; Ren, W.; Lu, Z.; Li, Z. TDLAS Monitoring of carbon dioxide with temperature compensation in power plant exhausts. *Appl. Sci.* **2019**, *9*, 442. [CrossRef]
24. Wang, X.; Ouyang, J.; Wei, Y.; Liu, F.; Zhang, G. Real-time vision through haze based on polarization imaging. *Appl. Sci.* **2019**, *9*, 142. [CrossRef]
25. Zhou, X.; Yu, J.; Wang, L.; Zhang, Z. Investigating the relation between absorption and gas concentration in gas detection using a diffuse integrating cavity. *Appl. Sci.* **2018**, *8*, 1630. [CrossRef]

© 2020 by the authors. Licensee MDPI, Basel, Switzerland. This article is an open access article distributed under the terms and conditions of the Creative Commons Attribution (CC BY) license (http://creativecommons.org/licenses/by/4.0/).

Communication

Watt-Level Continuous-Wave Single-Frequency Mid-Infrared Optical Parametric Oscillator Based on MgO:PPLN at 3.68 um

Jiaqun Zhao [1,*], Ping Cheng [2,*], Feng Xu [3], Xiaofeng Zhou [1], Jun Tang [1], Yong Liu [1] and Guodong Wang [1]

1. College of Science, Hohai University, Nanjing 211100, China; 20150057@hhu.edu.cn (X.Z.); 1510020115@hhu.edu.cn (J.T.); liuy@hhu.edu.cn (Y.L.); gdwang@hhu.edu.cn (G.W.)
2. College of Computer and Information, Hohai University, Nanjing 211100, China
3. College of Science, Nanjing University of Aeronautics and Astronautics, Nanjing 211106, China; fengxu@nuaa.edu.cn
* Correspondence: zhaojq@hhu.edu.cn (J.Z.); chengping1219@hhu.edu.cn (P.C.); Tel.: +86-25-8378-6640 (J.Z.)

Received: 13 July 2018; Accepted: 5 August 2018; Published: 10 August 2018

Abstract: We report a continuous-wave single-frequency singly-resonant mid-infrared optical parametric oscillator (OPO). The OPO is based on 5 mol % MgO-doped periodically poled lithium niobate (MgO:PPLN) pumped by a continuous-wave single-frequency Nd:YVO$_4$ laser at 1064 nm. A four-mirror bow-tie ring cavity configuration is adopted. A low-finesse intracavity etalon is utilized to compress the linewidth of the resonant signal. A single-frequency idler output power higher than 1 W at 3.68 µm is obtained.

Keywords: mid-infrared; single-frequency; optical parametric oscillator (OPO); MgO:PPLN crystal; continuous-wave (CW)

1. Introduction

Tunable laser sources in the mid-infrared range are widely used in laser spectroscopy, atmospheric pollution monitoring, remote detection, and differential absorption lidar. In particular, continuous-wave (CW) single-frequency mid-infrared laser sources with broad wavelength tunability are more suitable for high-resolution spectral analysis [1–4] and atom physics [5]. Different techniques have been applied to obtain a mid-infrared laser source. Quantum cascade lasers have been proved to be a method to generate mid-infrared radiation and Razeghi et al. have done extensive work in this area [6,7]. A solid-state laser based on metal-ion-doped crystals can directly generate mid-infrared radiation; for example, research on a Fe:ZnSe laser has been reported [8,9]. An alternative method to reach the mid-infrared wavelength range is to utilize nonlinear frequency downconversion devices such as optical parametric oscillators (OPOs). OPOs with wide wavelength-tunability and a narrow linewidth have become a very important mid-infrared laser source. Compared with birefringent phase-matched (BPM) OPOs, quasi-phase-matched (QPM) OPOs can utilize the largest nonlinear optical tensor element of nonlinear crystals, and make the three interacting waves (pump ω_p, signal ω_s, and idler ω_i) collinearly propagate in nonlinear crystals so that the distance of nonlinear interaction is largely enhanced. Many QPM nonlinear materials such as periodically poled LiTaO$_3$ (PPLT), periodically poled LiNbO$_3$ (PPLN), periodically poled KTiOPO$_4$ (PPKTP), periodically poled RbTiOAsO$_4$ (PPRTA), periodically poled GaAs, and periodically poled GaP have been studied. Among these materials, PPLN is an excellent nonlinear crystal for QPM OPOs, having a relatively high nonlinear coefficient (d_{33} ~27.2 pm/V) with a wide transparent range (0.35–5 µm). Compared with PPLN, MgO-doped periodically poled lithium niobate (MgO:PPLN) has a much higher photorefractive damage threshold. Therefore, MgO:PPLN is widely used as a QPM nonlinear crystal in mid-infrared OPOs [10–15].

To obtain a narrow linewidth idler output from an OPO, a narrow linewidth pump laser source is necessary in the OPO system. In addition, additional wavelength-selective elements are generally utilized to suppress the linewidth of the oscillated signal in the OPO cavity. Peng et al. presented a narrow linewidth PPMgLN OPO, and the linewidth of the 2.98 μm idler was within 0.30–0.63 nm by theoretical analysis [16]. Henderson et al., demonstrated a singly-resonant CW OPO pumped by an all-fiber pump source, with a 3.17 μm idler linewidth of 1 MHz [17]. Vainio et al., demonstrated a singly-resonant CW OPO operating without mode hops for several hours due to the good thermal control of the MgO:PPLN crystal [18]. Reflecting volume Bragg gratings (VBGs) have been widely used in laser devices to obtain a narrow linewidth output. For example, Zeil et al., reported a singly-resonant CW OPO with optimum extraction efficiency, in which a single-longitudinal-mode signal output was obtained by employing a variable-reflectivity VBG as the output coupler of a ring cavity [19]. In addition, Xing et al., devised self-seeding dual etalon-coupled cavities in the OPO system pumped by a single-longitudinal-mode pulsed Yb-fiber laser. The linewidth of the oscillated signal was suppressed and the linewidth of the idler was efficiently narrowed [20].

In this paper, we report our experimental work on a CW single-frequency MgO:PPLN OPO pumped by a CW single-frequency Nd:YVO$_4$ laser at 1064 nm. We obtained a CW single-frequency 3.68 μm idler laser with an output power higher than 1 W. Wavelength tuning can be achieved through thermal control of the nonlinear crystal and use of the different grating periods.

2. Experimental Setup

The experimental configuration of the CW single-frequency MgO:PPLN OPO is shown schematically in Figure 1. The nonlinear medium used for the OPO is 5 mol % MgO-doped periodically poled lithium niobate (MgO:PPLN, HC Photonics) with a length of 50 mm and a laser aperture of 8 mm × 1 mm. The MgO:PPLN crystal contains seven domain grating periods from 28.5 μm to 31.5 μm with 0.5-μm increments. The crystal is antireflection-coated for the signal wavelength (R < 1%@1.4–1.7 μm), idler wavelength (R < 1%@3–4 μm), and pump wavelength (R < 1%@1.064 μm). The crystal is mounted in a temperature-controlled oven, in which the crystal temperature can be adjusted in the range of 25–200 °C with a temperature stability of ± 0.1 °C. A simple bow-tie ring cavity is used in the OPO system. The ring cavity consists of two identical curved cavity mirrors (M_1 and M_2) and two flat mirrors (M_3 and M_4), which are all made of CaF$_2$ and are antireflection-coated at the pump wavelength (T > 98%@1064 nm) and idler wavelength (T > 95%@3–5 μm), and have high reflectivity at the signal wavelength (R > 99.8%@1.4–1.7 μm). The OPO configuration gives a singly resonant OPO, which is resonant for the signal frequency. The two identical curved cavity mirrors (M_1 and M_2), enclosing the MgO:PPLN crystal, have a 75-mm radius of curvature and are separated by a distance of 120 mm. The nonlinear crystal is placed at the center between the two curved mirrors (M_1 and M_2). The other two flat cavity mirrors (M_3 and M_4) are separated by 35 mm. The total resonator length is about 325 mm.

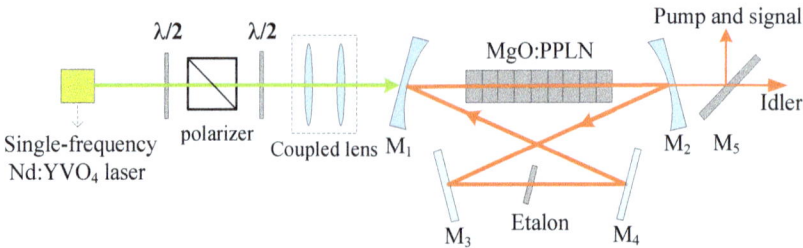

Figure 1. Configuration of the continuous-wave (CW) single-frequency MgO-doped periodically poled lithium niobate (MgO:PPLN) optical parametric oscillator (OPO).

The pump source is a continuous-wave single-frequency Nd:YVO$_4$ laser that produces over 10 W of radiation at 1064 nm. The Nd:YVO$_4$ laser has an excellent beam quality of M^2 ~1.1 and an output of linear polarization, which was described in Reference [21]. When the power is changed, the laser beam characteristics change significantly. In order to maintain a stable output, the Nd:YVO$_4$ laser is operated at maximum output power in this experiment. A combination of a half-wave plate and a polarizing beam splitter is used as a power attenuator to change the incident pump power. By using the second half-wave plate, the pump polarization is aligned along the crystallographic z-axis of the MgO:PPLN crystal to utilize the largest nonlinear coefficient d_{33}. The pump beam is mode-matched to the OPO cavity with a series of convex lenses, producing a 1/e^2 waist radius of 60 µm at the center of the MgO:PPLN crystal. Its waist yields a focusing parameter ξ_p ~1.1. With the current OPO cavity, the signal beam waist at the center of the MgO:PPLN crystal is about 70 µm, resulting in optimum mode-matching to the pump (ξ_s ~ξ_p). To enhance the single-frequency operation of the OPO, an uncoated 0.5-mm-thick yttrium aluminium garnet (YAG) plate is used as an intracavity etalon with a free spectral range of 120 GHz. A 45° flat dichroic mirror M$_5$ is utilized as a filter to separate the idler from the output beams.

3. Experimental Results and Discussion

The wavelengths of the OPO signal and idler are recorded with a laser spectrum analyzer (EXFO WA-650) combined with a wavelength meter (EXFO WA-1500). When the pump beam passes through a 29.5 µm grating period of the MgO:PPLN crystal and the crystal temperature is controlled at 120.0 ± 0.1 °C, the idler wavelength is 3.68 µm (Figure 2) and the corresponding signal wavelength is 1.49 µm (Figure 3). The wavelengths of the pump (λ_p), signal (λ_s), and idler (λ_i) waves are in accord with the conservation of energy ($1/\lambda_p = 1/\lambda_s + 1/\lambda_i$).

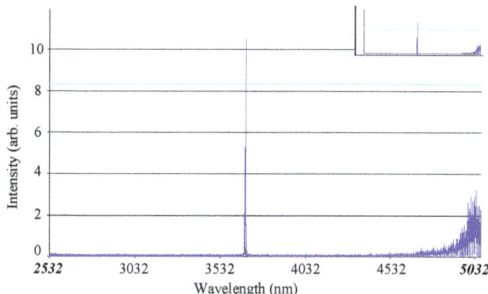

Figure 2. Idler wavelength of the MgO:PPLN OPO at temperature T = 120 °C for the grating period Λ = 29.5 µm.

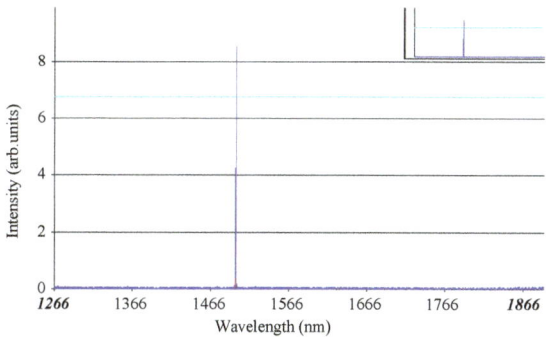

Figure 3. Signal wavelength of the MgO:PPLN OPO at temperature T = 120 °C for the grating period Λ = 29.5 µm.

To investigate its longitudinal mode structure, the signal spectral information is monitored by a 1.5 μm scanning confocal Fabry-Perot (F-P) interferometer, with a free spectral range of 1.5 GHz. As shown in Figure 4, the upper trace is the F-P ramp voltage and the lower trace is the voltage of the signal transmission through the F-P interferometer. Figure 4a shows the F-P spectrum of the signal from the MgO:PPLN OPO without the YAG etalon. To lock the cavity mode and reduce the spectral noise, an uncoated 0.5-mm-thick YAG plate is used as an intracavity etalon inserted in the cavity. By adjusting the angle of the etalon carefully, the signal spectral noise can be reduced. Figure 4b shows the F-P spectrum of the signal from the MgO:PPLN OPO with the YAG etalon. As can be seen, a single-frequency operation of the signal is presented in Figure 4b. According to the energy conservation condition $\omega_i + \omega_s = \omega_p$, the single-frequency operation of the idler from the MgO:PPLN OPO can be confirmed.

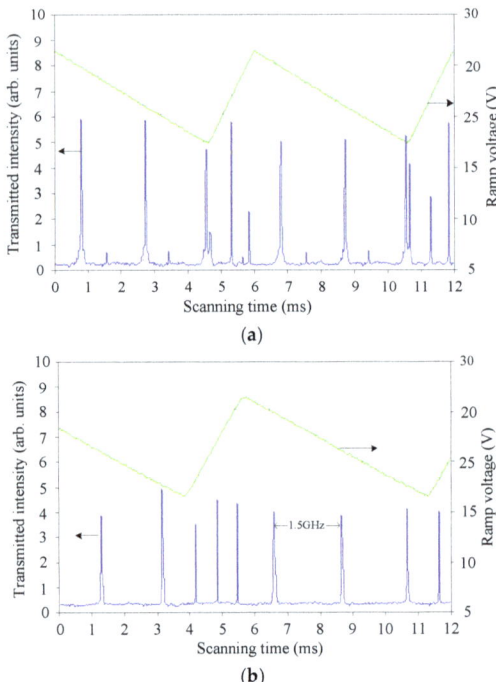

Figure 4. F-P spectrum of the signal from the MgO:PPLN OPO: (**a**) cavity without etalon; (**b**) cavity with an uncoated yttrium aluminium garnet (YAG) etalon.

The idler output power, as a function of the incident pump power, is measured by a power meter (Coherent PM2). Figure 5 shows the measured idler power versus incident pump power. When the YAG etalon is inserted in the four-mirror ring cavity, the oscillated threshold of the OPO is increased from 2 W to 5 W. Without the etalon in the OPO cavity, with a pump power of 10 W from the single-frequency Nd:YVO$_4$ laser, the 3.68 μm idler power is 1.3 W emitting from mirror M$_2$. With an intracavity etalon, the 3.68 μm idler power is 1.1 W, corresponding to an optical efficiency of 11%.

The idler beam quality is also measured as a function of the idler power. By making use of the knife-edge method, the idler beam radius as a function of the distance from mirror M$_2$ is achieved. By using a nonlinear fitting method, the beam quality factor M^2 can be obtained. For a single-frequency idler output power of 1 W at 3.68 μm, the values of M^2 are measured to be about 1.3 and 1.2 in the horizontal and vertical directions, respectively.

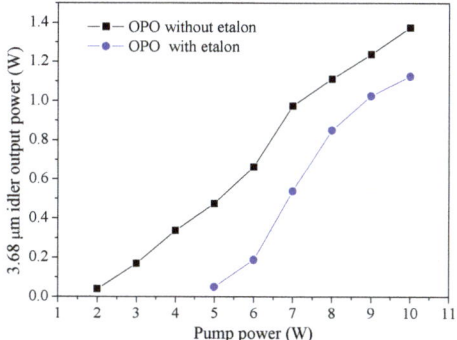

Figure 5. Output power of the 3.68 µm idler versus incident pump power.

We can tune the signal and idler wavelengths by changing the temperature of the MgO:PPLN crystal. According to the Sellmeier equations [22], the theoretical tuning curves for the seven available grating periods are shown in Figure 6. By shifting the MgO:PPLN crystal to keep the pump beam passing through the different grating periods, and changing the nonlinear crystal temperature between 20 °C and 200 °C, the OPO is able to generate idler wavelengths ranging from 2.9 to 4.1 µm.

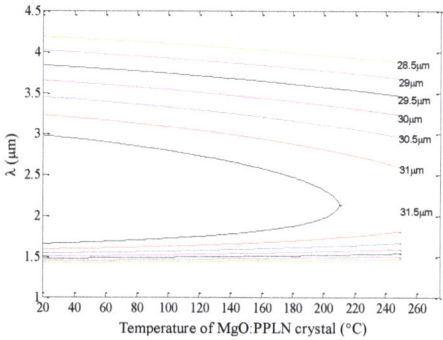

Figure 6. Theoretical tuning curves for eight periods of the MgO:PPLN crystal.

4. Conclusions

In conclusion, we have demonstrated a continuous-wave single-frequency mid-infrared MgO:PPLN OPO pumped by a continuous-wave single-frequency Nd:YVO$_4$ laser at 1064 nm. The symmetrical design of the system can easily achieve mode-matching. We used an uncoated 0.5-mm-thick YAG etalon to enhance the single-frequency operation of the MgO:PPLN OPO. With the etalon in the cavity, the OPO produced a single-frequency output of 1.1 W at 3.68 µm. By using different grating periods and adjusting the nonlinear crystal temperature between 20 °C and 200 °C, the idler wavelength of the OPO can be continuously tuned in the range of 2.9–4.1 µm.

Author Contributions: J.Z., P.C., and F.X. conceived and designed the experiment; J.T. and Y.L. performed the experiment; X.Z. and G.W. supervised the entire work; J.Z. and P.C. wrote the paper.

Funding: This work was supported by the Fundamental Research Funds for the Central Universities (2016B02014, 2016B12114, and 2016B01914), National Natural Science Foundation of China (NSFC) (61378027, 61301199), and A Project Funded by the Priority Academic Program Development of Jiangsu Higher Education Institutions.

Acknowledgments: The authors thank Yuezhu Wang for providing technical support.

Conflicts of Interest: The authors declare no conflict of interest.

References

1. Kovalchuk, E.V.; Dekorsy, D.; Lvovsky, A.I.; Braxmaier, C.; Mlynek, J.; Peters, A.; Schiller, S. High-resolution Doppler-free molecular spectroscopy with a continuous-wave optical parametric oscillator. *Opt. Lett.* **2001**, *26*, 1430–1432. [CrossRef] [PubMed]
2. Verbraak, H.; Ngai, A.K.Y.; Persijn, S.T.; Harren, F.J.M.; Linnartz, H. Mid-infrared continuous wave cavity ring down spectroscopy of molecular ions using an optical parametric oscillator. *Chem. Phys. Lett.* **2007**, *442*, 145–149. [CrossRef]
3. Zaske, S.; Lee, D.-H.; Becher, C. Green-pumped cw singly resonant optical parametric oscillator based on MgO:PPLN with frequency stabilization to an atomic resonance. *Appl. Phys. B* **2010**, *98*, 729–735. [CrossRef]
4. Ricciardi, I.; Tommasi, E.D.; Maddaloni, P.; Mosca, S.; Rocco, A.; Zondy, J.J.; Natale, P.D. A narrow-bandwidth, frequency-stabilized OPO for sub-Doppler molecular spectroscopy around 3 µm. *Proc. SPIE* **2012**, *8434*, 84341Z.
5. Mickelson, P.G.; Martinez de Escobar, Y.N.; Anzel, P.; De Salvo, B.J.; Nagel, S.B.; Traverso, A.J.; Yan, M.; Killian, T.C. Repumping and spectroscopy of laser-cooled Sr atoms using the $(5s5p)^3P_2$-$(5s4d)^3D_2$ transition. *J. Phys. B Mol. Opt. Phys.* **2009**, *42*, 235001. [CrossRef]
6. Bandyopadhyay, N.; Slivken, S.; Bai, Y.; Razeghi, M. High power, continuous wave, room temperature operation of λ~3.4 µm and λ~3.55 µm InP-based quantum cascade lasers. *Appl. Phys. Lett.* **2012**, *100*, 212104–212107. [CrossRef]
7. Razeghi, M. High-performance InP-based mid-IR quantum cascade lasers. *IEEE J. Sel. Top. Quantum Electron.* **2009**, *15*, 941–951. [CrossRef]
8. Jelínková, H.; Němec, M.; Šulc, J.; Miyagi, M.; Iwai, K.; Takaku, H.; Doroshenko, M.; Basiev, T.T.; Komar, V.K.; Gerasimenko, A.S. Transfer of Fe:ZnSe laser radiation by hollow waveguide. *Laser Phys. Lett.* **2011**, *8*, 613–616. [CrossRef]
9. Evans, J.W.; Berry, P.A.; Schepler, K.L. 840 mW continuous-wave Fe:ZnSe laser operating at 4140 nm. *Opt. Lett.* **2012**, *37*, 5021–5023. [CrossRef] [PubMed]
10. Murray, R.T.; Runcorn, T.H.; Guha, S.; Taylor, J.R. High average power parametric wavelength conversion at 3.31–3.48 µm in MgO:PPLN. *Opt. Express* **2017**, *25*, 6421–6430. [CrossRef] [PubMed]
11. Kemlin, V.; Jegouso, D.; Debray, J.; Segonds, P.; Boulanger, B.; Menaert, B.; Ishizuki, H.; Taira, T. Widely tunable optical parametric oscillator in a 5 mm thick 5% MgO:PPLN partial cylinder. *Opt. Lett.* **2013**, *38*, 860–862. [CrossRef] [PubMed]
12. Zhao, J.Q.; Yao, B.Q.; Zhang, X.L.; Li, L.; Ju, Y.L.; Wang, Y.Z. An efficient, compact intra-cavity continuous-wave mid-infrared SRO with a narrow line width. *Laser Phys. Lett.* **2013**, *10*, 045801. [CrossRef]
13. Wei, X.; Peng, Y.; Wang, W.; Chen, X.; Li, D. High-efficiency mid-infrared laser from synchronous optical parametric oscillation and amplification based on a single MgO:PPLN crystal. *Appl. Phys. B* **2011**, *104*, 597–601. [CrossRef]
14. Koch, P.; Ruebel, F.; Nittman, M.; Bauer, T.; Bartschke, J.; L'huillier, J.A. Narrow-band, tunable 2 µm optical parametric oscillator based on MgO:PPLN at degeneracy with a volume Bragg grating output coupler. *Appl. Phys. B* **2011**, *105*, 715–720. [CrossRef]
15. Das, R.; Kumar, S.C.; Samanta, G.K.; Ebrahim-Zadeh, M. Broadband, high-power, continuous-wave, mid-infrared source using extended phase-matching bandwidth in MgO:PPLN. *Opt. Lett.* **2009**, *34*, 3836–3838. [CrossRef] [PubMed]
16. Peng, Y.F.; Wei, X.B.; Xie, G.; Gao, J.R.; Li, D.M.; Wang, W.M. A high-power narrow-linewidth optical parametric oscillator based on PPMgLN. *Laser Phys.* **2013**, *23*, 055405. [CrossRef]
17. Henderson, A.; Stafford, R. Low threshold, singly-resonant CW OPO pumped by an all-fiber pump source. *Opt. Express* **2006**, *14*, 767–772. [CrossRef] [PubMed]
18. Vainio, M.; Peltola, J.; Persijn, S.; Harren, F.J.; Halonen, L. Singly resonant cw OPO with simple wavelength tuning. *Opt. Express* **2008**, *16*, 11141–11146. [CrossRef] [PubMed]
19. Zeil, P.; Thilmann, N.; Pasiskevicius, V.; Laurell, F. High-power, single-frequency, continuous-wave optical parametric oscillator employing a variable reflectivity volume Bragg grating. *Opt. Express* **2014**, *22*, 29907–29913. [CrossRef] [PubMed]

20. Xing, T.; Wang, L.; Hu, S.; Cheng, T.; Wu, X.; Jiang, H. Widely tunable and narrow-bandwidth pulsed mid-IR PPMgLN-OPO by self-seeding dual etalon-coupled cavities. *Opt. Express* **2017**, *25*, 31810–31815. [CrossRef] [PubMed]
21. Zhao, J.Q.; Wang, Y.Z.; Yao, B.Q.; Ju, Y.L. High efficiency, single-frequency continuous wave Nd:YVO$_4$/YVO$_4$ ring laser. *Laser Phys. Lett.* **2010**, *7*, 135–138. [CrossRef]
22. Gayer, O.; Sacks, Z.; Galun, E.; Arie, A. Temperature and Wavelength Dependent Refractive Index Equations for MgO-doped Congruent and Stoichiometric LiNbO$_3$. *Appl. Phys. B* **2008**, *91*, 343–348. [CrossRef]

© 2018 by the authors. Licensee MDPI, Basel, Switzerland. This article is an open access article distributed under the terms and conditions of the Creative Commons Attribution (CC BY) license (http://creativecommons.org/licenses/by/4.0/).

Article

Passively Q-Switched Operation of a Tm,Ho:LuVO$_4$ Laser with a Graphene Saturable Absorber

Wei Wang [1], Linjun Li [1,2,*], Hongtian Zhang [2,*], Jinping Qin [2], Yuang Lu [3], Chong Xu [2], Shasha Li [2], Yingjie Shen [4], Wenlong Yang [5], Yuqiang Yang [5] and Xiaoyang Yu [1,*]

1. The Higher Educational Key Laboratory for Measuring & Control Technology and Instrumentations of Heilongjiang Province, Harbin University of Science and Technology, Harbin 150080, China; weiweisnoopy@126.com
2. Heilongjiang Province Key Laboratory of Optoelectronics and Laser Technology, Heilongjiang Institute of Technology, Harbin 150050, China; qjinping@sina.com (J.Q.); xucong_htc@163.com (C.X.); lsshljgx@163.com (S.L.)
3. College of Information Science and Technology, Northwest University, Xi'an 710069, China; lya970809@163.com
4. School of Opto-Electronic Information Science and Technology, Yantai University, Yantai 264005, China; yingjieyj@163.com
5. College of Sciences, Harbin University of Science and Technology, Harbin 150080, China; yangwenlong1983@163.com (W.Y.); yuqiangy110@sina.com (Y.Y.)
* Correspondence: llj7897@126.com (L.L.); zhonghongtian805@163.com (H.Z.); yuxiaoyang62@126.com (X.Y.); Tel.: +86-451-8862-7940 (L.L.)

Received: 21 April 2018; Accepted: 4 June 2018; Published: 9 June 2018

Abstract: A passively Q-switched (PQS) operation of Tm,Ho:LuVO$_4$ laser is experimentally demonstrated with a graphene saturable absorber (SA) mirror. An average output power of 1034 mW at 54.5 kHz is acquired with an 8% optical–optical conversion efficiency. The energy per pulse of 40.4 µJ and a peak power of 56.07 W are achieved; the narrowest pulse width of 300 ns is acquired, and the output wavelengths of Tm,Ho:LuVO$_4$ are 2075.02 nm in a continuous wave (CW) regime and 2057.03 nm in a PQS regime.

Keywords: Tm,Ho:LuVO$_4$ laser; PQS; graphene saturable absorber

---

1. Introduction

Solid-state lasers emitting at 2 µm with ultra-short pulses are used in many fields, such as remote sensing, medicine, and gas detection, as well as frequency conversion of mid-IR wavelengths, due to their high peak power and pulse energy [1–4]. A passively Q-switched (PQS) with a saturable absorber (SA) is a compact way, in which microsecond (µs) pulsed Q-switching operation in the mid-Mid-infrared wavelength range can be achieved. Recently, many SAs, such as carbon nanotubes, two-dimensional (2D) materials and ion-doped crystals, with broadband saturable absorption at 1–3 µm, have been used for passive Q-switching operations [5–8]. The carbon nanotubes are typically a one-dimensional (1D) SA material, which has been widely applied in fiber lasers emitting at wavelengths of 1–2 µm [5,9], but its performance is poor when used in solid-state lasers emitting at 2 µm, because its bandwidth is limited by the diameter of single-walled carbon nanotubes, and its broadband saturable absorption characteristics rely on mixing single-walled carbon nanotubes with different diameters [9]. 2D materials are low-cost materials offering excellent performance and are used as the SA for PQS and mode-locked mode operations, due to their ultrafast recovery time, moderate modulation depth, high nonlinear effects and broadband saturable absorption [10,11]. Compared with the 1D and 2D SAs, ion-doped crystals, such as the Cr:ZnS crystal, have inherent defects which affect the stability, electronic structures and optical properties of ion-doped crystals [8,12].

Among the 2D materials, graphene is a 2D sheet of sp2-bonded carbon atoms with a honeycomb lattice [13]: this has attracted attention from workers in many disciplines, especially as it has special capabilities with regard to nonlinear saturable absorption. Graphene has very good thermal conductivity and optical properties, due to its zero-bandgap structure, and pure non-defective single-layer graphene with a thermal conductivity of up to 5300 W/m·K and its optical absorption is decided, independent of optical frequencies and optical conductivity constants [14]. Therefore, graphene is used as an SA in fiber and solid-state laser devices at a 1–2 µm wavelength range to achieve stable PQS and mode-locked laser emissions, owning to its ultrafast recovery time and moderate modulation depths [14–18]. Loiko et al. have demonstrated that a PQS mode of Ytterbium lasers with a graphene SA possessing a 1.9-µJ pulse energy and a 190-ns pulse duration were achieved from Yb-doped calcium niobium gallium garnet disordered garnet crystal [15]. Li et al. demonstrated a PQS mode operation of Nd:LiF$_4$ laser with an SA of graphene in 2016. The output wavelengths of 1.31 and 1.32 µm and an average output power of 1.33 W were acquired with the largest pulse energy of 17.3 µJ [16]. Serres et al. demonstrated a Tm:KLu(WO$_4$)$_2$ laser with an SA in 2015, and a 310-mW average output power was achieved [17]. Duan et al. demonstrated a PQS mode Ho:YVO$_4$ laser at 2052.1 nm with a graphene SA in 2016, and a 265.2-ns pulse width at 131.6 kHz was acquired [14]. Zhang et al. demonstrated a fiber laser operating at 1.94 µm with a graphene SA in 2012, and a 3.6-ps pulse width and a 0.4-nJ energy at 6.46 MHz were obtained in a thulium-doped fiber laser [18].

Vanadate crystals are of particular interest as a potential laser host material, because they are suitable for rare-earth ions doped or co-doped lasers, such as Nd:GdVO$_4$, Nd:GdVO$_4$, Tm,Ho:YVO$_4$, Tm,Ho:GdVO$_4$, Ho:LVO$_4$ and Tm,Ho:LuVO$_4$ [8,19–25]. Nd:GdVO$_4$ and Nd:YVO$_4$ lasers have been demonstrated at 1 µm wavelength range [26,27]. Tm,Ho:YVO$_4$ and Tm,Ho:GdVO$_4$ have been used as the laser crystal in continuous wave (CW), PQS, and acousto-optically (AO) Q-switched mode operation, and PQS Tm,Ho:YVO$_4$ and Tm,Ho:GdVO$_4$ laser were demonstrated with an SA of a 2 mm-thick Cr:ZnS crystal with a pulse width of less than 100 ns and a pulse energy over 100 µJ [8,28]. However, 2D materials were rarely used as an SA for Tm^{3+} and Ho^{3+} co-doped vanadate crystals in PQS laser operations. Compared with other vanadate crystals, the LuVO$_4$ crystal has larger absorption and emission cross-sections in the vicinity of 800 nm and 1.064 µm, respectively [29]. In addition, various Ho:LuVO$_4$ lasers (e.g., CW, Q-switched, actively mode-locked and single-longitudinal mode Ho:LuVO$_4$ laser) have been demonstrated [14,21–24], which show that the LuVO$_4$ crystal is an attractive host material at 2 µm wavelength range. We demonstrated an AO Q-switched Tm,Ho:LuVO$_4$ laser in 2018. An average output power of 3.77 W was achieved at a pulse repetition frequency of 10 kHz with an incident pump power of 14.7 W, and a pulse energy of 2.54 mJ was obtained at 1 kHz with a pulse duration of 69.9 ns [25]. However, the Tm,Ho:LuVO$_4$ crystal is a new laser crystal at 2 µm wavelength range, and its use therein has rarely been reported, and in particular, it has never been reported in use in PQS-mode operation.

In this paper, We demonstrated a PQS-mode operation of Tm,Ho:LuVO$_4$ laser with a graphene SA at 77 K. A Tm,Ho:LuVO$_4$ crystal was dual end-faces pumped by a laser diode (LD) with a center wavelength of 798.6 nm and the output power of 13 W. A pulse energy of 40.4 µJ and an average output power of 1034 mW were acquired at 2057.03 nm, and the narrowest pulse width was 300 ns.

2. Experimental Setup

The experimental setup for the PQS mode operation of a Tm,Ho:LuVO$_4$ laser is shown in Figure 1. A $^5I_7 \rightarrow ^5I_8$ laser transition of Ho^{3+} in a Tm,Ho:LuVO$_4$ crystal was used to achieve a laser emission with a 2 µm wavelength range. A Tm,Ho:LuVO$_4$ crystal was used in a resonator cavity with a 150-mm physical cavity length, and the physical length from a dichroic mirror (M5) to a 300-mm concave radius mirror (M6) was 60 mm. An LD (nLight Corp., Vancouver, WA, USA, NL-PPS50-10027) with a center output wavelength of 798.6 nm, corresponding to the output power of 13 W, was used as the pump source for the Tm,Ho:LuVO$_4$ laser. The temperature of the LD was selected at 298.15 K in the experiment, and the output power of LD was coupled to a fiber (a core diameter was 400 µm and numerical aperture (NA) was 0.22). Tm,Ho:LuVO$_4$ with the mass percentages of 5% and 0.5% for Tm and Ho was cut along a-axis, of which the dimension was 4 mm×4 mm×8 mm. The end-faces of the laser crystal were coated at pump wavelengths from 790 to 810 nm and its laser wavelength range from 1.9 to 2.2 µm with a transmission efficiency over 99.5% was used. The laser crystal was cooled at 77 K. In addition, the laser crystal was located in the middle of M5 and M6.

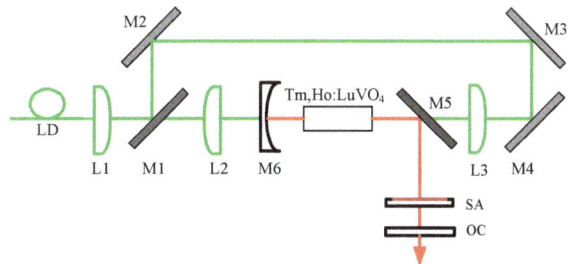

Figure 1. Experimental setup of the passively Q-switched Tm,Ho:LuVO$_4$ laser.

A configuration with double-end pumping was chosen to relieve the thermal loading of the laser crystal. The pumped laser from the LD was divided into two beams after being collimated by a collimation lens (L1) and a split light mirror (M1). Then, One of the divided pumped laser was refocused on one end-face of the laser crystal by a focusing lens (L2), and the other pumped laser was reflected by the mirrors (M2, M3 and M4) and refocused on one end-face of the laser crystal by a focusing lens (L3). The focal lengths of L1, L2 and L3 were 25, 50 and 50 mm, respectively, and the pump spots with 800 µm in diameter were placed at the input surfaces of the laser crystal. A flat 45° mirror was used as a splitting light mirror (M1), which was coated at 790–810 nm with a 50%-transmission-efficiency and 50%-reflection-efficiency material on one face and another high-transmission material ($T > 99.0\%$) on the other face. Three flat 45° mirrors were used as reflection mirrors (M2, M3 and M4), which were coated at 790–810 nm with a high-reflectivity ($R > 99.5$) material on one face. A dichroic mirror (M5) was a plano mirror placed at 45° with respect to the incident light, and it was coated with a high-transmission material working at 798.6 nm on two faces and another high-reflection material working at 2000–2100 nm on one face. A concave mirror with a radius of 300 mm (M6) was coated with a high-transmission material working at 798.6 nm on two faces, and was coated at a high-reflectivity material working at 2000–2100 nm on its concave face. A plane mirror coated with a 2% or 5% transmittance at 2000–2100 nm was used as an output coupler (OC) mirror of the laser. The mirror made from a CaF$_2$ crystal was used as the substrate of the SA, and a graphene crystal was chosen as the material for the SA. The graphene material dissolved in ethyl alcohol was coated onto the surface of one face at the CaF$_2$ mirror with a spin coating machine (KW-4A, Chinese Academy of Sciences, Beijing, China).

3. Experimental Results and Discussion

Two kinds of OC mirrors (with transmittances of 2% and 5%) were chosen in CW and PQS mode operation of an a-cut Tm,Ho:LuVO$_4$ laser to achieved optimal output performances. In CW and PQS mode operations, the output power and the average output power of the laser are shown in Figure 2. Under the CW mode operation, output powers of 2310 and 2398 mW were achieved with 2% and 5% transmittances of the OC mirror, and optical–optical conversion efficiencies were 17.8% and 18.5%, respectively. In the PQS mode operation, a graphene SA mirror placed between M5 and OC mirror was located near the OC mirror. The transmittance of a few-layer graphene material at 2075.72 nm was measured and found to have been approximately 88.6%, and the modulation depths were 69.0% ($T = 2\%$) and 58.3% ($T = 5\%$). The laser spot radii on the surface of M6, left end face of laser crystal, right end face of laser crystal, M5, graphene SA and the OC mirror were calculated to be approximately 445, 416, 395, 367, 310 and 310 µm, respectively, by the ABCD matrix without considering the thermal lens effect. The average output powers of 604 and 1034 mW were achieved with 2% and 5% transmittances of the OC mirror, respectively, corresponding to optical–optical conversion efficiencies of 6.0% and 8.0%. In the PQS mode operation of Tm,Ho:LuVO$_4$ laser, the pump threshold power was approximately 5 W with a 2% or 5% transmittance of the OC mirror. The laser was operated under a stable condition, and the pump power of 10.03 W or 13 W was injected into the laser cavity with a 2% or 5% transmittance of the OC mirror to avoid the damage of the graphene SA. Compared with the high transmittance of the OC mirror, a low transmittance of OC was chosen to acquire a stable PQS mode operation, due to more energy being stored in the crystal, but the damage threshold of the SA was easily reached, and the graphene SA was damaged at a peak power density of approximately 0.019 MW/cm^2 at 2057.03 nm with a 2% transmittance of the OC mirror in the experiment.

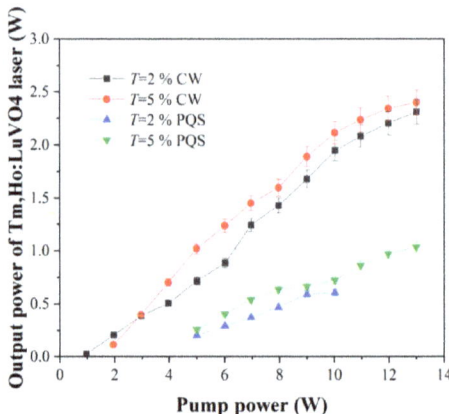

Figure 2. Output power of Tm,Ho:LuVO$_4$ laser in CW and PQS mode operations verse the pump power.

Under the PQS mode operation, the pulse repetition frequency (PRF) of the Tm,Ho:LuVO$_4$ laser is shown in Figure 3. As can be seen, the PRF increased with increasing pump power, and the highest PRFs were 35.92 kHz (T = 2%) and 54.5 kHz (T = 5%), respectively, corresponding to energies per pulse of 16.82 and 18.97 µJ. In addition, the maximum pulse energy of 40.4 µJ was achieved with a PRF of 24.01 kHz (T = 5%) and an average output power of 970 mW (T = 5%), corresponding a pulse width of 1.429 µs. The pulse widths of Tm,Ho:LuVO$_4$ laser are shown in Figure 4, and the narrowest pulse width of 300 ns was acquired under a 2% transmittance of the OC mirror. The pulse widths were 2.521 and 8.250 µs with the pump powers of 6.96 and 5.01 W, respectively for a 5% transmittance of the OC mirror. With a detector (Thorlabs, Newton, NJ, USA, PDA10PT-EC) and an oscilloscope with a bandwidth of 1 GHz (Tektronix, Beaverton, OR, USA, DPO4104), the typical Q-switched pulse trains were recorded in 40 and 200 µs time scales (Figure 5). Therefore, the highest pulse energy and the peak power of the Tm,Ho:LuVO$_4$ laser were calculated to be 40.4 µJ and 56.07 W based on the findings in Figure 2.

A 721A IR laser wavelength meter with a measuring range of 1300–5000 nm (Bristol Instruments Inc., Victor, NY, USA) was used to obtain an output wavelength of Tm,Ho:LuVO$_4$ laser in CW and PQS modes operations (Figure 6). An output wavelength of 2075.72 nm was achieved in the CW mode operation, and 2057.03 nm was achieved in the PQS mode operation.

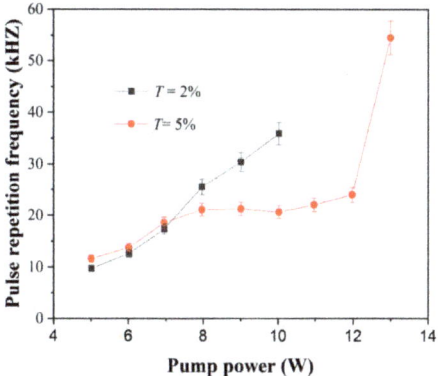

Figure 3. PRF of the PQS Tm,Ho:LuVO$_4$ laser verse the pump power.

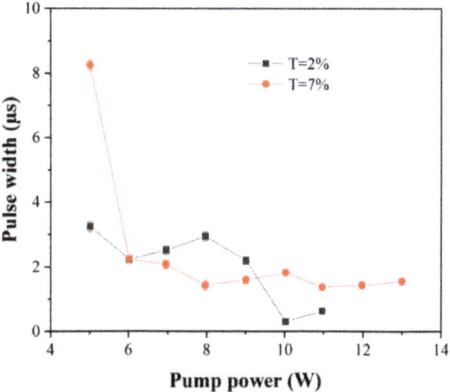

Figure 4. Pulse widths of Tm,Ho:LuVO$_4$ laser verse the pump power.

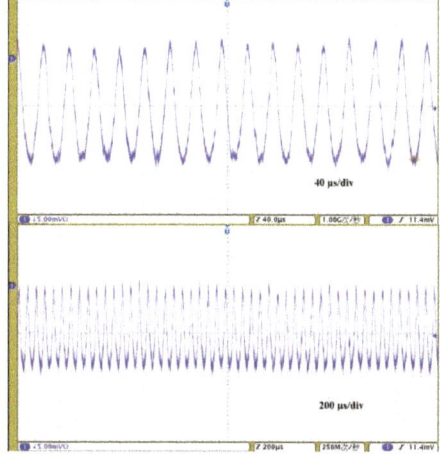

Figure 5. The pulse trains in 40 μs and 200 μs time scales.

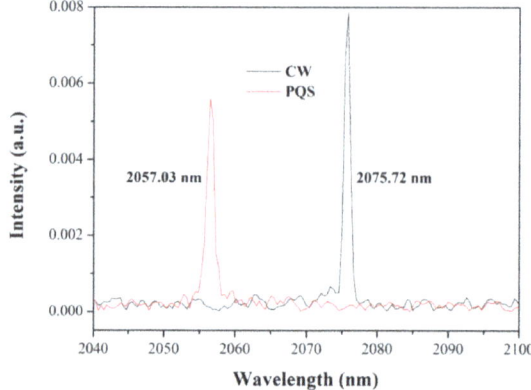

Figure 6. The output wavelengths of the Tm,Ho:LuVO$_4$ laser in CW and PQS mode Operations.

A slit scanning beam profiler (BP109-IR2, Thorlabs Inc., Newton, NJ, USA) was used to acquire beam quality factors of Tm,Ho:LuVO$_4$ laser. Output beam profiles of Tm,Ho:LuVO$_4$ laser were measured, and 2D and 3D graphics are shown in Figure 7. In addition, beam quality factors M_x^2 and M_y^2 were measured to be 1.19 and 1.19, respectively, at approximately 1-W average output power from the PQS Tm,Ho:LuVO$_4$ laser.

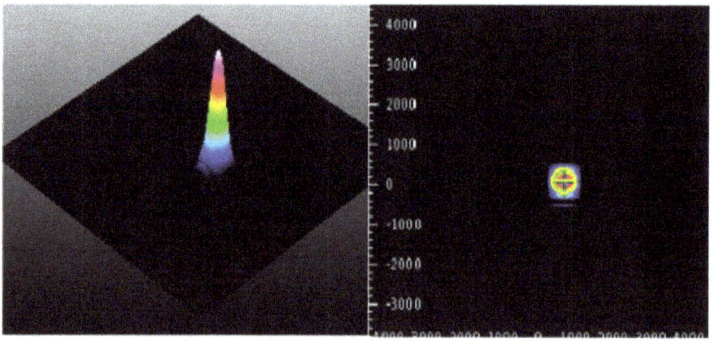

Figure 7. 2D and 3D output beam profiles of Tm,Ho:LuVO$_4$ laser.

4. Conclusions

In conclusion, we have experimentally demonstrated a Tm,Ho:LuVO$_4$ laser under the PQS mode operation with a graphene SA for the first time. A pulse energy of 40.4 µJ and an average output power of 1034 mW were acquired at 2057.03 nm. In addition, the beam quality factors ($M_x^2 = 1.19$ and $M_y^2 = 1.19$) were obtained at about 1-W average output power from the PQS Tm,Ho:LuVO$_4$ laser.

Author Contributions: In this paper, W.W., L.L., H.Z. and X.Y. conceived and designed the experiments; J.Q., Y.L. and Y.S. performed the experiments; W.W. and L.L. analyzed the data; W.Y. and Y.Y. contributed reagents, materials and analysis tools; S.L. and C.X. wrote the paper.

Acknowledgments: This work was supported by the National Natural Science Foundation of China (No. 61378029, 61775053, 51572053 and 51777046), the Science Foundation for Outstanding Youths of Heilongjiang Province (JC2016016), the Science Foundation for Youths of Heilongjiang Province (QC2017078), and the Program for Equipment Pre-research Field Funds (No. 6140414040116CB01012).

Conflicts of Interest: We declare that this article does not have any conflicts of interest because the author's order is sorted by actual contribution for the paper.

References

1. Scholle, K.; Heumann, E.; Huber, G. Single mode Tm and Tm, Ho:luag lasers for LIDAR applications. *Laser Phys. Lett.* **2004**, *1*, 285–290. [CrossRef]
2. Nishioka, N.S.; Domankevitz, Y. Comparison of tissue ablation with pulsed holmium and thulium lasers. *IEEE J. Quantum Electron.* **1990**, *26*, 2271–2275. [CrossRef]
3. Lagatsky, A.A.; Fusari, F.; Kurilchik, S.V.; Kisel, V.E.; Yasukevich, A.S.; Kuleshov, N.V. Optical spectroscopy and efficient continuous-wave operation near 2 μm for a Tm, Ho:KYW laser crystal. *Appl. Phys. B* **2009**, *97*, 321–326. [CrossRef]
4. Spiers, G.D.; Menzies, R.T.; Jacob, J.; Christensen, L.E.; Phillips, M.W.; Choi, Y. Atmospheric CO_2 measurements with a 2 μm airborne laser absorption spectrometer employing coherent detection. *Appl. Opt.* **2011**, *50*, 2098–2111. [CrossRef] [PubMed]
5. Liu, J.; Wang, Y.G.; Qu, Z.S.; Fan, X.W. 2 μm passive Q-switched mode-locked Tm^{3+}:YAP laser with single-walled carbon nanotube absorber. *Opt. Laser Technol.* **2012**, *44*, 960–962. [CrossRef]
6. Loiko, P.A.; Serres, J.M.; Mateos, X.; Liu, J.; Zhang, H.; Yasukevich, A.S.; Yumashev, K.V.; Petrov, V.; Griebner, U.; Aguiló, M.; et al. Passive Q-switching of Yb bulk lasers by a graphene saturable absorber. *Appl. Phys. B* **2016**, *122*, 105. [CrossRef]
7. Fan, M.Q.; Li, T.; Zhao, S.Z.; Li, G.Q.; Gao, X.C.; Yang, K.J.; Li, D.C.; Kränkel, C. Multilayer black phosphorus as saturable absorber for an $Er:Lu_2O_3$ laser at 3 μm. *Photonics Res.* **2016**, *4*, 181–186. [CrossRef]
8. Du, Y.Q.; Yao, B.Q.; Liu, W.; Cui, Z.; Duan, X.M.; Ju, Y.L.; Yu, H. Highly efficient passively Q-switched Tm, $Ho:GdVO_4$ laser with kilowatt peak power. *Opt. Eng.* **2016**, *55*, 046112. [CrossRef]
9. Schmidt, A.; Rivier, S.; Steinmeyer, G.; Yim, J.H.; Cho, W.B.; Lee, S.; Rotermund, F.; Maria, C.A.; Mateos, P.X.; Aguiló, M.; et al. Passive mode locking of Yb: KLuW using a single-walled carbon nanotube saturable absorber. *Opt. Lett.* **2008**, *33*, 729–731. [CrossRef] [PubMed]
10. Guo, B. 2D noncarbon materials-based nonlinear optical devices for ultrafast photonics. *Chin. Opt. Lett.* **2018**, *16*, 020004. [CrossRef]
11. Wang, C.; Zhao, S.Z.; Li, T.; Yang, K.J.; Luan, C.; Xu, X.D.; Xu, J. Passively Q-switched Nd:LuAG laser using few-layered MoS_2 as saturable absorber. *Opt. Commun.* **2018**, *406*, 249–253. [CrossRef]
12. Karhu, E.A.; Ildstad, C.R.; Poggio, S.; Furtula, V.; Tolstik, N.; Sorokina, I.T.; Belbruno, J.J.; Gibson, U.J. Vapor deposited Cr-doped ZnS thin films: Towards optically pumped mid-infrared waveguide lasers. *Opt. Mater. Express* **2016**, *6*, 2947–2955. [CrossRef]
13. Seyller, T.; Bostwick, A.; Emtsev, K.V.; Horn, K.; Ley, L.; Mcchesney, J.L.; Ohta, T.; Riley, J.D.; Rotenberg, E.; Speck, F. Epitaxial Graphene: A New Material. *Phys. Status Solidi B* **2008**, *245*, 1436–1446. [CrossRef]
14. Lin, W.M.; Duan, X.M.; Cui, Z.; Yao, B.Q.; Dai, T.Y.; Li, X.L. A passively Q-switched $Ho:YVO_4$ Laser at 2.05 μm with Graphene Saturable Absorber. *Appl. Sci.* **2016**, *6*, 128. [CrossRef]
15. Loiko, P.; Serres, J.M.; Mateos, X.; Yu, H.H.; Zhang, H.J.; Liu, J.H.; Yumashev, K.; Griebner, U.; Petrov, V.; Magdalena, A.F.D. Q-Switching of Ytterbium Lasers by A Graphene Saturable Absorber. *Nano-Opt.: Princ. Enabl. Basic Res. Appl.* **2017**, 533–535. [CrossRef]
16. Li, S.X.; Li, T.; Zhao, S.Z.; Li, G.Q.; Hang, Y.; Zhang, P.X. 1.31 and 1.32 μm dual-wavelength $Nd:LuLiF_4$ laser. *Opt. Laser Technol.* **2016**, *81*, 14–17. [CrossRef]
17. Serres, J.M.; Loiko, P.; Mateos, X.; Yumashev, K.; Griebner, U.; Petrov, V.; Aguiló, M.; Díaz, F. $Tm:KLu(WO_4)_2$ microchip laser Q-switched by a graphene-based saturable absorber. *Opt. Express* **2015**, *23*, 14108–14113. [CrossRef] [PubMed]
18. Zhang, M.; Kelleher, E.J.R.; Torrisi, F.; Sun, Z.; Hasan, T.; Popa, D.; Wang, F.; Ferrari, A.C.; Popov, S.V.; Taylor, J.R. Tm-doped fiber laser mode-locked by graphene-polymer composite. *Opt. Express* **2012**, *20*, 25077–25084. [CrossRef] [PubMed]
19. Sato, Y.; Taira, T. Comparative Study on the Spectroscopic Properties of $Nd:GdVO_4$ and $Nd:YVO_4$ with Hybrid Process. *IEEE J. Select. Top. Quantum Electron.* **2005**, *11*, 613–620. [CrossRef]
20. Ma, Y.F.; Yu, X.; Li, X.D.; Fan, R.W.; Yu, J.H. Comparison on performance of passively Q-switched laser properties of continuous-grown composite $GdVO_4/Nd:GdVO_4$ and $YVO_4/Nd:YVO_4$ crystals under direct pumping. *Appl. Opt.* **2011**, *50*, 3854–3859. [CrossRef] [PubMed]
21. Yao, B.Q.; Cui, Z.; Duan, X.M.; Du, Y.Q.; Han, L.; Shen, Y.J. Resonantly pumped room temperature $Ho:LuVO_4$ laser. *Opt. Lett.* **2014**, *39*, 6328–6330. [CrossRef] [PubMed]

22. Cui, Z.; Yao, B.Q.; Duan, X.M.; Xu, S.; Du, Y.Q.; Yuan, J.H. Output characteristics of actively q-switched Ho:LuVO$_4$ laser at room temperature. *Opt. Express* **2015**, *23*, 13482–13487. [CrossRef] [PubMed]
23. Duan, X.; Zhang, P.; Cui, Z.; Yao, B.; Wang, Y. Actively mode-locked Ho:LuVO$_4$ laser at 2073.8 nm. *Opt. Eng.* **2016**, *55*, 126104. [CrossRef]
24. Xu, L.W.; Ju, L.; Wu, J.; Li, Y.Y.; Zhang, Z.G.; Ju, Y.L. Tm-fiber pumped single-longitudinal-mode Ho:LuVO$_4$ laser. *Opt. Quantum Electron.* **2017**, *49*, 12. [CrossRef]
25. Wang, W.; Yang, X.G.; Shen, Y.J.; Li, L.J.; Zhou, L.; Yang, Y.Q.; Bai, Y.F.; Xie, W.Q.; Ye, G.C.; Yu, X.Y. High beam quality of a Q-switched 2-μm Tm, Ho:LuVO$_4$ laser. *Appl. Phys. B* **2018**, *124*, 82. [CrossRef]
26. Sato, Y.; Taira, T.; Pavel, N.; Lupei, V. Laser operation with near quantum-defect slope efficiency in Nd:YVO$_4$ under direct pumping into the emitting level. *Appl. Phys. Lett.* **2003**, *82*, 844–846. [CrossRef]
27. Chen, F.; Yu, X.; Zhang, K.; He, Y.; Zheng, C.B.; Wang, C.R.; Guo, J. Diode-pumped acousto-optical Q-switched 912 nm Nd:GdVO$_4$ laser and extra-cavity frequency-doubling of 456 nm deep-blue light emission. *Opt. Laser Technol.* **2015**, *68*, 36–40. [CrossRef]
28. Du, Y.Q.; Yao, B.Q.; Li, J.; Pan, Y.B.; Duan, X.M.; Cui, Z.; Shen, Z.C.; Ju, Y.L. Stable passively Q-switched Tm, Ho YVO$_4$ laser with near 100 ns pulse duration at 2 μm. *Laser Phys. Lett.* **2014**, *11*, 115817. [CrossRef]
29. Zhang, H.; Liu, J.; Wang, J.; Xu, X.; Jiang, M. Continuous-wave laser performance of Nd:LuVO$_4$ crystal operating at 1.34 micron. *Appl. Opt.* **2005**, *44*, 7439–7441. [CrossRef] [PubMed]

© 2018 by the authors. Licensee MDPI, Basel, Switzerland. This article is an open access article distributed under the terms and conditions of the Creative Commons Attribution (CC BY) license (http://creativecommons.org/licenses/by/4.0/).

Article

A Tunable Mid-Infrared Solid-State Laser with a Compact Thermal Control System

Deyang Yu [1,2], Yang He [1,2], Kuo Zhang [1], Qikun Pan [1], Fei Chen [1,*] and Lihong Guo [1]

[1] Changchun Institute of Optics, Fine Mechanics and Physics, Chinese Academy of Sciences, Changchun 130033, China; yudeyang830@163.com (D.Y.); heyang_3g@126.com (Y.H.); cole_fx@163.com (K.Z.); panqikun2005@163.com (Q.P.); guolh@ciomp.ac.cn (L.G.)
[2] University of Chinese Academy of Sciences, Beijing 100039, China
* Correspondence: feichenny@126.com; Tel.: +86-431-8617-6197

Received: 3 May 2018; Accepted: 20 May 2018; Published: 26 May 2018

Abstract: Tunable mid-infrared lasers are widely used in laser spectroscopy, gas sensing and many other related areas. In order to solve heat dissipation problems and improve the environmental temperature adaptability of solid-state laser sources, a tunable all-fiber laser pumped optical parametric oscillator (OPO) was established, and a compact thermal control system based on thermoelectric coolers, an automatic temperature control circuit, cooling fins, fans and heat pipes was integrated and designed for the laser. This system is compact, light and air-cooling which satisfies the demand for miniaturization of lasers. A mathematical model and method was established to estimate the cooling capacity of this thermal control system under different ambient environments. A finite-element model was built and simulated to analyze the thermal transfer process. Experiments in room and high temperature environments were carried out and showed that the substrate temperature of a pump module could be maintained at a stable value with controlled precision to 0.2 degrees, while the output power stability of the laser was within ±1%. The experimental results indicate that this compact air-cooling thermal control system could effectively solve the heat dissipation problem of mid-infrared solid-state lasers with a one hundred watts level pump module in room and high temperature environments.

Keywords: tunable mid-infrared solid-state laser; thermal control; all-fiber laser; thermoelectric cooling; finite-element analysis; optical parametric oscillator

1. Introduction

Since mid-infrared lasers have minimum attenuation in atmospheric transmission and cover the absorption peaks of many atoms and molecules, they are widely used in laser spectroscopy, atmosphere monitoring, photoelectric detection, remote sensing survey and many other fields [1–4]. The pump laser is one of the most important key modules in tunable mid-infrared solid-state lasers. It generates a large amount of waste heat in the working process and needs to be dissipated in time, otherwise it will cause wavelength shift, decrease of output power and even damage the laser [5,6]. Thus, it is of vital importance to dissipate the waste heat and maintain the temperature of pump laser at a suitable value. Various conventional cooling methods are used for lasers, such as water cooling, forced air cooling, thermoelectric cooling, heat pipe cooling, micro-channel cooling, compressor refrigeration and so on [7–9].

High power lasers mainly use the water cooling method to dissipate heat due to its advantages of high heat transfer coefficient and heat flux density [10]. However, it requires the connection of an external water cooler which causes the laser to have a large. Forced air cooling, heat pipe cooling, micro-channel cooling and other passive cooling methods can only ensure the temperature of the laser is higher than the ambient environment [11], while in some situations, the temperature of the laser needs to be controlled to be lower than the environment. Thermoelectric cooling, usually considered using the thermoelectric cooler (TEC), has advantages over conventional cooling devices, including being compact in size, having high reliability, no mechanical moving parts, no working liquid, light in weight, being powered by direct current, and easily switching between cooling and heating modes [12]. Due to these advantages, TECs are widely used in many fields including electronic device cooling, diode laser (LD) cooling and temperature control, domestic refrigeration, scientific equipment and so on [13–15].

In recent years, many studies have reported TEC temperature control systems for laser cooling. Jian Dong et al. [16] investigated a high power diode-side-pumped Q-switched Nd:YAG solid-state laser under 808nm side-pumping without a water cooler. Two TECs were adopted to maintain the temperature of the LD module, and it was observed that precision of the TEC controller impacted the relative stability of the output energy. Wei Zhang et al. [11] used a finite-element analysis method to study the performance of a micro semiconductor laser with TEC; the thermal cooling methods on the hot side of TEC were also investigated. Hao Wang et al. [17] developed a dynamic thermal model for a tunable laser module with a proportion-integration-differentiation (PID) temperature controller based on finite-element analysis. Temperature variation and wavelength shift of the laser were simulated by changing PID parameters and discussed. Limei Shen et al. [18] investigated a miniature thermoelectric module for pulse laser cooling, and the effects of pulse and step variable voltage on the miniature TEC module (TEM) were numerically and experimentally studied.

Recent studies of thermal management using TECs for lasers have mainly focused on single diode lasers with medium and low output powers, while high output power pump lasers always use the water cooling method to dissipate heat [11], which cannot satisfy the demand for miniaturization. In this paper, an all-fiber laser pumped optical parametric oscillator (OPO) laser was established with a compact thermal control system based on an automatic temperature control circuit, TEMs, cooling fins, axial fans and heat pipes. This thermal control system was designed and analyzed in finite-element method. A mathematical model to estimate the cooling capacity of the system under different ambient environments was established. Experiments in room and high temperature environments were carried out to validate the performance of the mid-infrared solid-state laser and the thermal control system.

2. Module Design

2.1. Design of a Tunable Mid-Infrared Solid-State Laser

The tunable mid-infrared solid-state laser that was investigated in the current study was an all-fiber laser pumped OPO, made up of an all-fiber laser pump module, beam control module and OPO module. The schematic diagram of the laser source is shown in Figure 1. The all-fiber laser pump module is based on the structure of the master oscillator power amplifier (MOPA), and includesmaster-oscillation and power-amplifier two parts. The master-oscillation stage uses a linearly polarized DFB laser to produce seed light at 1064 nm.The power-amplifier is made up of three-stages: linear, polarized and Yb doped fiber amplifier (YDFA) and can produce a 1064 nm laser at maximum output power of 50 W. The facula and polarization-state of the fiber pump laser are adjusted and isolated through a beam control module and then poured into OPO module on the MgO:PPLN crystal. The optical parametric oscillation process happens through the feedback effect of resonant cavity and obtains a near-infrared signal light and a mid-infrared idler. We used a dichroicmirror to filter out the signal light and finally achieved a mid-infrared laser at maximum output power of 5 W.

There is a certain relationship between the polarization period and temperature of MgO:PPLN crystal with the output wavelength of mid-infrared laser. The variation in the curve of the mid-infrared laser wavelength with the temperature of the crystal during polarization periods of 29 μm, 29.5 μm and 30 μm is shown in Figure 2. In our design, the wavelength of the mid-infrared laser was tuned by changing the temperature of the crystal. The MgO:PPLN crystal was fixed in the temperature control furnace, a film heater was used to control the temperature of the crystal and a PT1000 temperature sensor was used to measure the temperature of the crystal. We were able to obtain a certain wavelength from the mid-infrared laser by controlling the temperature of MgO:PPLN crystal at a specific value.

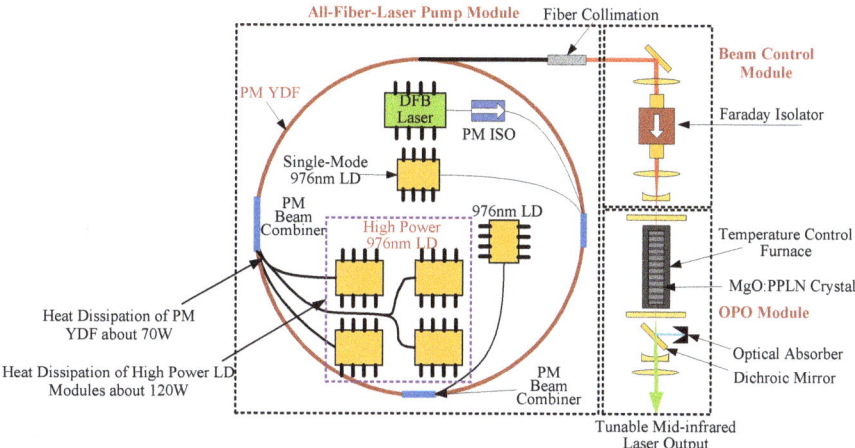

Figure 1. The schematic diagram of the designed tunable all-fiber laser pumped optical parametric oscillator (OPO).

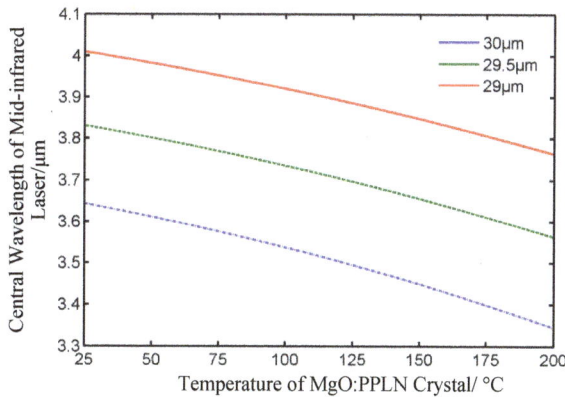

Figure 2. The variation in the curve of the laser wavelength with changes in the temperature of the crystal at different polarization periods.

2.2. Design of the Thermal Control System

There were four high power 976 nm LD modules in the pump module and each had a maximum output power of 30 W. The electro-optic conversion efficiency of this LD module is about 50%; thus, the total heat generated from the LD modules was about 120 W. In addition, the maximum output power of the fiber laser pump module was over 50W; thus, the maximum heat generated from the fiber was no more than 70 W. The total heat generated in the all-fiber laser pump module was no more than 200 W.

Each LD module was tightly fixed on the upper surface of a thin heat sink and fully contacted through grease. The fiber was encircled into a round shape and tightly pasted on the heat sink. The thin heat sink was made from molybdenum–copper material which has advantages of low density and high thermal conductivity. Thus, we were able to dissipate the waste heat generated from LD modules and fibers by cooling and keeping the upper surface temperature of the heat sink at a suitable value. A schematic diagram of the laser and thermal control system is shown in Figure 3.

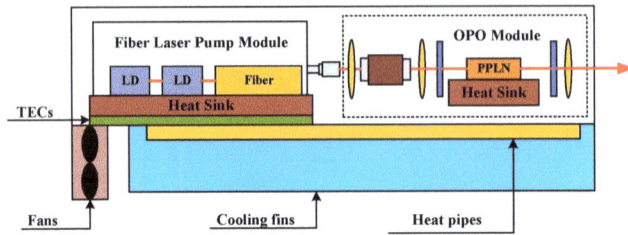

Figure 3. Schematic diagram of the laser and thermal control system.

2.2.1. Constitution of the Thermal Control System

To achieve a better automatic cooling ability for the mid-infrared laser, the thermal control system was composed of an automatic temperature control circuit, TEMs, and an aluminum cabinet with cooling fins, axial fans and heat pipes. The heat sink was stuck to the cold side of TECs through grease and compressed tightly, allowing full contact and consequently, better heat transfer. The hot sides of the TECs were positioned close to the bottom of the cabinet. Thus, when the laser was working, the large amount of waste heat generated from the pump source module was absorbed by TECs and dissipated to the bottom of cabinet. The cooling fins were distributed outside the bottom of the cabinet to increase the heat transfer area. The axial fans were fixed on one side of the fins, increasing the heat convection coefficient through forced convection cooling. There were also two heat pipes fixed at the bottom of laser cabinet to improve heat transfer performance of the system.

The TEM used here was MCTE1-12712L-S, manufactured by Multicomp (Warsaw, Poland) with parameters shown in Table 1. There were 16 pieces of TECs divided into 4 groups to cool the heat sink. Each group was made up of 4 TECs with two pieces electrically connected in series and then electrically connected in parallel together. As shown in Figures 4 and 5, two groups of TECs in a line were used to cool the LD modules and the remaining two groups were used to cool the fiber. This electrical connection and configuration mode improved the heat transfer efficiency by increasing the contact area and simplified the circuit structure through one output channel to drive 4 TECs. The axial fan used here was a pulse width modulation (PWM) speed controlled fan with 24 V direct current (DC) power supply. The max rated speed of this fan was 18,000 r/min and the maximum airflow was 0.011 m^3/s. There were 7 fans parallel connected in a line to increase airflow.

Table 1. TEC parameters.

Parameters		
Internal resistance	$1\,\Omega \pm 10\%$	
I_{max}	12 A	
V_{max}	15.4 V	
-	$T_h = 27\,°C$	$T_h = 50\,°C$
Q_{max}	110 W	134 W
ΔT_{max}	68 °C	75 °C
Solder melting point	138 °C	

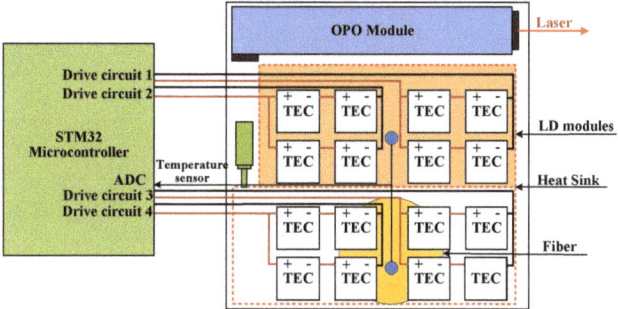

Figure 4. Diagrammatic sketch of the thermoelectric cooler (TEC) layout.

Figure 5. Picture of TEC layout.

2.2.2. Hardware Circuit Design of the Thermal Control System

The main function of the system hardware circuit was to realize automated thermal control for the laser through exporting suitable drive power on the TECs. The diagram of the hardware circuit is shown in Figure 6. The circuit consisted of a STM32 microcontroller as control core, two PT1000 platinum resistor sensors to measure the upper surface temperature of heat sink, one PT1000 to measure the temperature of the MgO:PPLN crystal, an LCD12864 display to show the measuring temperature, setting temperature and operating state of the laser, a 24V DC power supply for the PWM fans, four H-bridges to drive the TECs, and a universal synchronous/asynchronous receiver/transmitter (USART) interface to communicate with a personal computer. The platinum resistor is a kind of temperature sensor with high stable performance and wide temperature measure ranging from −200 to +650 °C. The PT1000 temperature characteristics can be approximately expressed with the following equation:

$$R_t = 1000 + 3.9 \times t. \tag{1}$$

R_t is the resistance value (Ω) at temperature of t (°C). An 0.1 mA precise constant current source flows into the PT1000 resistor. The voltage on two sides of the resistor is amplified by an instrumentation amplifier INA128 before entering the analog-to-digital converter (ADC) interface on STM32. Then, the STM32 chip calculates the temperature value according to the conversion formula.

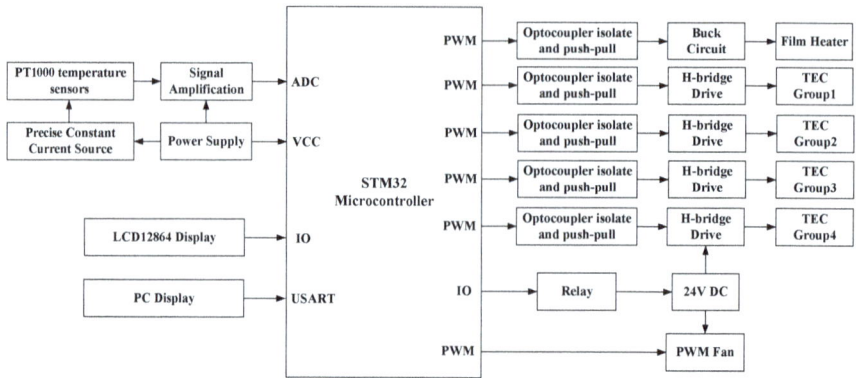

Figure 6. Diagram of the hardware circuit.

2.2.3. Software Design of the Thermal Control System

The software designed for the automated system was embedded in the STM32 microcontroller and programmed in C programming language based on a Keil development environment. It was made up of six function modules: a signal acquisition and process module for temperature measurement, an LCD12864 display module, a PID temperature control module, a PWM fan control module, a serial communicating module and a system monitoring and protecting module.

In the temperature control process, due to the time-varying, nonlinearity and uncertainty factors of the laser and external environment, the voltage exerted on the TEMs was calculated based on a parameter self-tuning incremental PID algorithm. The design of this PID algorithm is shown as below:

$$\Delta U(k) = K_p \times [e(k) - e(k-1)] + K_i \times e(k) + K_d[e(k) - 2 \times e(k-1) + e(k)] \qquad (2)$$

(1) $|e| > E_{a1} : K_p = K_{p1}, K_d = K_{d2}, K_i = 0$;
(2) $E_{a2} > |e| > E_{a1} : K_p = K_{p2}, K_d = K_{d1}, K_i = K_{i2}$;
(3) $|e| < E_{a2} : K_p = K_{p1}, K_d = K_{d1}, K_i = K_{i1}$.

The relationships among the PID parameters are as follows: $K_{p1} > K_{p2} > 0$, $K_{d1} > K_{d2} > 0$, $K_{i1} > K_{i2} > 0$, $E_{a1} > E_{a2} > 0$. $K_{p1}, K_{p2}, K_{d1}, K_{d2}, K_{i1}, K_{i2}, E_{a1}$ and E_{a2} are constants. When $|e|$ is larger, we choose a larger K_p and smaller K_d to promote the tracking performance of the system, setting $K_i = 0$ to avoid larger overshoot when the system responds. When $|e|$ is of medium size, we set smaller K_p and K_i values to decrease the system response overshoot. When $|e|$ is smaller, we set larger K_p and K_i values to decrease the steady error and a medium-sized K_d to avoid system vibration near the set value.

3. Mathematical Model and Numerical Simulation

3.1. Thermal Analysis of TEM

The commercialized TEM is normally composed of a series of thermocouples electrically connected in series and thermally converged in parallel. When direct current flows through a thermocouple which is composed of an N semiconductor element and a P semiconductor element, the phenomenon of heat absorption and release happens at the contact surface. The current flows from the N element to the P element above the contact surface and the surface absorbs heat from the outside, becoming the cold side. The current flows from the P element to the N element below the contact surface and the surface releases heat to the outside, becoming the hot side [12]. The cooling capacity (Q_c, heat absorption power on the cold side) of the TEM is

$$Q_c = S_m I T_c - \frac{1}{2} I^2 R_m - K_m(T_h - T_c). \tag{3}$$

The quantity of heat production, Q_h, on the hot side of the TEM is

$$Q_h = S_m I T_h + \frac{1}{2} I^2 R_m - K_m(T_h - T_c). \tag{4}$$

In Equations (3) and (4), S_m is the Seebeck coefficient of TEM, T_c is the temperature of the cold side of the TEM and T_h is the temperature of the hot side of the TEM, I is the direct current through the module, R_m is the electrical resistance of the TEM and K_m is the thermal conductance of the TEM.

The performance parameters, S_m, R_m and K_m, can be figured out by the TEM parameters of I_{max}, ΔT_{max}, T_h and V_{max} given by the manufacturer through Equations (5)–(7) [19,20], of which I_{max} is the direct current flowing through TEM that causes the maximum temperature difference, ΔT_{max}. ΔT_{max} is the maximum temperature difference between the cold and hot sides of the TEM at a certain temperature, T_h. V_{max} is the direct voltage that diverges the ΔT_{max}.

$$R_m = \frac{(T_h - \Delta T_{max}) \times V_{max}}{T_h \times I_{max}} \tag{5}$$

$$K_m = \frac{(T_h - \Delta T_{max}) \times V_{max} \times I_{max}}{2 \times T_h \times \Delta T_{max}} \tag{6}$$

$$S_m = \frac{V_{max}}{T_h} \tag{7}$$

3.2. Mathematical Model of the Thermal Control System

In order to estimate the performance of the thermal control system, a mathematical model of the system was established to calculate the cooling capacity. The thermal resistance network diagram is shown in Figure 7.

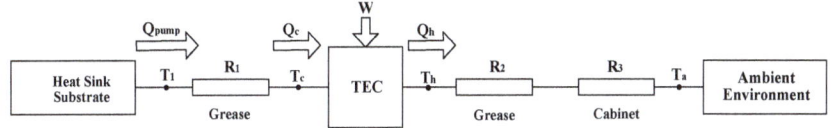

Figure 7. The thermal resistance network diagram.

R_1 is the thermal resistance of the thermally conductive grease between the substrate of the heat sink and the cold side of the TEMs. R_2 is the thermal resistance of the thermally conductive grease between the hot side of the TEMs and the bottom inside of the cabinet. R_1 is approximately equal to R_2. R_3 is the thermal resistance from the bottom inside of the cabinet to the ambient environment. T_1 is the upper surface temperature of heat sink, since the 4 LD modules and fibers are closely fixed on the heat sink and they are the main heat sources. Thus, it is important to maintain T_1 at a suitable and stable temperature to ensure the pump laser module works reliably. T_c is the cold side temperature of the TEMs, T_h is the hot side temperature of TEMs and T_a is the temperature of the ambient environment. Q_{pump} is the heat generated by the pump laser module and Q_c is the cooling capacity of the TEM, since there are 16 pieces of TECs used in the system, and assuming each TEC hasthe same Q_c, it can be deduced that the equation at the steady heat transfer state is

$$Q_{pump} = 16 \times Q_c. \tag{8}$$

The one-dimensional steady heat transfer equation of the thermal resistance, R_1, can be carried out by thermal analysis as shown below:

$$Q_c = \frac{T_1 - T_c}{R_1}. \tag{9}$$

Q_h is the heat release output at the hot side of TEM. Thus, we can determine the one-dimensional steady heat transfer equation:

$$Q_h = \frac{T_h - T_a}{R_2 + R_3}. \tag{10}$$

From thermodynamic Equations (3), (4) and (8)–(10), the TEC cooling capacity (Q_c) can be written as the following expression:

$$Q_c(T_1, T_a, I) = \frac{[S_m{}^2 I^2 (R_2 + R_3) - S_m I - K_m] T_1 + K_m T_a + \frac{1}{2} I^2 R_m [(2K_m - S_m I)(R_2 + R_3) + 1]}{[S_m I (R_2 + R_3) - 1](S_m I R_1 + 1) - K_m (R_1 + R_2 + R_3)}. \tag{11}$$

From Equation (11), it can be deduced that Q_c mainly depends on the TEM performance parameters: S_m, R_m, K_m, TEC drive current I, the system thermal resistances (R_1, R_2 and R_3), the upper surface temperature of the heat sink (T_1) and the ambient environment temperature (T_a). If it is assumed that T_1 is 293.15 K at which the pump laser module could work steadily and normally, I and T_a are set to regular values, and S_m, R_m and K_m are calculated according to the TEM parameters of I_{max}, ΔT_{max}, T_h and V_{max} (which are given by the manufacturer), then we just need to know the system's thermal resistances (R_1, R_2 and R_3) to calculate Q_c. If Q_c multiplied by 16 is greater than Q_{pump}, we can make the conclusion that the pump laser module can work at a suitable and stable temperature. The temperature of heat sink T_1 can also be written as the following expression:

$$T_1(Q_c, T_a, I) = \frac{\{[S_m I(R_2+R_3)-1](S_m I R_1+1) - K_m(R_1+R_2+R_3)\}Q_c - K_m T_a - \frac{1}{2}I^2 R_m[(2K_m - S_m I)(R_2+R_3)+1]}{S_m{}^2 I^2 (R_2+R_3) - S_m I - K_m}. \tag{12}$$

If assuming Equation (8) is established and Q_{pump} is at its maximum value, then we can obtain Q_c as a fixed constant, setting I and T_a at regular values, and calculate S_m, R_m and K_m, allowing the temperature (T_1) to be calculated out by the expression. Finally, it can be concluded whether the pump laser module is working at a suitable temperature by comparing T_1 with the normal operating temperature range.

3.3. Numerical Simulation

A finite-element model was established using COMSOL 4.4 software to estimate the cooling capacity of the thermal control system, and the thermal transfer process was mainly analyzed to simulate the temperature rise and the thermal distribution in the laser cabinet. The heat dissipation of the mid-infrared OPO laser is a multi-physics field coupled process with heat conduction, heat convection and gas turbulent flow.

The material properties and size of the finite-element model are shown in Table 2. The whole laser cabinet structure including fins was made by aluminum material. The aluminum 6063-T83 with thermal conductivity 201 W/(m·K) was defined as the cabinet material in the software. When the thermal control system works, the cold side of the TEMs absorbs the heat produced by the pump laser module, and releases heat at the hot side simultaneously. This heat quantity was taken as the heat source in the model and delivered through thermal conduction to the laser cabinet. A 500 W boundary heat source was defined to simulate the quantity of heat produced by the TEMs. There were 14 cooling fins at the bottom of the laser cabinet. The shape of each fin was $300 \times 50 \times 2$ mm^3, and the fin spacing was 20 mm. Seven fans were defined in the software, and the flow rate was set at 0.011 m^3/s. The thermal conductivity of the heat pipe was set in segmented mode to make a closer equivalent simulation of its physical features. The thermal conductivity of the evaporation section was set at 2×10^4 W/(m·K), while the thermal conductivity of the condensation section was set at 500 W/(m·K).

Table 2. Material properties and sizes in the finite-element model.

	Material	Thermal Conductivity (W/m·K)	Size	Number	Flow Rate (m³/s)
Cabinet	Aluminum 6063-T83	201	330 × 330 × 115 mm³ (L × W × H) Wall thickness: 10mm	/	/
Fin	Aluminum 6063-T83	201	300 × 50 mm² (L × W) Fin thickness: 2mm Fin spacing: 20mm	14	/
Heat pipe	/	Evaporation section: 2×10^4 Condensation section: 500	280 × 12 × 5 mm³ (L × W × H)	2	/
Fan	/	/	40 × 40 × 10 mm³	6	0.011

The finite-element mesh model is shown in Figure 8 and the size unit of this model is in millimeters, the results for the temperature analysis of the model at ambient environment temperatures of 25 degrees and 45 degrees are shown in Figure 9, and the temperature units are degrees. It can be deduced that the highest temperature rise occurred at the left edge of the laser cabinet and with almost the same value at the different ambient temperatures. It is assumed that the two groups of TEMs were close to the left edge of the cabinet with the highest temperature at the hot side, since the simulated heat source was 500 W and the temperature rise was almost 20 K, thus the thermal resistance (R_3) for the TEMs was 0.04 K/W. The other two groups of TEMs in the middle of the cabinet almost had a 15 K temperature rise at the hot side and the thermal resistance (R_3) for them was 0.03 K/W.

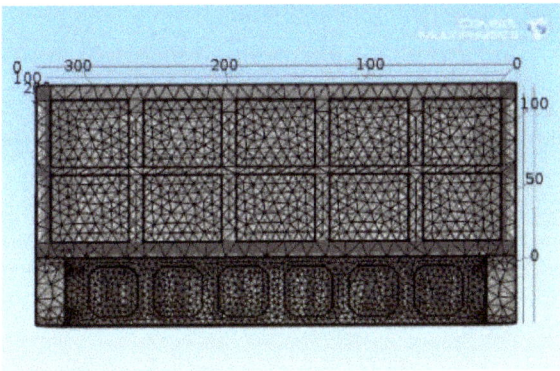

Figure 8. Finite-element mesh model of the system.

The cooling capacity of the system is discussed in two situations—at 25 °C (room temperature) and 45 °C (high temperature). From the temperature rising simulation results in Figure 9, we approximately calculated the TEC performance parameters, R_m, K_m and S_m, for $T_h = 50$ °C and the following parameters were obtained: $R_m = 0.985$ Ω, $K_m = 0.946$ K/W and $S_m = 0.048$ V/K. Using Equation (11), in the first situation, $T_a = 298.15$ K (25 °C), the TEC drive current (I) was defined as 5A, and it was assumed that $R_1 = R_2 = 0.1$ K/W and $T_1 = 293.15$ K (20 °C) at which the pump source can work stably. For the two groups of TEMs with $R_3 = 0.04$ K/W, the cooling output of each TEM was $Q_{c1} = 39.5$ W, and for the other two groups of TEMs with $R_3 = 0.03$ K/W, the cooling output of each TEM was $Q_{c2} = 40$ W. Consequently, the total cooling output (Q_c) of the 16 TEMs was about 636W.

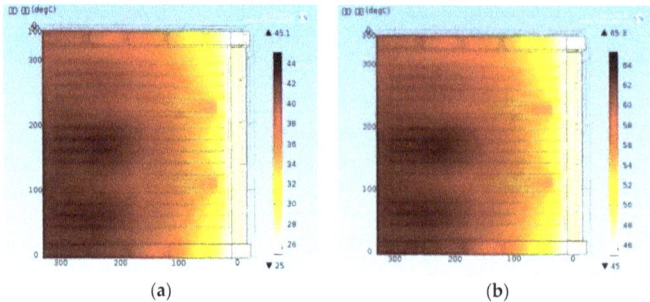

Figure 9. Finite-element analysis results: (**a**) in 25 degrees and (**b**) in 45 degrees.

When T_a = 318.15 K (45 °C) with the same other parameters, the two groups of TEMs with R_3 = 0.04 K/W had Q_{c1} = 24 W, and the other two groups of TEMs with R_3 = 0.03 K/W had Q_{c2} = 24.4 W. Thus, the total cooling output (Q_c) of the 16 TEMs was about 387.2W. The calculated results indicate that the temperature of the LD modules and the fiber in the pump source can be controlled at the setting value in the ambient temperatures of 25 °C and 45 °C. The estimated cooling capacity of the system in different ambient environments is shown in Table 3.

Table 3. The estimated cooling capacity of the system in different ambient environments.

Ambient Environment Temperature (T_a)	T_1	TEC Driving Current	Position of TEMs	R_3	TEM Cooling Output (Q_{c1})	System Cooling Capacity (Q_c)
298.15 K (25 °C)	293.15 K (20 °C)	5 A	TEMs near the edge TEMs in the middle	0.04 K/W 0.03 K/W	39.5 W 40 W	636 W
318.15 K (45 °C)	293.15 K (20 °C)	5 A	TEMs near the edge TEMs in the middle	0.04 K/W 0.03 K/W	24 K/W 24.4 W	387.2 W

4. Experiment Results and Discussion

Experiments in room and high temperature environments were conducted to verify the actual cooling capacity of the thermal control system and the performance of the mid-infrared OPO laser. The output power of OPO laser was measured using the laser power meter in 25 °C and 45 °C temperature environments. The laser controlling temperature was set to 20 °C. The near-field pattern of the laser was analyzed through the laser beam quality analyzer (PY3) in room temperature. The picture of the mid-infrared OPO laser is shown in Figure 10. The black case on the left side is the electrical control unit, and the case with fans on the right side is the optical unit.

Figure 10. Picture of the mid-infrared OPO laser.

In the room temperature environment, when the electrical control unit was powered on, the thermal control system began to control the refrigeration of the TEMs to reduce the temperature of the laser. About 4 min later, the upper surface temperature of heat sink was reduced and stabilized at about 20.1 °C. Then the pump source was turned on and the laser began to output light. Since the pump source generated heat and transferred it to the heat sink, the temperature rose to about 20.6 °C. The temperature control circuit acquired rose in temperature and calculated the PWM control variable based on the PID algorithm. Then the output voltage exerted on TEMs was increased appropriately and the cooling output was amplified. About 1 min later, the temperature was reduced and stabilized at about 20.1 °C. The temperature variation of the pump laser over time is shown in Figure 11.

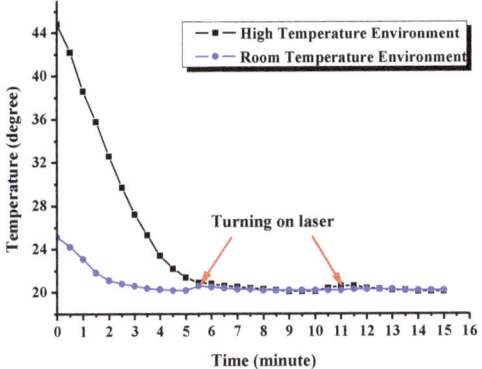

Figure 11. Temperature variation of the pump laser.

In the high temperature experiment, the mid-infrared OPO laser was put in the high and low temperature test boxes, as shown in Figure 12, and the environment temperature was set to 45 °C. About 10 min after the electrical control unit was powered on, the temperature was reduced and stabilized at about 20.1 °C. Then, the pump source was turned on and the laser began to output light, and the temperature rose to about 20.7 °C. About 1 min later, the temperature was reduced and stabilized at about 20.1 °C, as shown in Figure 11.

Figure 12. Experiment in the high temperature environment.

The output power variation of the mid-infrared OPO laser with 70% maximum pump power in room and high temperature environments is shown in Figure 13. The near-field pattern of the OPO laser at room temperature was also measured and is shown in Figure 13. The following formula was used to calculate the output power stability (S) of the OPO laser:

$$S = \pm \frac{P_{max} - P_{min}}{2 \times P_{avr}} \quad (13)$$

P_{max}, P_{min} and P_{avr} are the maximum, minimum and average output powers of the OPO laser with the measured values. The calculated output power stability values of the laser in room and high temperature environment were ±0.7% and ±1%, respectively. The one-hour output power stability of the laser in room temperature environment was within ±1%. The experimental results above show that the thermal control system has a strong cooling capacity to keep the substrate temperature of the pump laser module at the setting value. The laser was able to work normally in both the room and high temperature environments, and when pump power of the laser changed, the temperature control circuit was able to adjust the output drive voltage on TEMs according to the measured temperature value, keeping the temperature of the pump laser at a stable and suitable value.

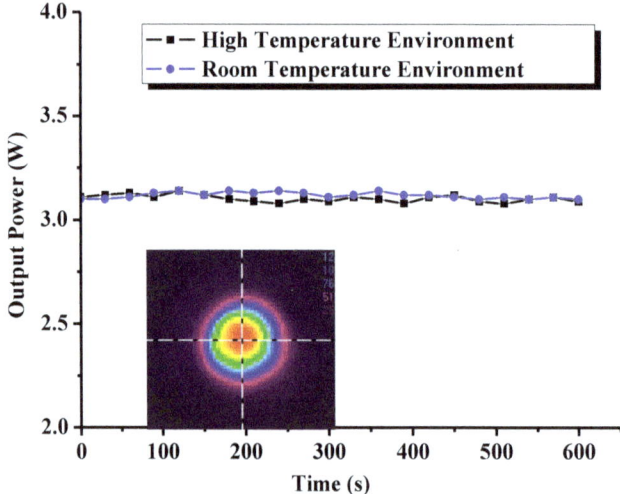

Figure 13. Output power variation and near-field pattern of the OPO laser.

5. Conclusions

In this paper, a tunable all-fiber laser pumped OPO was established, and a thermal control system based on TEMs, automatic temperature control circuit, PID algorithm, cooling fins, fans and heat pipes was integrated and designed for the laser. This system is compact, light, air-cooling and has automatic temperature control. A finite-element model of the cooling system was built and simulated in the COMSOL 4.4 software to analyze the thermal transfer process. A mathematical model and method were established to estimate the cooling capacity of the system under different ambient environments. The estimated maximum cooling capacity of this system was approximately 636 W in 25 °C and 387 W in 45 °C. The substrate temperature of the pump laser module was able to be maintained at a stable value with a steady state error of no more than 0.2 °C in room and high temperatures when the OPO laser was working. The output power stability of the OPO laser in room and high temperatures were ±0.7% and ±1% respectively. The experimental results indicate that this compact air-cooling thermal control system can effectively solve the heat dissipation problem of mid-infrared solid-state lasers with a one hundred watts level pump module in room and high temperature environments.

Author Contributions: F.C., Y.H. and D.Y. conceived, designed and performed the experiment; Y.H. and Q.P. contributed the all-fiber laser pump module design and the OPO module design, K.Z. established the finite-element model of the cooling system, D.Y. contributed to the thermal control system design and wrote the paper, F.C. and L.G. supervised the whole work.

Acknowledgments: The authors acknowledge the financial support of the Open Fund Project of the State Key Laboratory of Laser and Material Interaction (No.SKLLIM1713), the National Natural Science Foundation of China (61705219), the National Defense Science and Technology Innovation Fund of the Chinese Academy of Sciences (No. CXJJ-16M228), the Major Science and Technology Biding Project of Jilin Province (No. 20160203016GX), the Young and Middle-Aged Science and Technology Innovation Leader and Team Project of Jilin Province (No. 20170519012JH), and the Youth Innovation Promotion Association of CAS.

Conflicts of Interest: The authors declare no conflict of interest.

References

1. Jin, Y.W.; Cristescu, S.M.; Harren, F.J.; Mandon, J. Two-crystal mid-infrared optical parametric oscillator for absorption and dispersion dual-comb spectroscopy. *Opt. Lett.* **2014**, *39*, 3270–3273. [CrossRef] [PubMed]
2. Liu, J.; Tang, P.H.; Chen, Y.; Zhao, C.J.; Shen, D.Y.; Wen, S.C.; Fan, D.Y. Highly efficient tunable mid-infrared optical parametric oscillator pumped by a wavelength locked, Q-switched Er:YAG laser. *Opt. Express.* **2015**, *23*, 20812–20819. [CrossRef] [PubMed]
3. Steinle, T.; Kumar, V.; Floess, M.; Steinmann, A.; Marangoni, M.; Koch, C.; Wege, C.; Gerullo, G.; Giessen, H. Synchronization-free all-solid-state laser system for stimulated Raman scattering microscopy. *Light Sci. Appl.* **2016**, *5*, e16149. [CrossRef]
4. Yan, M.; Luo, P.L.; Iwakuni, K.; Millot, G.; Hansch, T.W.; Picque, N. Mid-infrared dual-comb spectroscopy with electro-optic modulators. *Light Sci. Appl.* **2017**, *6*, e17076. [CrossRef]
5. Darvish, G.; Moravvej-Farshi, M.K.; Zarifkar, A.; Saghafi, K. Pre-compensation techniques to suppress the thermally induced wavelength drift in tunable DBR lasers. *IEEE J. Quantum Electron.* **2008**, *44*, 958–965. [CrossRef]
6. Andreoni, E.; Xu, J.H.; Cartaleva, S.; Celli, R.M.; Mango, F.; Gozzini, S. A simple system of thermal control and frequency stabilization of solitary diode lasers. *Rev. Sci. Instrum.* **2000**, *71*, 3648–3652. [CrossRef]
7. Datta, M.; Choi, H.W. Microheat exchanger for cooling high power laser diodes. *Appl. Therm. Eng.* **2015**, *90*, 266–273. [CrossRef]
8. Shu, S.L.; Hou, G.Y.; Wang, L.J.; Tian, S.C.; Vassiliev, L.L.; Tong, C.Z. Heat dissipation in high-power semiconductor lasers with heat pipe cooling system. *J. Mech. Sci. Technol.* **2017**, *31*, 2607–2612. [CrossRef]
9. Jia, G.N.; Qiu, Y.T.; Yan, A.R.; Yao, S.; Wang, Z.Y. Laser three-dimensional printing microchannel heat sink for high-power diode laser array. *Opt. Eng.* **2016**, *55*, 096105. [CrossRef]
10. Mu, J.; Feng, G.J.; Yang, H.M.; Zhang, H.; Zhou, S.H. Research on multiannular channel liquid cooling method for thin disk laser. *Opt. Eng.* **2014**, *53*, 056110. [CrossRef]
11. Zhang, W.; Shen, L.M.; Yang, Y.X.; Chen, H.X. Thermal management for a micro semiconductor laser based on thermoelectric cooling. *Appl. Therm. Eng.* **2015**, *90*, 664–673. [CrossRef]
12. Zhao, D.L.; Tan, G. A review of thermoelectric cooling: Materials, modeling and applications. *Appl. Therm. Eng.* **2014**, *66*, 15–24. [CrossRef]
13. Li, C.; Jiao, D.; Jia, J.Z.; Guo, F.; Wang, J. Thermoelectric cooling for power electronics circuits: Modeling and active temperature control. *IEEE Trans. Ind. Appl.* **2014**, *50*, 3995–4005. [CrossRef]
14. Li, J.H.; Zhang, X.R.; Zhou, C.; Zheng, J.G.; Ge, D.S.; Zhu, W.H. New application of an automated system for high-power LEDs. *IEEE/ASME Trans. Mechatron.* **2016**, *21*, 1035–1042. [CrossRef]
15. Noguchi, N.; Okuchi, T. A peltiercooling diamond anvil cell for low-temperature Raman spectroscopic measurements. *Rev. Sci. Instrum.* **2016**, *87*, 125107. [CrossRef] [PubMed]
16. Dong, J.; Liu, X.S.; Peng, C.; Liu, Y.Q.; Wang, Z.Y. High Power Diode-Side-Pumped Q-Switch Nd:YAG Solid-State Laser with a Thermoelectric Cooler. *Appl. Sci.* **2015**, *5*, 1837–1845. [CrossRef]
17. Wang, H.; Yu, Y.L. Dynamic modeling of PID temperature controller in a tunable laser module and wavelength transients of the controlled laser. *IEEE J. Quantum Electron.* **2012**, *48*, 1424–1431. [CrossRef]

18. Shen, L.M.; Chen, H.X.; Xiao, F.; Yang, Y.X.; Wang, S.W. The step-change cooling performance of miniature thermoelectric module for pulse laser. *Energy Convers. Manag.* **2014**, *80*, 39–45. [CrossRef]
19. Lineykin, S.; Ben-Yaakov, S. Analysis of thermoelectric coolers by a spice-compatible equivalent-circuit model. *IEEE Power Electron. Lett.* **2005**, *3*, 63–66. [CrossRef]
20. Lineykin, S.; Ben-Yaakov, S. Modeling and analysis of thermoelectric modules. *IEEE Trans. Ind. Appl.* **2007**, *43*, 505–512. [CrossRef]

 © 2018 by the authors. Licensee MDPI, Basel, Switzerland. This article is an open access article distributed under the terms and conditions of the Creative Commons Attribution (CC BY) license (http://creativecommons.org/licenses/by/4.0/).

Review

Review of Recent Advances in QEPAS-Based Trace Gas Sensing

Yufei Ma

National Key Laboratory of Science and Technology on Tunable Laser, Harbin Institute of Technology, Harbin 150001, China; mayufei@hit.edu.cn; Tel.: +86-451-86413161

Received: 5 September 2018; Accepted: 28 September 2018; Published: 4 October 2018

Abstract: Quartz-enhanced photoacoustic spectroscopy (QEPAS) is an improvement of the conventional microphone-based photoacoustic spectroscopy. In the QEPAS technique, a commercially available millimeter-sized piezoelectric element quartz tuning fork (QTF) is used as an acoustic wave transducer. With the merits of high sensitivity and selectivity, low cost, compactness, and a large dynamic range, QEPAS sensors have been applied widely in gas detection. In this review, recent developments in state-of-the-art QEPAS-based trace gas sensing technique over the past five years are summarized and discussed. The prospect of QEPAS-based gas sensing is also presented.

Keywords: quartz-enhanced photoacoustic spectroscopy; quartz tuning fork; gas sensing; detection limit; laser spectroscopy; practical applications

1. Introduction

Photoacoustic spectroscopy (PAS) is an indirect absorption spectroscopy. It is based on the photoacoustic effect, which was first discovered by Alexander Graham Bell in 1880 [1]. The principle of PAS is shown in Figure 1. When the laser output is absorbed by a gas sample, the absorbed laser energy is converted to heat energy by the non-radiative relaxation processes of gas molecules, and will subsequently result in an increase in the local temperature and pressure in the sample. If the laser is modulated, the absorption of laser energy in a gas sample leads to the generation of an acoustic wave. The intensity of the produced acoustic wave is related to the sample concentration. The signal amplitude S of PAS is expressed as in Equations (1) and (2) [2]:

$$S \sim \frac{\alpha Q P}{f} \quad (1)$$

$$\alpha = \sigma N \quad (2)$$

where α is the absorption coefficient, Q is the quality factor of spectrophone, P is the optical power of excitation source, f is the resonance frequency, σ is the absorption cross section, and N is the gas molecular concentration. From Equation (1), it can be seen that the signal amplitude of PAS sensor is inversely proportional to the resonance frequency of the photo-acoustic cell. However, if the resonance frequency is too low, it makes the PAS sensor more sensitive to $1/f$ noise, environmental noise, and sample gas flow noise. Finally, it will result in a low signal-to-noise ratio (SNR) for the PAS sensor [3].

Unlike an optical detector used in direct absorption spectroscopy, in traditional PAS, a sensitive microphone is employed to detect the acoustic wave. PAS has the advantages of low cost and signal enhancement on increasing the laser excitation power. However, for the microphone-based PAS, the low value of Q factor (<100) and the large size of the photoacoustic cell limit its performance and actual applications [4–8]. Quartz-enhanced photoacoustic spectroscopy (QEPAS) is a modification of the conventional microphone-based PAS, which was first reported in 2002 [9].

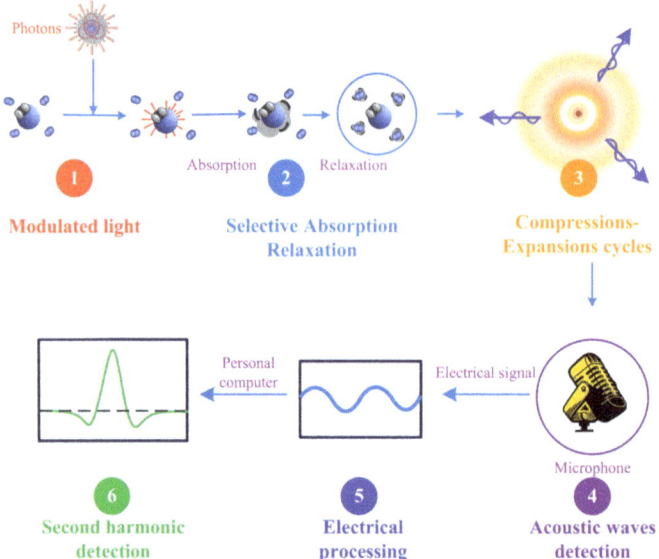

Figure 1. The principle of the PAS technique.

In the QEPAS technique, a quartz tuning fork (QTF) is used as an acoustic wave transducer. The QTF is a commercially available millimeter-sized piezoelectric element, which is usually used in watches, clocks, and electronic circuits to provide clock rate [10–12]. The high Q-factor (~100,000 in a vacuum and ~10,000 in a standard atmosphere pressure) and narrow resonance frequency band (<1 Hz) of QTF improve the QEPAS selectivity and immunity to environmental acoustic noise [13–19]. Due to the merits of high selectivity and sensitivity, low cost, compactness, and a large dynamic range, QEPAS sensors have been widely applied in gas detection for atmospheric monitoring [20–27], chemical analysis [28–32], biomedical diagnostics [33–36], and trace gas sensing [37–44]. Different QEPAS sensor architectures were developed to meet the requirements of a large number of applications.

2. Standard QEPAS-Based Gas Sensor System

A standard QEPAS-based gas sensor system is shown in Figure 2. A diode laser or quantum cascade laser (QCL) is often used as a gas sample excitation source. For a diode laser, the emission wavelength is less than 3 µm. Usually it is a fiber-coupled output. The gas analytes flow into an acoustic detection module (ADM) with a slow flow rate to avoid flow noise. Usually the rate is smaller than 200 mL/min. In order to increase the Q factor of QTF and narrow the gas absorption line width, a pressure controller is adopted to reduce the gas pressure. The laser wavelength is modulated through employing a sinusoidal dither at frequency f ($f = f_0/2$, where f_0 is the resonance frequency of QTF). The QTF in QEPAS sensor converts the acoustic wave into an electrical signal due to the piezoelectric effect, and the resonant property of QTF enhances the signal amplitude. Commercially available QTFs with a f_0 of ~32.76 kHz are typically employed. An obvious enhancement of the QTF current signal can be obtained after adding two metallic tubes (mRs) to the QTF sensor architecture. The mRs act as micro-resonators [45]. To obtain a strong acoustic wave coupling, the length L of mR should be optimized. Typically, the optimum L is in the range of $\lambda_s/4 < L < \lambda_s/2$, where λ_s is the acoustic wave wavelength [46]. The weak current signal of QTF (in the scale of nanoampere) is amplified and converted into voltage by a transimpedance amplifier (TA) with a resistance of ~MΩ. The lock-in amplifier is used to demodulate the second harmonic component ($2f$) generated by the QTF. To improve the slow vibrational-translational (V-T) relaxation processes of the target analyte, H_2O vapor is usually added to the target trace gas. The presence of H_2O vapor in the analyzed gas increases the V-T relaxation process, which results in a higher detected amplitude of the QEPAS signal. Furthermore, the excitation laser wavelengths and absorption lines of the target analyte should avoid overlap with the metastable vibrational transitions of atmospheric nitrogen (N_2) and oxygen (O_2) to ensure efficient V-T relaxation of the target molecule, thus contributing to the QEPAS signal.

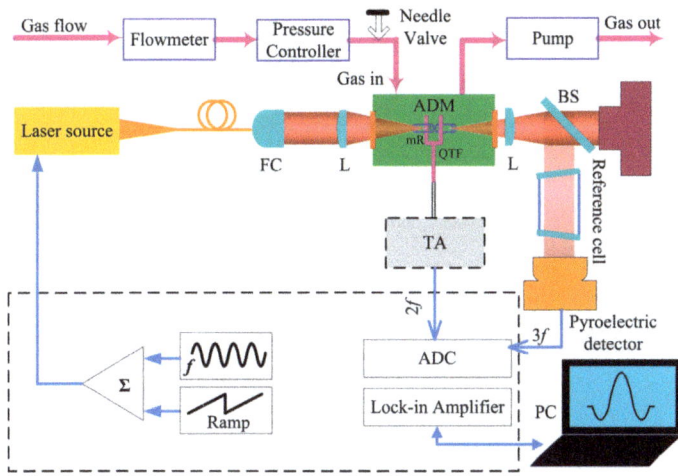

Figure 2. Schematic of the QEPAS sensor. TA: transimpedance amplifier; mR: micro-resonator; ADM: acoustic detection module; L: plano-convex lens; FC: fiber collimator.

3. QEPAS Sensor with High-Power Excitation Source

When a laser source with a higher optical power is used in the QEPAS sensor, more molecules are excited, which results in a stronger acoustic wave. Therefore, different from tunable diode laser absorption spectroscopy (TDLAS) and other laser absorption spectroscopy, a distinct merit of the QEPAS sensor is that its performance can be improved when the excitation laser power is increased. As shown in Equation (1), QEPAS detection sensitivity is proportional to the excitation laser power P.

3.1. QEPAS Sensor Based on an EDFA-Amplified Diode Laser

Single-mode diode lasers with a butterfly package or transistor outline (TO) package are usually used in QEPAS sensor system due to their low cost and compactness [47–49]. However, the output power of the diode laser is low, typically in the range of several milliwatts (<50 mW), which significantly limits the QEPAS sensor sensitivity. An optical fiber amplifier is usually used for optical signal amplification. With the advantage of high-power output (which easily reaches watt level), it is used extensively in optical communications. In a commercially available optical fiber amplifier, an optical fiber with a single-mode structure (and doped with rare-earth ions such as erbium and ytterbium) is used as the active medium. The fiber is pumped by diode lasers. The optical fiber amplifier usually used is an erbium-doped fiber amplifier (EDFA). EDFA has many obvious advantages, such as fiber compatibility, low noise, high gain, and polarization independence [50,51]. When an appropriate seed diode laser is injected, an EDFA can be used to obtain an amplification gain of more than 30 decibel (dB). Three operating wavelength bands (L band: 1565–1610 nm, C band: 1520–1570 nm, and S band: 1450–1550 nm) of EDFA are commercially available.

Recently, a QEPAS sensor system based on an EDFA-amplified diode laser was successfully used in acetylene (C_2H_2), hydrogen sulfide (H_2S), and ammonia (NH_3) detection [52–54]. In [52], a near-infrared, continuous-wave diode laser with emitting power of 6.7 mW served as the seed laser. The output of this laser was sent to an EDFA for power amplification. The maximum output power of the EDFA was 1500 mW. The emission spectra for the seed laser and the EDFA-amplified diode laser are shown in Figure 3. The amplified diode laser with high output power of 1500 mW had the same signal-to-noise ratio as the seed laser (~30 dB). Due to the fact that with high power intensity it is easy to attain high gain, compared to the seed laser, the line width of the amplified diode laser was compressed (see Figure 3a,b). Narrower line width represents better spectral resolution and is beneficial for improving the sensor's selectivity.

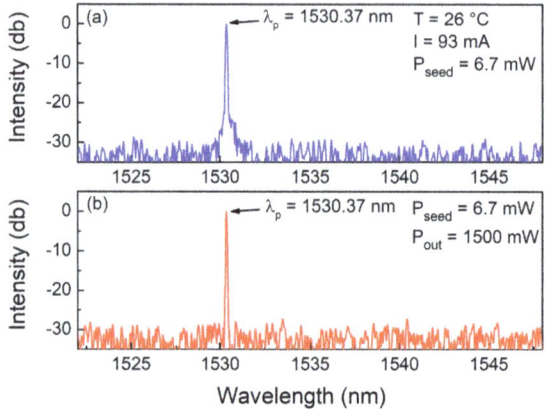

Figure 3. Emission spectra for a seed laser and an EDFA-amplified diode laser.

The schematic of an EDFA-based QEPAS sensor is shown in Figure 4. After being amplified by an EDFA, the laser beam of a diode laser was injected into an acoustic detection module (ADM). For C_2H_2, H_2S, and NH_3 sensing, the detection limits of EDFA-QEPAS sensors were 33.2 ppb (parts per billion by volume) [52], 142 ppb [53], and 418.4 ppb [54], respectively. No signal saturation was observed in these investigations, which means that an EDFA with higher output power can be adopted to further improve the QEPAS sensor performance.

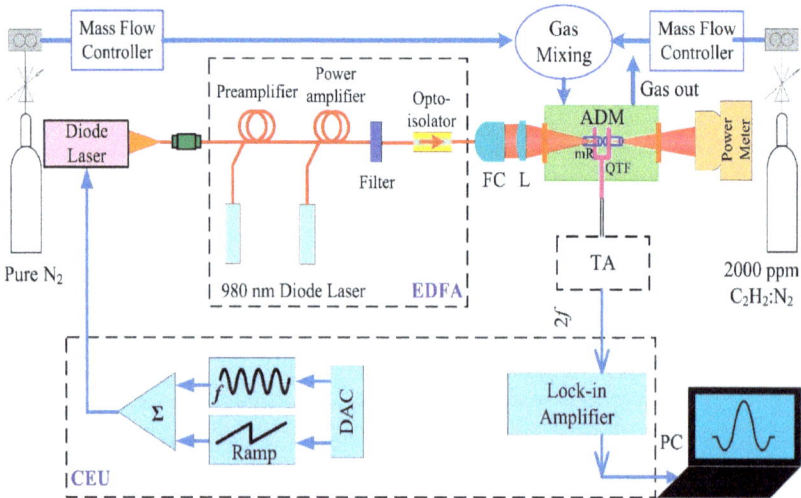

Figure 4. EDFA-based QEPAS sensor system.

3.2. Intra-Cavity QEPAS Sensor

An intra-cavity QEPAS (I-QEPAS) sensor was first reported by Vincenzo Spagnolo [55] and was further developed for sensitive carbon dioxide (CO_2) and nitric oxide (NO) detection [56–58]. Due to the output coupling loss, the laser power intensity in the intra-cavity is much higher than outside. In [55], a bow-tie type optical resonator with a high finesse (>1500) was developed, which is shown in Figure 5. A distributed feedback quantum cascade laser (QCL) emitting at 4.33 µm was used as the excitation source for carbon dioxide (CO_2) detection. Compare to the incident laser power, the intra-cavity optical power was boosted by a factor of ~250. Finally, a ppt (parts per trillion) level detection limit and a normalized noise equivalent absorption (NNEA) of 3.2×10^{-10} cm^{-1} $W/\sqrt{Hz}$ were achieved.

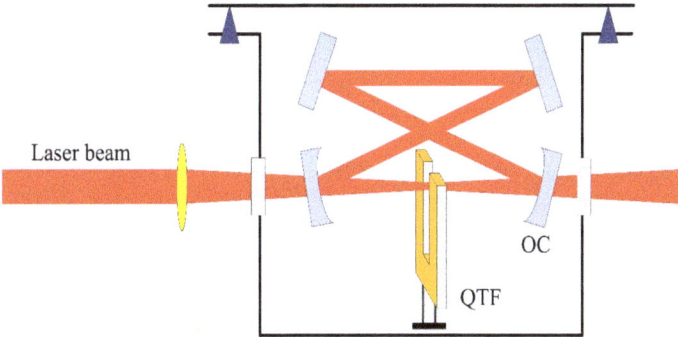

Figure 5. Schematic of the optical resonator.

An I-QEPAS sensor based on fiber-ring laser was reported by Ren [59]. The schematic of the experimental setup is shown in Figure 6. An erbium-doped fiber laser with a ring cavity structure was adopted. A fiber coupler #1 with a ratio of 1:9 was used to ensure that the laser power intensity within the cavity was much higher than outside. Hence, it would make maximum use of the laser power for the I-QEPAS. A fiber Bragg grating (FBG) was used to modulate the laser wavelength. C_2H_2 detection at the wavelength of 1531.6 nm was performed to evaluate the fiber-ring laser based I-QEPAS sensor performance. A minimum detection limit of C_2H_2 of 29 ppb at 300 s integration time was achieved when the optimized laser modulation depth and gas pressure were selected. The linear dynamic range of the fiber-ring laser-based I-QEPAS sensor was on the order of 10^5.

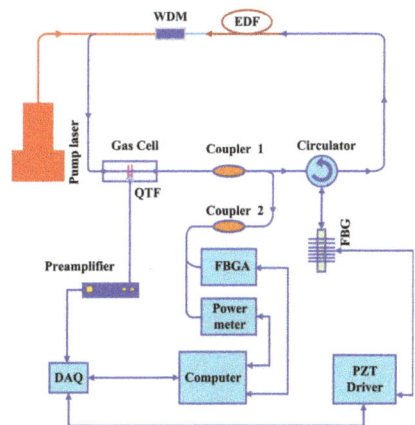

Figure 6. Schematic of the I-QEPAS sensor based on fiber-ring laser.

4. QEPAS Sensor Based on Custom QTF

A commercially available, low-cost QTF (<$1) with a resonance frequency f_0 of ~32.76 kHz is usually used in the QEPAS sensor system [60–62]. Typically, the dimension of the two prongs of the QTF is 3.6 mm in length, 0.6 mm in width, and 0.36 mm in thickness, and the gap between the two prongs is 300 µm [63]. As shown in Equation (1), the signal amplitude of the QEPAS sensor is inversely proportional to the QTF resonance frequency. A QTF with a smaller f_0 will result in a longer integration time, which is an advantage for increasing the QEPAS sensor signal level [64–66].

4.1. QEPAS Sensor Employing a Custom QTF with a Large Prong Gap

When excitation sources with bad beam quality such as light-emitting diodes (LEDs), vertical cavity surface emitting lasers (VCSELs), and terahertz (Thz) QCLs are used in QEPAS sensors, the large beam diameter and divergence angle make the light beam unable to pass through the micro-resonators and the 300 µm gap between the two prongs of usually used standard QTF totally, resulting in obvious optical noise. Custom QTFs with a large prong gap can solve this issue [67–69]. The first custom-made QTF employed in a QEPAS sensor combining a 3.93 THz QCL was reported by Spagnolo [70]. The gap between the two QTF prongs was 1 mm. The fundamental resonance frequency and Q factor at atmospheric pressure of this custom QTF were 4246 Hz and 9930, respectively. The comparison of parameters of geometries for a standard QTF and the custom QTF is shown in Table 1. Methanol detection was performed to verify the performance of this THz-QEPAS sensor with custom QTF. The obtained detection limit was 7 ppm at 4 s integration time and the corresponding NNEA was 2×10^{-10} cm^{-1}W/$\sqrt{\text{Hz}}$.

Table 1. Parameters for standard and custom QTFs.

QTF	f_0 (kHz)	Length (mm)	Width (mm)	Thickness (mm)	Gap (mm)
Standard	32.768	3.6	0.6	0.36	0.3
Custom	4.246	20	1.4	0.8	1

4.2. QEPAS Sensor Employing Overtone Flexural Mode of Custom QTF

If the fundamental resonance frequency of custom QTF is small (a few kHz), the resonance frequency of the first overtone flexural mode should not be too high. Therefore, the first overtone flexural mode of QTF can be used in a QEPAS sensor [71–74]. The schematic of the fundamental flexural mode and the first overtone flexural mode of QTF is shown in Figure 7. In [71], a custom QTF with fundamental mode resonance ~2.87 kHz was adopted and the first overtone at ~17.7 kHz was utilized. The measured Q factor for the first overtone resonance was 31,373.81, which is ~2.6 times higher than when it operated in fundamental flexural resonance mode (12,098.97). A high Q factor is beneficial to improve the QEPAS signal level. To evaluate the sensor performance, a near infrared (IR) diode laser emitting at 1.37 µm and water vapor (H_2O) as the target gas were selected. The comparison of the first overtone mode and the fundamental mode was carried out at the same conditions, in terms of a 1.7% water concentration, a pressure of 75 Torr, an optimized laser wavelength modulation depth, and laser beam focusing point. The normalized peak value of measured second harmonics for the first overtone mode is ~5.3 times higher than that achieved using the fundamental mode.

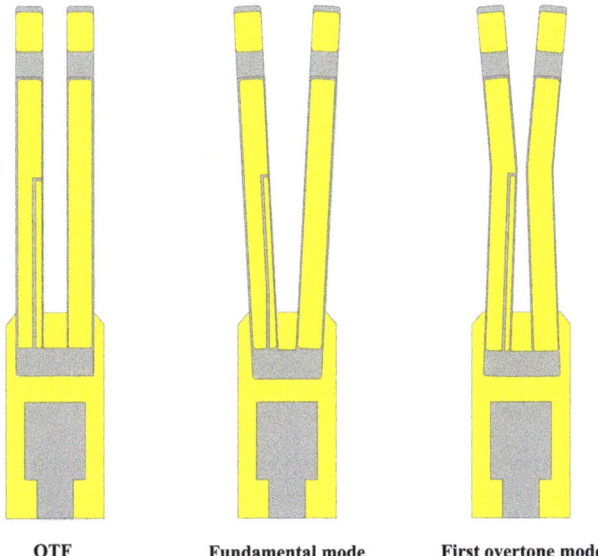

Figure 7. Schematic of the custom QTF, fundamental flexural mode, and the first overtone flexural mode.

4.3. QEPAS Sensor Employing a Custom Small-Gap QTF

In the QEPAS technique, the acoustic wave generated by the reaction between gas molecules and laser at the gap of the two prongs can be regarded as spherical wave oscillation. The spherical wave propagation decreases with the cube of the distance, which means that the energy of the acoustic wave will attenuate rapidly as the distance from the generation point of the acoustic wave increases. Based on the above analysis, employing a QTF with a small gap will increase the strength of the acoustic wave acting on the prongs of QTF. Therefore, the signal level of QEPAS would be enhanced. The schematic of the laser beam, a QTF, and the generated acoustic wave is shown in Figure 8. The first custom QTF with a small gap of 200 µm used in the QEPAS sensor was reported by Ma [75]. Using the COMSOL software, a finite element modeling (FEM) was constructed, and the displacement of QTF with different micro-resonators (mR) was calculated theoretically. For comparison, a standard QTF with a 300-µm gap was also investigated. The calculated results are shown in Figure 9. An experimental evaluation was carried out by employing a distributed feedback diode laser with emitting wavelength of 1.39 µm and choosing water vapor (H_2O) as the target analyte. The detailed parameters of QTF and experimental results are displayed in Table 2. It can be seen that using a custom QTF with a small gap of 200 µm improved the QEPAS sensor performance significantly. This improvement can be explained as follows. Firstly, due to a lower energy loss of propagation, the QTF with a small gap of 200 µm allowed the acoustic wave to push the fork prongs more efficiently. Secondly, a decrease of QTF mass compared to that of a standard QTF would result in bigger displacement.

Table 2. Dimensions and parameters of two QTFs.

Parameter	Unit	Small-Gap QTF	Standard QTF [63]
Length	mm	3.42	3.6
Width	mm	0.3	0.6
Thickness	mm	0.44	0.36
Gap	µm	200	300
Detection limit	ppm	1.85	5.9
NNEA	$cm^{-1}W/\sqrt{Hz}$	2.02×10^{-8}	7.73×10^{-8}

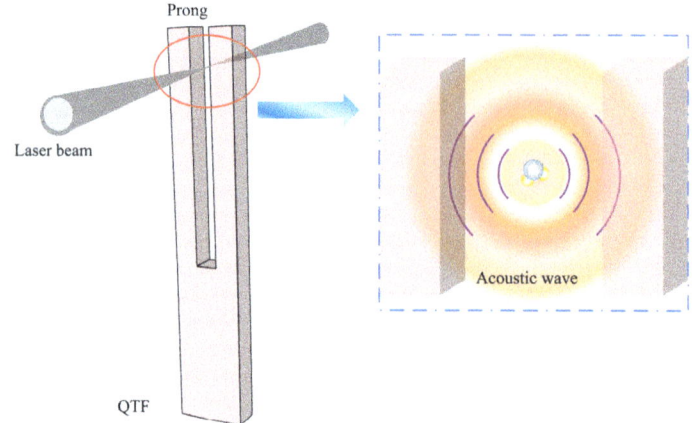

Figure 8. Schematic of the QTF, the laser beam, and the generated acoustic wave.

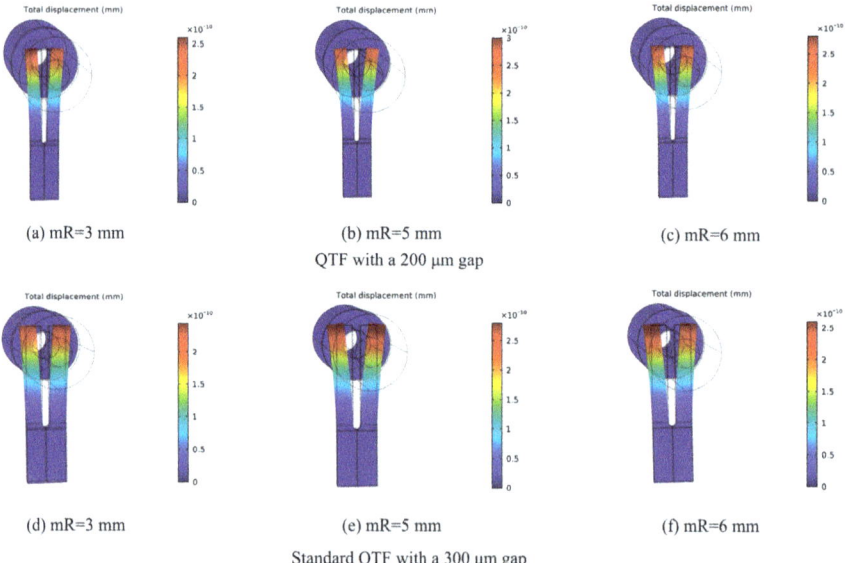

Figure 9. Calculated displacement for two different QTFs with various lengths of micro-resonators (mRs): (**a–c**) for the QTF with a 200 µm gap; (**d–f**) for the standard QTF with a 300 µm gap.

5. FEW-QEPAS Sensor for Long-Distance and Distributed Gas Detection

Usually the QEPAS sensor is applied to gas concentration measurement for one point. It is hard for it to realize distributed gas sensing. Fiber evanescent wave (FEW) technology has the advantages of anti-adverse environment, anti-electromagnetic interference, low transmission loss, flexible in operation, and on-line measurement. In 2017, Ma combined the QEPAS and FEW techniques and proposed a new FEW-QEPAS method for long-distance and distributed gas sensing [76]. Due to the fact that the power of the FEW increases with decreasing diameter of the micro-nano fiber, the diameter of the tapered fiber should be small enough (<2 µm) to get a strong FEW. Using the flame-brushing method, the tapered fibers were fabricated from a standard single-mode fiber (SMF). To evaluate the FEW sensor performance, a single mode fiber with a length of 3 km was used to deliver the laser, and three tapers were fabricated as sensing units. The measured diameters of the three tapered fibers were ~1.67 µm, 1.77 µm, and 1.12 µm, respectively. The power percentage of FEW for the three tapers was calculated by a finite element method (FEM) of COMSOL Multiphysics (COMSOL Inc., Stockholm, Sweden) and is depicted in Figure 10.

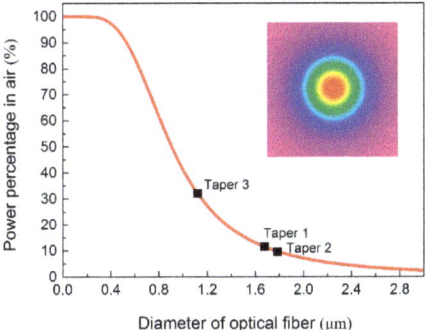

Figure 10. FEW power as a function of fiber diameter for three tapers (inset: optical field distribution for fiber taper).

The schematic of the experimental setup for FEW-QEPAS sensor is depicted in Figure 11. An experimental evaluation was carried out by employing a distributed feedback diode laser with emitting wavelength of 1.53 µm and choosing acetylene (C_2H_2) as the target analyte. The detection limit of 30 ppm, 51 ppm and 13 ppm at 1 s integration time was obtained for the three tapers, respectively. The corresponding NNEA were 3.55×10^{-6}, 3.52×10^{-6}, and 3.66×10^{-6} cm^{-1} W/Hz$^{-1/2}$, respectively. Further improvement of the performance of FEW-sensor can be achieved by using a fiber taper with a smaller diameter and lower transmission loss.

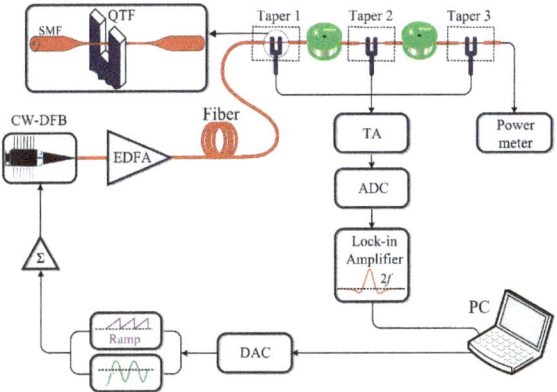

Figure 11. Schematic of the FEW-QEPAS sensor.

6. 3D-Printed QEPAS Sensor

To be useful in practical applications, a QEPAS sensor should be portable and small in terms of size, weight, and power consumption. Typically, a group of block-shaped lenses is used for laser beam propagating and transforming [77,78]. As a consequence, it results in unstable configuration and large size and therefore limits the sensor's performance and real applications. The 3D printing technique does not need traditional machine tools and multi-machining process. It holds the merits of "free manufacturing", reduction in processing time and procedure, high integration and stability, and so on. A 3D-printed QEPAS acoustic detection module (ADM) was first demonstrated by Ma [79,80].

The designed 3D model of ADM and the 3D-printed product are displayed in Figure 12. A UV-curable resin was used as the processing material for the 3D printing. The 3D-printed ADM had dimensions of 29 mm in length, 15 mm in width, and 8 mm in thickness, and was sealed by a quartz glass window. A fiber-coupled grin lens with a diameter of 1.8 mm was adopted for laser focusing. The optical components of fiber and grin lens and the photoacoustic detection components of QTF and a pair of micro-resonators (mRs) were assembled together. The total weight of the 3D-printed ADM, including all the above components, was ~5 g.

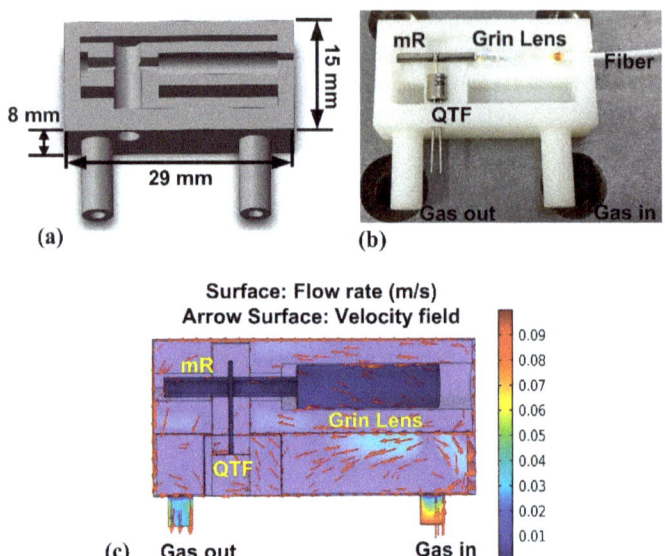

Figure 12. Schematic of the 3D-printed ADM (**a**) Designed 3D model; (**b**) 3D-printed ADM with optical and photoacoustic detection components; (**c**) the flow field within the ADM.

A portable QEPAS sensor was further developed based on the above 3D-printed ADM. The QEPAS system contains two floors, in which Floor I is for the circuitry section, including a diode laser driver, a custom lock-in amplifier, and a transimpedance amplifier. In order to facilitate its usage outside, a rechargeable battery with running time of 8 h was used as the power supply. Floor II is for the optical section, in which a fiber beam splitter (FBS) with a splitting ratio of 1:9, an integrated ADM, and a reference cell were assembled. The portable QEPAS sensor had a dimension of 250 × 250 × 50 mm^3. The weight of it was ~3.5 kg.

7. Conclusions and Future Outlook

In this review, recent developments of the QEPAS-based trace gas sensing technique in the past five years are summarized and discussed. Compared to other absorption spectroscopy such as tunable diode laser absorption spectroscopy (TDLAS), a distinct advantage of the QEPAS sensor is that its performance can be improved when the output power of the excitation laser source is increased. Based on this feature, the near-infrared light source of diode laser was power amplified by an erbium-doped fiber amplifier in the communications bands to improve the QEPAS sensor performance. Due to the output coupling loss, the laser power intensity in the intra-cavity is much higher than outside. Therefore, placing a QTF into the intra-cavity can enhance the signal level of QEPAS. When excitation sources such as LEDs, VCSELs, and terahertz (Thz) QCLs are used in QEPAS sensors, the bad beam quality of large beam diameter and divergence angle make the light beam unable to pass through the micro-resonators and the 300 μm gap between the two prongs of standard QTFs, resulting in obvious optical noise. A custom QTF with a large prong gap of ~1 mm can solve this issue well. If the fundamental resonance frequency of the custom QTF is small (a few kHz), the resonance frequency of the first overtone flexural mode should not be too high (<20 kHz). Therefore, the first overtone flexural mode of QTF can be used in the QEPAS sensor. The several-fold increase in the Q factor of the first overtone mode improved the QEPAS sensor detection sensitivity significantly. Unlike the custom QTF with a large prong gap, a custom QTF with a small gap of 200 μm with the merit of lower energy loss of propagation allowed the acoustic wave to push the fork prongs more efficiently. A decrease of QTF mass compared to a standard QTF would result in bigger displacement. Therefore, employing a QTF with a small gap of 200 μm in QEPAS is another way to improve the detection limit. Fiber evanescent wave (FEW) has the advantages of being flexible in operation and on-line measurement. By combining FEW and QEPAS techniques together, a long distance of 3 km and distributed gas sensing were achieved. The 3D printing technique has the merits of "free manufacturing", reduction in processing time and procedure, high integration, and stability. A portable, small QEPAS sensor was obtained using a 3D-printed QEPAS acoustic detection module with a weight of ~5 g. Further improvement of the QEPAS sensor detection sensitivity can be obtained by using a THz laser source with high output power in order to achieve the minimum detectable concentration at the ppb or sub-ppb level. Furthermore, the intra-cavity QEPAS sensor can be realized in the novel mid-infrared solid-state laser oscillations with the merits of high laser power intensity and excellent beam quality [81,82]. Unlike optical detectors used in other laser spectroscopy such as TDLAS, QTF is immune to the laser wavelength. This feature is especially attractive for QEPAS gas sensing in the >10 μm region, where the availability of optical detectors is limited and the strong absorption band of gas molecules exist. The overall size of a typical QEPAS sensor platform can be significantly reduced to a size that makes this device suitable for applications requiring lightweight and compact sensors such as balloon- or miniature unmanned aerial vehicle (UAV)-based atmospheric measurements or portable low-cost sensors for first responders, such as in mining accidents or forest fires near urban centers. The QEPAS-based sensor has a wide range of applications in environmental monitoring, atmospheric chemistry, industrial chemical analysis, and medical and biomedical diagnostics, as well as in law enforcement.

Funding: This research was funded by the National Natural Science Foundation of China (Grant No. 61505041 and 61875047), the Natural Science Foundation of Heilongjiang Province of China (Grant No. F2015011), the Fundamental Research Funds for the Central Universities, and the Application Technology Research and the Development Projects of Harbin (No. 2016RAQXJ140).

Conflicts of Interest: The author declares no conflict of interest.

References

1. Bell, A.G. On the production and reproduction of sound by light: The photophone. *Am. J. Sci.* **1880**, *20*, 305–324. [CrossRef]
2. Kosterev, A.A.; Tittel, F.K.; Serebryakov, D.V.; Malinovsky, A.L.; Morozov, I.V. Applications of quartz tuning forks in spectroscopic gas sensing. *Rev. Sci. Instrum.* **2005**, *76*, 043105. [CrossRef]
3. Elia, A.; Lugarà, P.M.; Franco, C.D.; Spagnolo, V. Photoacoustic techniques for trace gas sensing based on semiconductor laser sources. *Sensors* **2009**, *9*, 9616–9628. [CrossRef] [PubMed]
4. Manninen, A.; Sand, J.; Saarela, J.; Sorvajärvi, T.; Toivonen, J.; Hernberg, R. Electromechanical film as a photoacoustic transducer. *Opt. Express* **2009**, *17*, 16994–16999. [CrossRef] [PubMed]
5. Bozóki, Z.; Pogany, A.; Szabo, G. Photoacoustic instruments for practical applications: Present, potentials, and future challenges. *Appl. Spectrosc. Rev.* **2011**, *46*, 1–37. [CrossRef]
6. Elia, A.; Franco, C.D.; Lugarà, P.M.; Scamarcio, G. Photoacoustic spectroscopy with quantum cascade lasers for trace gas detection. *Sensors* **2006**, *6*, 1411–1419. [CrossRef]
7. Schmohl, A.; Miklós, A.; Hess, P. Detection of ammonia by photoacoustic spectroscopy with semiconductor lasers. *Appl. Opt.* **2002**, *41*, 1815–1823. [CrossRef] [PubMed]

8. Zheng, H.; Lou, M.; Dong, L.; Wu, H.; Ye, W.; Yin, X.; Kim, C.S.; Kim, M.; Bewley, W.W.; Merritt, C.D.; et al. Compact photoacoustic module for methane detection incorporating interband cascade light emitting device. *Opt. Express* **2017**, *25*, 16761–16770. [CrossRef] [PubMed]
9. Kosterev, A.A.; Bakhirkin, Y.A.; Curl, R.F.; Tittel, F.K. Quartz-enhanced photoacoustic spectroscopy. *Opt. Lett.* **2002**, *27*, 1902–1904. [CrossRef] [PubMed]
10. Mordmüller, M.; Köhring, M.; Schade, W.; Willer, U. An electrically and optically cooperated QEPAS device for highly integrated gas sensors. *Appl. Phys. B* **2015**, *119*, 111–118. [CrossRef]
11. Petra, N.; Zweck, J.; Kosterev, A.A.; Minkoff, S.E.; Thomazy, D. Theoretical analysis of a quartz-enhanced photoacoustic spectroscopy sensor. *Appl. Phys. B* **2009**, *94*, 673–680. [CrossRef]
12. Gong, P.; Xie, L.; Qi, X.Q.; Wang, R. A QEPAS-based central wavelength stabilized diode laser for gas sensing. *IEEE Photonics Technol. Lett.* **2015**, *27*, 545–548. [CrossRef]
13. Liu, Y.N.; Chang, J.; Lian, J.; Liu, Z.J.; Wang, Q.; Qin, Z.G. Quartz-enhanced photoacoustic spectroscopy with right-angle prism. *Sensors* **2016**, *16*, 214. [CrossRef] [PubMed]
14. Kosterev, A.A.; Bakhirkin, Y.A.; Tittel, F.K. Ultrasensitive gas detection by quartz-enhanced photoacoustic spectroscopy in the fundamental molecular absorption bands region. *Appl. Phys. B* **2005**, *80*, 133–138. [CrossRef]
15. Lassen, M.; Lamard, L.; Feng, Y.; Peremans, A.; Petersen, J.C. Off-axis quartz-enhanced photoacoustic spectroscopy using a pulsed nanosecond mid-infrared optical parametric oscillator. *Opt. Lett.* **2016**, *41*, 4118–4121. [CrossRef] [PubMed]
16. Rück, T.; Bierl, R.; Matysik, F. NO_2 trace gas monitoring in air using off-beam quartz enhancedphotoacoustic spectroscopy (QEPAS) and interference studies towards CO_2, H_2O and acoustic noise. *Sens. Actuators B* **2018**, *255*, 2462–2471. [CrossRef]
17. Dong, Y.J.; Chen, J.; Luo, L.Q.; Forsberg, E.; He, S.L.; Yan, C.S. Modeling and implementation of a fiber-based quartz-enhanced photoacoustic spectroscopy system. *Appl. Opt.* **2015**, *54*, 4202–4206. [CrossRef]
18. Starecki, T.; Wieczorek, P.Z. A high sensitivity preamplifier for quartz tuning forks in QEPAS (quartz enhanced photoacoustic spectroscopy) applications. *Sensors* **2017**, *17*, 2528. [CrossRef] [PubMed]
19. Ngai, A.K.Y.; Persijn, S.T.; Lindsay, I.D.; Kosterev, A.A.; Groß, P.; Lee, C.J.; Cristescu, S.M.; Tittel, F.K.; Boller, K.J.; Harren, F.J.M. Continuous wave optical parametric oscillator for quartz-enhanced photoacoustic trace gas sensing. *Appl. Phys. B* **2007**, *89*, 123. [CrossRef]
20. Ma, Y.F.; Lewicki, R.; Razeghi, M.; Tittel, F.K. QEPAS based ppb-level detection of CO and N_2O using a high power CW DFB-QCL. *Opt. Express* **2013**, *21*, 1008–1019. [CrossRef] [PubMed]
21. Li, Z.L.; Wang, Z.; Wang, C.; Ren, W. Optical fiber tip-based quartz-enhanced photoacoustic sensor for trace gas detection. *Appl. Phys. B* **2016**, *122*, 147. [CrossRef]
22. Yi, H.M.; Maamary, R.; Gao, X.M.; Sigrist, M.W.; Fertein, E.; Chen, W.D. Short-lived species detection of nitrous acid by external-cavity quantum cascade laser based quartz-enhanced photoacoustic absorption spectroscopy. *Appl. Phys. Lett.* **2015**, *106*, 101109. [CrossRef]
23. Dong, L.; Spagnolo, V.; Lewicki, R.; Tittel, F.K. Ppb-level detection of nitric oxide using an external cavity quantum cascade laser based QEPAS sensor. *Opt. Express* **2011**, *19*, 24037–24045. [CrossRef] [PubMed]
24. Gray, S.; Liu, A.P.; Xie, F.; Zah, C. Detection of nitric oxide in air with a 5.2 µm distributed-feedback quantum cascade laser using quartz-enhanced photoacoustic spectroscopy. *Opt. Express* **2010**, *18*, 23353–23357. [CrossRef] [PubMed]
25. Zheng, H.D.; Dong, L.; Ma, Y.; Wu, H.P.; Liu, X.L.; Yin, X.K.; Zhang, L.; Ma, W.G.; Yin, W.B.; Xiao, L.T.; et al. Scattered light modulation cancellation method for sub-ppb-level NO_2 detection in a LD-excited QEPAS system. *Opt. Express* **2016**, *24*, A752–A761. [CrossRef] [PubMed]
26. Jiang, M.; Feng, Q.L.; Wang, C.Y.; Wei, Y.F.; Liang, T.L.; Wang, X.F. Ammonia sensor based on QEPAS with HC-PBF as reference cell. *Proc. SPIE* **2015**, *9620*, 96200F.
27. Triki, M.; Nguyen, B.T.; Vicet, A. Compact sensor for methane detection in the mid infrared region based on quartz enhanced photoacoustic spectroscopy. *Infrared Phys. Technol.* **2015**, *69*, 74–80. [CrossRef]
28. Waclawek, J.P.; Moser, H.; Lendl, B. Compact quantum cascade laser based quartz-enhanced photoacoustic spectroscopy sensor system for detection of carbon disulfide. *Opt. Express* **2016**, *24*, 6559–6571. [CrossRef] [PubMed]
29. Wang, Z.; Li, Z.; Ren, W. Quartz-enhanced photoacoustic detection of ethylene using a 10.5 µm quantum cascade laser. *Opt. Express* **2016**, *24*, 4143–4154. [CrossRef] [PubMed]

30. Nguyen, B.T.; Triki, M.; Desbrosses, G.; Vicet, A. Quartz-enhanced photoacoustic spectroscopy sensor for ethylene detection with a 3.32 μm distributed feedback laser diode. *Rev. Sci. Instrum.* **2015**, *86*, 023111. [CrossRef] [PubMed]
31. Helman, M.; Moser, H.; Dudkowiak, A.; Lendl, B. Off-beam quartz-enhanced photoacoustic spectroscopy-based sensor for hydrogen sulfide trace gas detection using a mode-hop-free external cavity quantum cascade laser. *Appl. Phys. B* **2017**, *123*, 141. [CrossRef]
32. Jahjah, M.; Belahsene, S.; Nähle, L.; Fischer, M.; Koeth, J.; Rouillard, Y.; Vicet, A. Quartz enhanced photoacoustic spectroscopy with a 3.38 μm antimonide distributed feedback laser. *Opt. Lett.* **2012**, *37*, 2502–2504. [CrossRef] [PubMed]
33. Pohlkötter, A.; Köhring, M.; Willer, U.; Schade, W. Detection of molecular oxygen at low concentrations using quartz enhanced photoacoustic spectroscopy. *Sensors* **2010**, *10*, 8466–8477. [CrossRef] [PubMed]
34. Milde, T.; Hoppe, M.; Tatenguem, H.; Mordmüller, M.; Ogorman, J.; Willer, U.; Schade, W.; Sacher, J. QEPAS sensor for breath analysis: A behavior of pressure. *Appl. Opt.* **2018**, *57*, C120–C127. [CrossRef] [PubMed]
35. Lewicki, R.; Kosterev, A.A.; Thomazy, D.M.; Risby, T.H.; Solga, S.; Schwartz, T.B.; Tittel, F.K. Real time ammonia detection in exhaled human breath using a distributed feedback quantum cascade laser based sensor. *Proc. SPIE* **2011**, *10*, 709–716.
36. Köhring, M.; Böttger, S.; Willer, U.; Schade, W. LED-absorption-QEPAS sensor for biogas plants. *Sensors* **2015**, *15*, 12092–12102. [CrossRef] [PubMed]
37. Liu, K.; Zhao, W.X.; Wang, L.; Tan, T.; Wang, G.S.; Zhang, W.J.; Gao, X.M.; Chen, W.D. Quartz-enhanced photoacoustic spectroscopy of HCN from 6433 to 6613 cm^{-1}. *Opt. Commun.* **2015**, *340*, 126–130. [CrossRef]
38. Gong, P.; Xie, L.; Qi, X.Q.; Wang, R.; Wang, H.; Chang, M.C.; Yang, H.X.; Sun, F.; Li, G.P. A quartz-enhanced photoacoustic spectroscopy sensor for measurement of water vapor concentration in the air. *Chin. Phys. B* **2015**, *24*, 014206. [CrossRef]
39. Dong, L.; Lewicki, R.; Liu, K.; Buerki, P.R.; Weida, M.J.; Tittel, F.K. Ultra-sensitive carbon monoxide detection by using EC-QCL based quartz-enhanced photoacoustic spectroscopy. *Appl. Phys. B* **2012**, *107*, 275–283. [CrossRef]
40. Cao, Y.C.; Jin, W.; Ho, L.H.; Liu, Z.B. Evanescent-wave photoacoustic spectroscopy with optical micro/nano fibers. *Opt. Lett.* **2012**, *37*, 214–216. [CrossRef] [PubMed]
41. Lewicki, R.; Wysocki, G.; Kosterev, A.A.; Tittel, F.K. QEPAS based detection of broadband absorbing molecules using a widely tunable, cw quantum cascade laser at 8.4 μm. *Opt. Express* **2007**, *15*, 7357–7366. [CrossRef] [PubMed]
42. Dong, L.; Kosterev, A.A.; Thomazy, D.; Tittel, F.K. Compact portable QEPAS multi-gas sensor. *Proc. SPIE* **2011**, *7945*, 631–640.
43. Ren, W.; Jiang, W.; Sanchez, N.P.; Patimisco, P.; Spagnolo, V.; Zah, C.; Xie, F.; Hughes, L.C.; Griffin, R.J.; Tittel, F.K. Hydrogen peroxide detection with quartz-enhanced photoacoustic spectroscopy using a distributed-feedback quantum cascade laser. *Appl. Phys. Lett.* **2014**, *104*, 041117. [CrossRef]
44. Bauer, C.; Willer, U.; Lewicki, R.; Pohlkötter, A.; Kosterev, A.A.; Kosynkin, D.; Tittel, F.K.; Schade, W. A mid-infrared QEPAS sensor device for TATP detection. *J. Phys. Conf. Ser.* **2009**, *157*, 012002. [CrossRef]
45. Ma, Y.F.; He, Y.; Yu, X.; Yu, G.; Zhang, J.B.; Sun, R. Research on high sensitivity detection of carbon monoxide based on quantum cascade laser and quartz-enhanced photoacoustic spectroscopy. *Acta Phys. Sin.* **2016**, *65*, 060701.
46. Dong, L.; Kosterev, A.A.; Thomazy, D.; Tittel, F.K. QEPAS spectrophones: Design, optimization, and performance. *Appl. Phys. B* **2010**, *100*, 627–635. [CrossRef]
47. Viciani, S.; de Cumis, M.S.; Borri, S.; Patimisco, P.; Sampaolo, A.; Scamarcio, G.; De Natale, P.; D'Amato, F.; Spagnolo, V. A quartz-enhanced photoacoustic sensor for H_2S trace-gas detection at 2.6 μm. *Appl. Phys. B* **2015**, *119*, 21–27. [CrossRef]
48. Ma, Y.F.; Yu, X.; Yu, G.; Li, X.D.; Zhang, J.B.; Chen, D.Y.; Sun, R.; Tittel, F.K. Multi-quartz-enhanced photoacoustic spectroscopy. *Appl. Phys. Lett.* **2015**, *107*, 021106. [CrossRef]
49. Liu, K.; Guo, X.Y.; Yi, H.M.; Chen, W.D.; Zhang, W.J.; Gao, X.M. Off-beam quartz-enhanced photoacoustic spectroscopy. *Opt. Lett.* **2009**, *34*, 1594–1596. [CrossRef] [PubMed]
50. Wu, M.C.; Olsson, N.A.; Sivco, D.; Cho, A.Y. A 970 nm strained-layer InGaAs/GaAlAs quantum well laser for pumping an erbium-doped optical fiber amplifier. *Appl. Phys. Lett.* **1990**, *56*, 221–223. [CrossRef]

51. Nakazawa, M.; Kimura, Y.; Suzuki, K. Efficient Er^{3+}-doped optical fiber amplifier pumped by a 1.48 μm InGaAsP laser diode. *Appl. Phys. Lett.* **1989**, *54*, 295–297. [CrossRef]
52. Ma, Y.F.; He, Y.; Zhang, L.G.; Yu, X.; Zhang, J.B.; Sun, R.; Tittel, F.K. Ultra-high sensitive acetylene detection using quartz-enhanced photoacoustic spectroscopy with a fiber amplified diode laser and a 30.72 kHz quartz tuning fork. *Appl. Phys. Lett.* **2017**, *110*, 031107. [CrossRef]
53. Wu, H.P.; Dong, L.; Zheng, H.D.; Liu, X.L.; Yin, X.K.; Ma, W.G.; Zhang, L.; Yin, W.B.; Jia, S.T.; Tittel, F.K. Enhanced near-infrared QEPAS sensor for sub-ppm level H_2S detection by means of a fiber amplified 1582 nm DFB laser. *Sens. Actuators B* **2015**, *221*, 666–672. [CrossRef]
54. Ma, Y.F.; He, Y.; Tong, Y.; Yu, X.; Tittel, F.K. Ppb-level detection of ammonia based on QEPAS using a power amplified laser and a low resonance frequency quartz tuning fork. *Opt. Express* **2017**, *25*, 29356–29364. [CrossRef]
55. Borri, S.; Patimisco, P.; Galli, I.; Mazzotti, D.; Giusfredi, G.; Akikusa, N.; Yamanishi, M.; Scamarcio, G.; de Natale, P.; Spagnolo, V. Intracavity quartz-enhanced photoacoustic sensor. *Appl. Phys. Lett.* **2014**, *104*, 091114. [CrossRef]
56. Patimisco, P.; Borri, S.; Galli, L.; Mazzotti, D.; Giufredi, G.; Akikusa, N.; Yamanishi, M.; Scamarcio, G.; Natale, P.D.; Spagnolo, V. High finesse optical cavity coupled with a quartz-enhanced photoacoustic spectroscopic sensor. *Analyst* **2015**, *140*, 736–743. [CrossRef] [PubMed]
57. Patimisco, P.; Sampaolo, A.; Tittel, F.K.; Spagnolo, V. Mode matching of a laser-beam to a compact high finesse bow-tie optical cavity for quartz enhanced photoacoustic gas sensing. *Sens. Actuators A* **2017**, *267*, 70–75. [CrossRef]
58. Wojtas, J.; Gluszek, A.; Hudzikowski, A.; Tittel, F.K. Mid-infrared trace gas sensor technology based on intracavity quartz-enhanced photoacoustic spectroscopy. *Sensors* **2017**, *17*, 513. [CrossRef] [PubMed]
59. Wang, Q.; Wang, Z.; Ren, W.; Patimisco, P.; Sampaolo, A.; Spagnolo, V. Fiber-ring laser intracavity QEPAS gas sensor using a 7.2 kHz quartztuning fork. *Sens. Actuators B* **2018**, *268*, 512–518. [CrossRef]
60. Ma, Y.F.; Tong, Y.; He, Y.; Yu, X.; Tittel, F.K. High power DFB diode laser based CO-QEPAS sensor: Optimization and performance. *Sensors* **2018**, *18*, 122. [CrossRef] [PubMed]
61. Ma, Y.F.; Yu, G.; Zhang, J.B.; Yu, X.; Sun, R. Sensitive detection of carbon monoxide based on a QEPAS sensor with a 2.3 μm fiber-coupled antimonide diode laser. *J. Opt.* **2015**, *17*, 055401. [CrossRef]
62. Li, Z.; Shi, C.; Ren, W. Mid-infrared multimode fiber-coupled quantum cascadelaser for off-beam quartz-enhanced photoacoustic detection. *Opt. Lett.* **2016**, *41*, 4095–4098. [CrossRef] [PubMed]
63. Ma, Y.F.; Yu, G.; Zhang, J.B.; Yu, X.; Sun, R.; Tittel, F.K. Quartz enhanced photoacoustic spectroscopy based trace gas sensors using different quartz tuning forks. *Sensors* **2015**, *15*, 7596–7604. [CrossRef] [PubMed]
64. Ma, Y.F.; He, Y.; Yu, X.; Chen, C.; Sun, R.; Tittel, F.K. HCl ppb-level detection based on QEPAS sensor using a low resonance frequency quartz tuning fork. *Sens. Actuators B* **2016**, *233*, 388–393. [CrossRef]
65. Ma, Y.F.; He, Y.; Yu, X.; Zhang, J.B.; Sun, R.; Tittel, F.K. Compact all-fiber quartz-enhanced photoacoustic spectroscopy sensor with a 30.72 kHz quartz tuning fork and spatially resolved trace gas detection. *Appl. Phys. Lett.* **2016**, *108*, 091115. [CrossRef]
66. Ma, Y.F.; He, Y.; Chen, C.; Yu, X.; Zhang, J.B.; Peng, J.B.; Sun, R.; Tittel, F.K. Planar laser-based QEPAS trace gas sensor. *Sensors* **2016**, *16*, 989. [CrossRef] [PubMed]
67. Patimisco, P.; Sampaolo, A.; Dong, L.; Giglio, M.; Scamarcio, G.; Tittel, F.K.; Spagnolo, V. Analysis of the electro-elastic properties of custom quartz tuning forks for optoacoustic gas sensing. *Sens. Actuators B* **2016**, *227*, 539–546. [CrossRef]
68. Wu, H.; Yin, X.; Dong, L.; Pei, K.; Sampaolo, A.; Patimisco, P.; Zheng, H.; Ma, W.; Zhang, L.; Yin, W.; et al. Simultaneous dual-gas QEPAS detection based on a fundamental and overtone combined vibration of quartz tuning fork. *Appl. Phys. Lett.* **2017**, *110*, 121104. [CrossRef]
69. Zheng, H.; Dong, L.; Patimisco, P.; Wu, H.; Sampaolo, A.; Yin, X.; Li, S.; Ma, W.; Zhang, L.; Yin, W.; et al. Double antinode excited quartz-enhanced photoacoustic spectrophone. *Appl. Phys. Lett.* **2017**, *110*, 021110. [CrossRef]
70. Borri, S.; Patimisco, P.; Sampaolo, A.; Beere, H.E.; Ritchie, D.A.; Vitiello, M.S.; Scamarcio, G.; Spagnolo, V. Terahertz quartz enhanced photo-acoustic sensor. *Appl. Phys. Lett.* **2013**, *103*, 021105. [CrossRef]
71. Sampaolo, A.; Patimsco, P.; Dong, L.; Geras, A.; Scamarcio, G.; Starecki, T.; Tittel, F.K.; Spagnolo, V. Quartz-enhanced photoacoustic spectroscopy exploiting tuning fork overtone modes. *Appl. Phys. Lett.* **2015**, *107*, 231102. [CrossRef]

72. Zheng, H.; Dong, L.; Sampaolo, A.; Wu, H.; Patimisco, P.; Yin, X.; Ma, W.; Zhang, L.; Yin, W.; Spagnolo, V.; et al. Single-tube on-beam quartz-enhanced photoacoustic spectroscopy. *Opt. Lett.* **2016**, *41*, 978–981. [CrossRef] [PubMed]
73. Zheng, H.; Dong, L.; Sampaolo, A.; Patimisco, P.; Ma, W.; Zhang, L.; Yin, W.; Xiao, L.; Spagnolo, V.; Jia, S.; et al. Overtone resonance enhanced single-tube on-beam quartz enhanced photoacoustic spectrophone. *Appl. Phys. Lett.* **2016**, *109*, 111103. [CrossRef]
74. Tittel, F.K.; Sampaolo, A.; Patimisco, P.; Dong, L.; Geras, A.; Starecki, T.; Spagnolo, V. Analysis of overtone flexural modes operation in quartz-enhanced photoacoustic spectroscopy. *Opt. Express* **2016**, *24*, A682–A692. [CrossRef] [PubMed]
75. Ma, Y.; Tong, Y.; He, Y.; Long, J.; Yu, X. Quartz-enhanced photoacoustic spectroscopy sensor with a small-gap quartz tuning fork. *Sensors* **2018**, *18*, 2047. [CrossRef] [PubMed]
76. He, Y.; Ma, Y.F.; Tong, Y.; Yu, X.; Peng, Z.F.; Gao, J.; Tittel, F.K. Long distance, distributed gas sensing based on micro-nano fiber evanescent wave quartz-enhanced photoacoustic spectroscopy. *Appl. Phys. Lett.* **2017**, *111*, 24110. [CrossRef]
77. Kosterev, A.A.; Mosely, T.S.; Tittel, F.K. Impact of humidity on quartz-enhancedphotoacoustic spectroscopy based detection of HCN. *Appl. Phys. B* **2006**, *85*, 295–300. [CrossRef]
78. Jahjah, M.; Vicet, A.; Rouillard, Y. A QEPAS based methane sensor with a 2.35 μm antimonide laser. *Appl. Phys. B* **2012**, *106*, 483–489. [CrossRef]
79. He, Y.; Ma, Y.; Tong, Y.; Yu, X.; Tittel, F.K. HCN ppt-level detection based on a QEPAS sensor with amplified laser and a miniaturized 3D-printed photoacoustic detection channel. *Opt. Express* **2018**, *26*, 9666–9675. [CrossRef] [PubMed]
80. Yang, X.; Xiao, Y.; Ma, Y.; He, Y.; Tittel, F.K. A Miniaturized QEPAS trace gas sensor with a 3D-printed acoustic detection module. *Sensors* **2017**, *17*, 1750. [CrossRef] [PubMed]
81. Wang, Y.C.; Lan, R.J.; Mateos, X.; Li, J.; Hu, C.; Li, C.Y.; Suomalainen, S.; Härkönen, A.; Guina, M.; Petrov, V.; et al. Broadly tunable mode-locked Ho:YAG ceramic laser around 2.1 μm. *Opt. Express* **2016**, *24*, 18003–18012. [CrossRef] [PubMed]
82. Yang, X.T.; Yao, B.Q.; Ding, Y.; Li, X.; Aka, G.; Zheng, L.H.; Xu, J. Spectral properties and laser performance of Ho:Sc$_2$SiO$_5$ crystal at room temperature. *Opt. Express* **2013**, *21*, 32566–32571. [CrossRef] [PubMed]

© 2018 by the author. Licensee MDPI, Basel, Switzerland. This article is an open access article distributed under the terms and conditions of the Creative Commons Attribution (CC BY) license (http://creativecommons.org/licenses/by/4.0/).

Review

A Review of Photothermal Detection Techniques for Gas Sensing Applications

Karol Krzempek

Laser & Fiber Electronics Group, Faculty of Electronics, Wroclaw University of Science and Technology, Wybrzeze Wyspianskiego 27, 50-370 Wroclaw, Poland; karol.krzempek@pwr.edu.pl

Received: 30 May 2019; Accepted: 3 July 2019; Published: 15 July 2019

Abstract: Photothermal spectroscopy (PTS) is a technique used for determining the composition of liquids, solids and gases. In PTS, the sample is illuminated with a radiation source, and the thermal response of the analyte (e.g., refractive index) is analyzed to gain information about its content. Recent advances in this unique method of detecting gaseous samples show that photothermal gas spectroscopy can be an interesting alternative to commonly used absorption techniques. Moreover, if designed properly, sensors using PTS detection technique can not only reach sensitivities comparable with other, more complex techniques, but can significantly simplify the design of the sensor. In this review, recent developments in photothermal spectroscopy of gases will be summarized and discussed.

Keywords: photothermal spectroscopy; gas sensing; detection limit; laser spectroscopy; practical applications; intracavity gas detection; interferometric gas detection

1. Introduction

Photothermal spectroscopy (PTS) is a group of spectroscopy techniques commonly used to measure thermal characteristics and optical absorption of samples (solids, liquids or gases) [1–3]. Photothermal techniques were investigated already in the 80s', mainly as a method for measuring the absorption of thin samples and liquids. Generally, in PTS, the analyte is illuminated with a radiation source (e.g., a laser beam) and the change in the thermal state of the sample (or the consequences of the change) is monitored. A portion of the light absorbed by the sample is lost to emission, the remaining part contributes to a temperature raise of the sample. This in turn influences the thermodynamic properties of the sample (or particles adjacent to it), which can be observed as temperature, density or pressure change. In PTS, those three parameters are most commonly monitored to determine the composition of the target analyte. PTS is in principle similar to photoacoustic spectroscopy (PAS) [4–6] and relies on an indirect method of measuring the absorption of the sample, which is the key distinction when compared to traditional absorption spectroscopy techniques. In absorption spectroscopy techniques e.g., tunable diode laser absorption spectroscopy (TDLAS), a basic setup consists of a tunable diode laser light source and a suitable detector [7,8]. The laser source used in the experiment is tuned across the characteristic absorption lines of the targeted gas in the path-length of the laser beam. Absorption of the molecules causes a reduction of the intensity, which is detected by a photodiode and signal-processed to determine the gas concentration or other properties of the gas (e.g., pressure, temperature, velocity) [9]. As this method usually relies on calculating the difference between the input light intensity and the light absorbed by the sample, the performance and accuracy of the sensor is limited especially in the case of small absorption and by reflections and scattering effects usually present in such systems. Moreover, accessing strong absorption features of numerous gas molecules requires using mid-infrared laser sources. In TDLAS, this implies incorporating expensive detectors (e.g., based on mercury cadmium telluride material). Unlike traditional absorption spectroscopy techniques, in PTS and PAS, the sensor does not rely on quantifying the signal directly involved in the absorption, but rather on the influence of the absorbed energy on the gas particles. Therefore, the amplitude of the registered PTS signal is proportional to the absorption, but the measurement does not require monitoring the parameters of the laser radiation exciting the gas molecules. Moreover, performed in a properly designed sensor, PTS is a zero-background measurement—no absorption equals to zero registered signal, thus PTS sensors require minimal calibration. Photothermal spectroscopy has been successfully employed in commercial lab-grade apparatus commonly used for analyzing solid samples. In such devices, photothermal deflection (PDS) or photothermal lens spectroscopy (TLS) effects are used [10–13].

PTS of gas samples is usually performed in a scheme as presented in Figure 1.

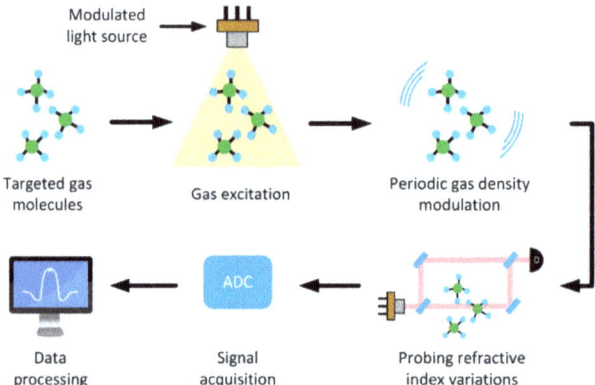

Figure 1. The principle of the photothermal spectroscopy techniques.

To perform PTS, the gas molecules have to be excited by an energy source. This is commonly accomplished by illuminating the gas sample with a laser source—*pump* laser. Similarly to traditional absorption spectroscopy, the wavelength of the light has to be appropriately chosen to target strong transitions of the molecules; thus, a significant portion of the light will be absorbed. Part of the absorbed energy will locally heat the sample (due to non-radiative relaxation of the molecules), causing a change in the density, which in turn modulates the refractive index (RI) of the gas. Because the variations of the RI are small ($\Delta n \sim 10^{-9}$), most PTS gas sensors rely on an interferometric signal retrieval sensor configuration (e.g., a Mach-Zehnder interferometer). In such a case, the interferometer is constructed to have one of the measurement beams interact with the excited gas sample. The induced RI modulation will result in a change in the optical paths of one arm of the interferometer and thus produce a signal which can be easily detected. The *pump* wavelength can be additionally modulated by a sinewave-function in the vicinity of the absorption line, causing a periodic RI modulation and can simplify the absorption signal retrieval by employing commonly used techniques e.g., wavelength modulation spectroscopy (WMS) [14,15].

The PTS-induced RI change in the gas sample can be estimated based on Equation (1) [16].

$$\Delta n = \frac{(n-1)}{T_0} \frac{\alpha P_{exc}}{4\pi a^2 \rho C_p f} \quad (1)$$

where α is the absorption coefficient of the targeted gas; f is the excitation beam modulation frequency; C_p is the specific heat of the gas mixture; ρ is the density of the gas mixture; P_{exc} is the *pump* beam power, T_0 is the absolute temperature and n is the refractive index of the gas mixture; and a is the *pump* beam radius. If required, the calculated photothermal-induced RI change can be converted to a phase shift of the *probe* beam, according to equation

$$\Delta\varnothing = (2\pi l \Delta n)/\lambda$$

where l is the laser-gas interaction length, and λ is the wavelength of the *probe* beam.

Although using multimode radiation in a Mach-Zehnder interferometer has been presented, this requires additional signal processing methods or using a bulky and expensive optical spectrum analyzer for calculating the induced phase shift. Therefore, in each of the presented PT gas spectrometers, a single mode and single frequency laser beam is used as the *probe*. From equation (1), it can be seen that the RI change and thus the resulting PTS signal is proportional to the absorption coefficient and the *pump* laser power. On the other hand, it is inversely proportional to the *pump* beam cross-section and the frequency of modulation. Hence, increasing the amplitude of the registered PTS signal can be conveniently achieved by utilizing a laser source capable of delivering high output powers and good beam quality (preferably with parameters M^2 close to 1, or with TEM00 mode for Gaussian beams, which permits tight focusing). Moreover, similarly to PAS, the PTS signal decreases with higher *pump* laser modulation frequencies [17]. This is a direct result of the limited thermal response time of the gas molecules and has been thoroughly explained in Reference [18]. This requires fine-tuning of the *pump* laser modulation frequency of each PTS sensor to optimize the signal-to-noise ratio (SNR), so a tradeoff between $1/f$ noise and registered signal is reached. Moreover, with a proper design, PTS gas sensors can have direct frequency readout of the induced RI modulation. The principle is similar to dispersion spectroscopy [19], where the gas concentration can be encoded into frequency deviation of the beat note. In such configurations, the PTS sensors represent excellent immunity to any variations of the signal amplitude. Moreover, PAS, PTS and CLADS are inherently baseline-free detection methods. If no targeted gas particles are present in the sensor, no signal is observed. This severely simplifies signal processing and limits the requirements for calibration and periodic re-adjustments of the sensor.

The main aim of this work is to present several interesting and unique approaches to photothermal gas sensing. The configurations will be discussed and their performance will be compared.

2. First Experimental Verification of PTS Gas Sensors

Because the PT-induced RI modulation is inherently small, researchers developed several methods to enable relatively non-complex retrieval of the signals. One of the methods uses interferometric signal retrieval. Although usually requiring several bulk optics-based components to set-up the interferometer, with careful design such PT sensors yield an acceptable SNR.

2.1. Optical Heterodyne Gas Spectrometer

One of the first attempts to use PT-induced effects for measuring highly diluted gas samples was proposed by Davis et al. in 1981 [18]. The method was adopted from PT techniques used in probing highly diluted liquid samples [20]. The Authors proposed to use a modified homodyne Mach-Zehnder (MZ) interferometer as a method of measuring the PT-induced RI modulation. The setup is depicted in Figure 2.

A single mode, single frequency beam of an HeNe laser was split into two arms to form an MZ interferometer (*probe*). A glass sample cell fitted with appropriate windows was inserted into the optical path of one of the interferometer arms. One of the mirrors (M on the schematic in Figure 2) was mounted on a piezoelectric transducer. By driving the position of the mirror, the MZ interferometer was stabilized in quadrature (in the middle between interference maximum and minimum), preventing long-term thermal variations affecting the signal readout. In this experiment, a CO_2 laser with a maximum output power of 20 W was used as the *pump* source. The sensor performance was evaluated by measuring methanol vapors (CH_3OH). Authors calculated the detection limit at a level of 3 ppb (part per billion). The achieved detection limit was excellent, proving that photothermal detection trace concentrations of gases can be an interesting alternative to commonly used methods. Although the sensors proposed by Davis reached sensitivity close to photon-noise-limited values (10^{-10} cm^{-1}), this was possible only by employing a very powerful *pump* laser (~20 W CW). Moreover, because the thermal response of the gas sample is limited, it is beneficial to work with low *pump* laser modulation frequencies. Davis used an optimal modulation frequency of 23 Hz, but found that due to the increased $1/f$ noise components, the sensor had to be secured on a highly stable optical table and enclosed in an additional enclosure, dampening acoustic noise. Although sensitive, this sensor layout is complex and implementing it in out-of-lab applications would be complicated. The publication by Davis et al. [18] is particularly interesting due to the extensive investigation of the thermal effects occurring in gases excited by optical beams. A detailed description is included in the appendix of the publication.

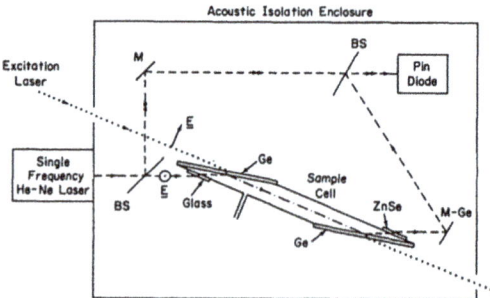

Figure 2. Schematic of the Mach-Zehnder interferometer-type photothermal gas sensor. BS—beam splitter, Ge—germanium window, M—mirror, ZnSe—zinc selenide window, E—beam polarization vector direction, M-Ge—germanium mirror. Reprinted with permission from [18] C The Optical Society.

2.2. Homodyne Hydrazine Detector with Folded Jamin Interferometer

One of the first attempts to use a modified Jamin interferometer [21] configuration to detect PT-induced RI changes was presented by Owens et al. in 1999 [22]. The schematic of the sensor is presented in Figure 3.

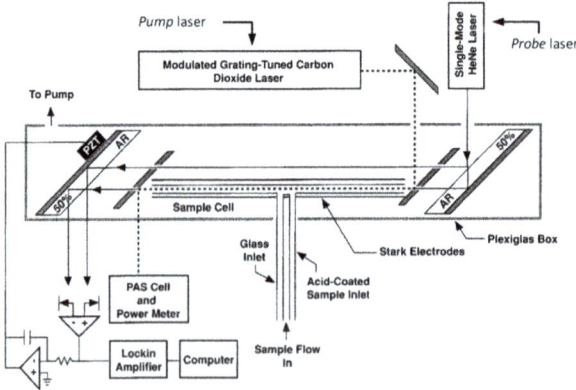

Figure 3. Schematic diagram of the unfolded Jamin interferometer PT ammonia detector. A single-mode He-Ne laser is used to form the interferometer. A 7 W modulated CO$_2$ laser beam is used as the *pump* source. Laser beam and gas interaction path length is 38 cm. Reprinted with permission from [22].

The proof-of-concept experiment used a single-mode He-Ne laser with an output power of 1 mW as the *probe* laser. The beam was split into two independent arms using two etalons oriented at a

45° angle. The etalons were appropriately coated to achieve a nearly even split ratio, forming an unfolded Jamin interferometer with a total optical path ~50 cm. Beams exiting the interferometer were focused onto two independent photodiodes. In this experiment, a static phase difference between the signal and the reference beam was maintained (at zero) by driving a piezoelectric transducer on which one of the etalons was mounted. This allowed for compensating any thermal drifts of the sensor (similarly to work by Davis). The core of the sensor was enclosed in plexiglass housing and the gas was delivered to the sensor via an inlet near the center of the interferometer. In this experiment, the detected gas sample was ammonia (NH_3). The gas particles were excited by an auxiliary, high-power (7 W), tunable CO_2 laser-*pump* laser. The emission of the *pump* laser was co-aligned with one of the arms of the interferometer, forming a 38 cm-long gas-laser interaction length. The laser was capable of reaching a strong NH_3 transition centered at 9.22 µm and was modulated at 1.2 kHz with a sinewave signal. By exciting the gas particles in one arm of the interferometer, a change in the refractive index was obtained. This imbalance was easily detected as a phase shift between both beams forming the interferometer. The signal analysis was simplified by using a lock-in amplifier (LIA) which was referenced by the 1.2 kHz modulation delivered to the *pump* laser. This NH_3 PTS detector was capable of reaching a detection limit of 250 ppt and 33 ppt for 1 s and 100 s integration time, respectively. Worth noting is the fact that the sensor had a linear response of over 5 orders of magnitude of the analyte concentration. Although reaching promising results, the proposed sensor configuration is rather bulky, requires precise alignment and reaching the detection limit is possible only in lab environment using a high-power excitation laser.

3. Fiber-Based Photothermal Gas Sensors

Although the preliminary experiments have proven that PT spectroscopy can with ease compete with traditional absorption gas detection techniques, the implementation of such sensors was significantly more complex compared to TDLAS or WMS approaches, even considering the fact that achieving results comparable to PTS requires implementing multipass gas absorption cells [23]. The limitations were mainly connected with the number of bulk optics required to set-up the sensor. The overall complexity limited out-of-lab applications and thus PT gas sensors were not extensively researched for years. This changed with the implementation of inexpensive, reliable and widely available fiber-based components.

3.1. Fiber-Based Heterodyne Detection of PTS Signals

The idea of using inexpensive and reliable fiber-based components in a PTS gas sensor was for example presented in Reference [24]. In this work, the PTS signal was registered in a Mach-Zehnder interferometer which was constructed based on telecom-wavelength (1550 nm) fiber technology. The schematic of the setup is presented in Figure 4.

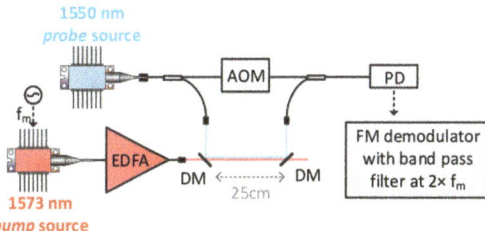

Figure 4. Schematic diagram of PAS measurement using a 2 µm tunable fiber laser as the pump source, and a Mach-Zehnder fiber interferometer at 1.55 µm for path-length modulation signal retrieval. TDFA—thulium doped fiber amplifier, DM—dichroic mirror, AOM—acousto-optical modulator (Ω = 40 MHz), PD—photodiode. Reprinted with permission from [24] C The Optical Society.

The base of the sensor was formed using a low-power singlemode 1.55 μm laser diode-*probe*. The radiation of the *probe* laser was split using a 3 dB fiber coupler to form two independent arms of the interferometer—reference and probe beams. The probe beam was launched into a 20 cm-long gas absorption cell filled with pure carbon dioxide (CO_2) gas at 740 Torr pressure. The reference beam was up-shifted by 40 MHz in frequency using an acousto-optic modulator (AOM). After passing the cell, the probe beam was combined by a second coupler with the reference beam and delivered to a photodiode to create a beatnote. The AOM frequency shift of the reference beam was necessary to move the beatnote to a higher frequency, limiting the 1/*f* noise of the sensor. In this experiment, a singlemode laser diode with emission wavelength of 1573 nm was used as a *pump* source, targeting a weak absorption line of CO_2 (R-branch of the 2v1+2v2+v3 combination band of CO_2). Additional erbium-doped fiber amplifier [25] boosted the *pump* radiation output power to approximately 300 mW. The *pump* beam was additionally modulated by a sinewave signal with frequency f_m and swept across the CO_2 absorption line using a triangular current ramp, similar to the traditional WMS gas detection technique. The effect of the PT-induced RI modulation was clearly observable as a frequency modulation of the beatnote. By incorporating an FM demodulator [26] and filtering the signal at $2 \times f_m$, the PTS signal could be resolved by techniques commonly used in WMS. This proof-of-concept PTS sensor reached an NEC of 400 ppm. Note, that the sensor had only a 20-cm-long gas-laser interaction length and a reasonable *pump* power intensity with a relatively large beam diameter (~3 mm).

3.2. Spatial Gas Sample Localization in an Fiber-Based PTS Sensor

Some practical applications of laser-based gas spectrometers e.g., at gas or oil processing plants, require to exactly localize the position of a gas leak [27–29]. This requires using open-path gas sensing techniques where the laser beam is transmitted in air and only the back reflected or scattered light is gathered for analysis [30,31]. Standard open-path absorption-based implementations of gas sensors do not allow to precisely define the position of the gas particles being measured, as the registered signal is averaged over the entire length of the interaction between the gas sample and the laser beam. This problem was also addressed in Reference [24]. The proposed configuration of a PTS spatially resolved gas detection is presented in Figure 5.

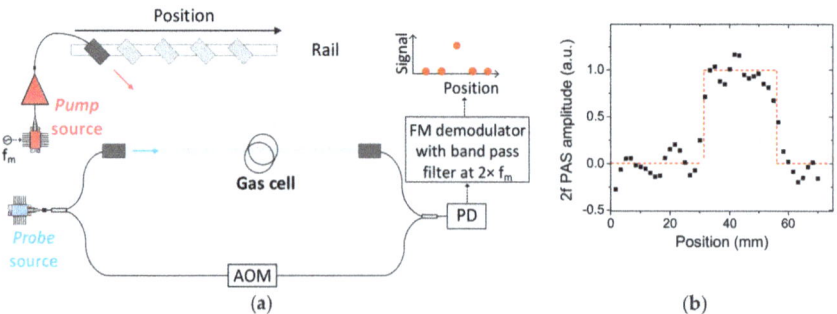

Figure 5. (a) PAS configuration for spatially resolved measurement. AOM—acousto optical modulator, PD—photodiode. Gas cell filled with pure CO_2 at atmospheric pressure was placed in the optical path of the fiber-based 1.55 μm Mach-Zehnder interferometer. (b) By scanning the position of the *probe* beam, a clear PT signal was registered only at a point, where the 2 μm beam intersected the gas-filled cell. Reprinted with permission from [24] C The Optical Society.

In this unique configuration, the PTS-induced RI modulation is probed similarly as in the configuration shown in Figure 4—using a fiber-based Mach-Zehnder interferometer working at 1.55 μm. Here, the *pump* laser was not collinearly aligned with the *probe* laser beam. Instead, the collimator outputting the 2 μm radiation was mounted on a mechanized optical rail and aligned to cross the *probe* beam at an angle (see Figure 5). By moving the stage, the *pump* beam could be positioned at a given point along the path-length of the *probe* beam with high precision, with only one intersection point. The proof-of-concept experiment was carried out by inserting a transparent gas cell filled with pure CO_2 into the path-length of the *probe* beam. In this configuration, the PTS signal was registered only at a certain position of the *pump* beam, where it coincided with the CO_2-filled gas cell. This experiment proved that appropriately designed PTS sensors are capable of resolving the position of the measured gas sample with reasonable precision.

3.3. PTS Signal Enhancement using Multipass Cells

According to the equation $\Delta\Phi = (2\pi l \Delta n)/\lambda$, the registered signal is linearly dependent on the length of the interaction between the gas sample and the beam that excites the molecules—similarly to standard absorption techniques. Therefore, a straightforward way to increase the sensitivity of a PTS sensor would be to maximize the this parameter. In Reference [32], a first demonstration of using commercially available multipass gas absorption cells to increase the gas-laser interaction length was presented. The layout of the setup is presented in Figure 6.

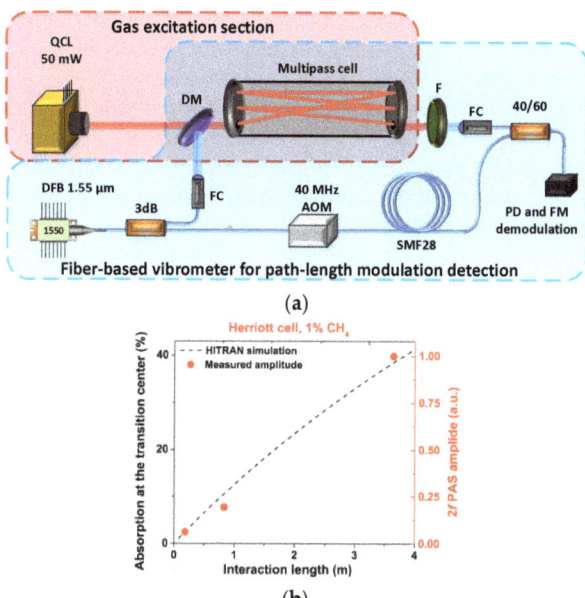

Figure 6. (**a**) Schematic of the experimental layout incorporating a multi-pass cell for interferometric detection of a PA signal. X0/Y0—fiber couplers with specified splitting ratio, DM—dichroic mirror, F—filter, FC—fiber collimator, PD—photodiode, QCL—quantum cascade laser, DFB—distributed feedback telecom laser diode. (**b**) 2*f* PTS signal amplitude in function of the laser-gas interaction length. Reprinted with permission from [32].

The base of the multipass-cell-assisted PTS sensor was a fiber-based Mach-Zehnder interferometer designed based on telecom wavelength components (1.55 μm). To increase the gas-laser interaction length, and thus achieve a higher PTS signal, a Herriot multipass cell was incorporated into the measurement arm of the interferometer. The Authors used a 50 mW quantum cascade laser (QCL) [33] as a *pump* source and targeted a strong absorption line of methane located in the v_4 band near 1389 cm^{-1}. The *pump* beam was spatially overlapped with the *probe* beam and launched into the multipass (MP) cell. The optical path in the MP cell was changed in steps, achieving 19 cm, 83 cm and 364 cm (1, 7 and 33 passes, respectively, obtained through axial rotation of the exit mirror). The results of the measurements presented as an inset in Figure 6 clearly demonstrated that the PTS signal scaled linearly with the increasing interaction length. Additionally, spherical mirrors used in the MP cell assured that both beams are traveling co-linearly (any misalignment would result in one of the beams not exiting the MP cell due to improper angular alignment) and are periodically re-focused in the center of the MP cell, thus keeping optimal beam spots of both the *pump* and the *probe* beams.

Interestingly, although the registered PTS signal was enhanced by the longer gas-laser interaction length, the sensitivity of the sensor estimated based on the Allan deviation plot was limited to an NEC of 10 ppmv of CH_4. According to the research, the main limitation of the MP cell-assisted PTS gas sensor was connected with mechanical and acoustic noise. In this configuration, the constructed MZ interferometer readout acts as a traditional vibrometer, picking-up all unwanted background signals. To fully exploit the multipass-cell-assisted enhancement of the PT signal, a different technique of detecting the RI modulation has to be adopted.

3.4. PTS in Hollow-Core Fibers

Hollow core fibers (HCF) have attracted much attention from researchers working on laser-based gas sensors [34]. Numerous experiments were published demonstrating the usefulness of such structures in gas sensors [34–40]. The problem of achieving reasonable PTS sensor performance was investigated by several Groups worldwide. Some researchers addressed the issue of PTS sensitivity by incorporating hollow-core fibers into the gas sensors. The appropriate design of an HCF-based PTS sensor addresses several crucial issues: optimized *pump* and *probe* beam overlap and beam size throughout the entire gas-laser path-length, limited gas volume required for the analysis and overall stability of the system. One of the most interesting experiments was presented by Jin et al. [41]. The schematic of the setup is presented in Figure 7.

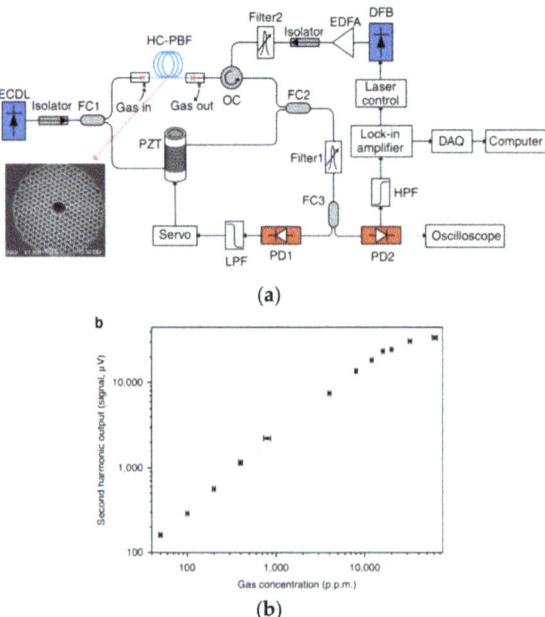

Figure 7. (a) HC-PCF-based PTS gas sensor. FC—fiber coupler, PZT—piezoceramic transducer, OC—optical circulator, EDFA—erbium doped fiber amplifier, LPF—low-pass filter, HPF—high-pass filter DAQ—digital acquisition card, DFB—distributed feedback laser diode. (b) Amplitude of the second harmonic signal in function of gas concentration. Reprinted with permission from [41].

The base of the presented sensor was a fiber-based MZ interferometer, similar to the ones explained in the previous experiments. A singlemode external cavity diode laser with a wavelength of 1556.59 nm was used as the *probe* laser. Here, the Authors decided to use phase detection for the PT-induced RI modulation. This method is significantly more sensitive compared to pure frequency demodulation, but requires using a piezo ceramic transducer (PZT) [42] in order to enable quadrature detection of the phase variations. The *probe* radiation was split into two arms—reference and measurement. The reference arm fiber was coiled around a PZT and stretched to maintain a 90 degree phase difference between the interfering beams. The measurement beam was coupled into a 0.62-m-long commercially available hollow-core photonic bandgap fiber (HC-PBF) [43,44], manufactured by NKT Photonics (HC-1550-02). The fiber had a core diameter of only 11 µm. Authors used but-coupling between the SMF28 standard fiber and the HC-PBF. A distributed feedback (DFB) semiconductor laser diode was used as the *pump* source which targeted the P (9) line of acetylene (C_2H_2) located at 1530.37 nm. The *pump* laser was modulated at f_m = 50 kHz and the resulting PT-induced phase-shift signal detected by the MZ interferometer was analyzed by an LIA at the second harmonic of the f_m. The Authors achieved an NEC of 2 ppb. Moreover, the sensor had a very wide dynamic range, nearly six orders of magnitude (5.3 × 10^5), which was superior to any previously reported fiber-based gas sensor at that time (inset in Figure 7). Nevertheless, the Authors admitted that the achieved results would be unobtainable in out-of-lab applications. The main identified issue was the feed-back loop controlling the quadrature operation of the MZ interferometer. In-lab experiments were conducted in an controlled environment, with limited mechanical and acoustic noise. Real-world applications would suffer from all sorts of disturbances influencing the MZ interferometer, and this would affect the stability of the quadrature phase feed-back loop and thus significantly increase the overall noise floor of the sensor.

3.5. Pulsed PTS in Hollow-Core Fibers

The HC-PBF-based PT gas layout was additionally investigated by Lin et al. in a an excellent paper published in 2016 [45]. This work not only presents a novel approach to PT gas sensing but also clearly describes the thermal properties of gas particles excited in an HCF. All experimental data were confronted with numerical models. Moreover, Authors analyzed drawbacks pointed out in the work by Jin et al. [41] and proposed an improved configuration of an HC-BPF-based photothermal gas sensor. The layout of the sensor is represented in Figure 8.

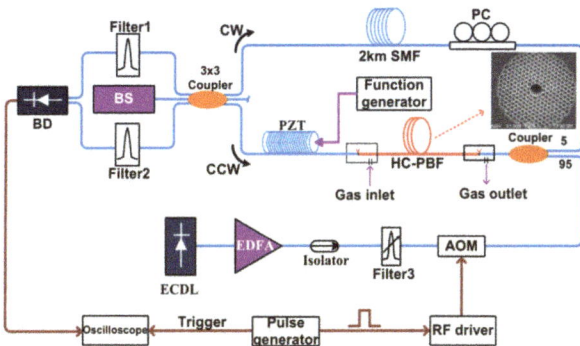

Figure 8. Pulsed PT gas sensor employing an HC-PCF as a gas cell. HC-PBF—hollow core photonic bandgap fiber, ECDL—external-cavity diode laser with wavelength around 1530.371 nm, EDFA—erbium doped fiber amplifier, AOM—acoustic-optic modulator, BD—balanced detector, BS—broadband source, PC—polarization controller. Reprinted with permission from [45].

Here, the Authors targeted the same gas as in the previous experiment using an ECDL exciting the P (9) absorption line of C_2H_2 at 1530.37 nm. The emission of the diode was amplified by a standard EDFA and additionally intensity modulated using an AOM. The pulse repetition rate was experimentally chosen and set to 500 Hz. Authors used the same 0.62-m-long HC-1550-02 fiber from NKT Photonics, which was filled with 7500 ppm C_2H_2 in nitrogen. The PT-induced phase modulation was detected using an all-fiber Sagnac interferometer which was probed by a super-luminescent light emitting diode (SLED) [46] with a bandwidth of 41 nm centered around 1545 nm. The Sagnac interferometer was configured so that the SLED radiation is divided into a 3 × 3 coupler and travels a common loop in opposite directions. The gas filled HC-PCF was inserted into the interferometer with an offset to its center. Authors used a 2-km-long single mode fiber as an optical delay line to ensure a constant, 10 µs delay between the light traveling clockwise (CW) and counterclockwise (CCW) through the HC-PCF. The CW and CCW beams are combined at the 3 × 3 coupler after traveling the interferometer and analyzed by a balanced photodetector. Compared to the experiment with the traditional MZ interferometer, both beams in the Sagnac interferometer travel through the same fiber-based loop. This limits the influence of environmental disturbances observed in previous experiments. The PT-induced phase shift was registered using a boxcar averager and analyzed for with different values of sample averaging. Ultimately, this configuration of a PT gas sensor reaches an NEC of 3.3 ppm of C_2H_2, but was mainly limited by the number of times the registered signal was averaged (in this case 10,000).

Both papers [41,45] show that integrating HCF into PT gas sensors gives excellent results and should be further explored as an interesting alternative for other commonly used detection techniques. Unfortunately, currently commercially available HCF's have two main drawbacks which severely limit their implementation in PTS sensors and laser-based gas sensors in general. The first limitation is connected with the wavelength-dependent attenuation of such fibers. Drawn from standard silica glass, HCF's suffer from tremendous attenuation in the mid-IR wavelength region, especially above 3.5 µm, mainly due to two photon absorption effects [47]. Transmission of a 1-cm-long fused silica glass has been plotted in Figure 9 in the function of a wavelength.

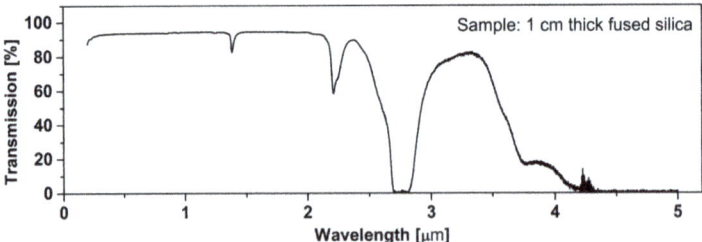

Figure 9. Transmission of 1-cm-thick fused silica glass in function of wavelength.

This limitation is clearly pointed out by most of the researchers working on the implementation of HCF in laser-based gas spectrometers. Moreover, currently available HCF's suffer from several additional drawbacks, severely limiting their out-of-lab application not only in PTS gas sensors, but any mid-IR laser-based sensors. The issues are connected with small inner core diameters and not purely single-mode operation. Multimode transmission of the light taking part in the sensing results in modal interferences, which are susceptible to bending of the fiber and mechanical and acoustic disturbances. In most cases, modal interferences significantly decrease the SNR. Moreover, small diameters of the hollow core drastically increase the time required to exchange the measured gas sample, and hence the response time of the sensor [48]. Fortunately, these issues have been perceived by researchers working on novel structures of HCFs. The result is the currently observed struggle to design and draw HCFs with low attenuation in the mid-IR wavelength region [49–51]. Preliminary experimental results have been published, validating the possibility of using e.g., antiresonant hollow-core fibers in the 3–4 µm wavelength region for various optical applications [52–54], which gives perspective for novel configurations of absorption and PT gas sensors.

4. Intra-Cavity PTS Gas Sensors

Several dissimilar configurations for registering the PT-induced RI modulation have been developed over the years. One of the approaches involves inserting the gas sample into the resonator of a *probe* laser. By exciting the gas sample via an auxiliary *pump* source, the resulting RI change would directly influence the parameters of the *probe* laser e.g., its output beam intensity or frequency.

4.1. PTS in an Intracavity He-Ne Laser Configuration

One of the first attempts to utilize an intracavity photothermal gas detection configuration was reported in the work by Fung et al. in 1986 [55]. The core of the sensors was a linear cavity HeNe laser operating at ~638.8 nm, with a layout as depicted in Figure 10.

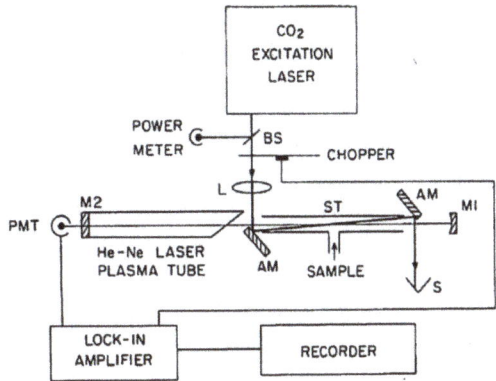

Figure 10. Schematic of an HeNe-based intracavity photothermal gas sensor. BS—beam splitter, AM—aluminum mirror, M1; M2—laser resonator mirrors, PMT—photomultiplier tube, S—beam stop, ST—gas sample cell. Reprinted with permission from [55] C The Optical Society.

The *probe* laser resonator was formed by two mirrors—M1 and M2, producing a cavity length of 50 cm. The gas sample was delivered with an open-ended gas cell (ST on the schematic) directly to the *probe* laser resonator. The cell was 20 cm long. In this experiment, a tunable 0.8 W CO_2 laser was used as a *pump* source to excite sulfur hexafluoride (SF_6). The beam of the *pump* source was aligned to coincide with the HeNe beam at a single point in the resonator. The PT effect induced by the *pump* laser was deflecting the beam of the HeNe laser and thus causing a change in the output power. The *pump* laser was chopper-modulated at 42.7 Hz and the PT signal was registered using an LIA. Authors calculated an NEC of 3 ppb. Although reaching a very good sensitivity, this configuration of a PT gas sensor would not have any particular applications in out-of-lab sensors, mainly due to the severe sensitivity to mechanical and acoustic noise.

4.2. All-Fiber Intracavity PT Gas Sensor

In the work by Zhao et al. [56], the authors used a very similar configuration to the one presented in Reference [45]. The schematic of the layout is presented in Figure 11.

The same 0.62-m-long HC-PCF was used as the gas cell and a Sagnac interferometer readout as proposed in Reference [45]. Here, the *pump* laser beam exciting the acetylene particles was additionally amplified in an Erbium-doped loop resonator built around the gas-filled hollow core fiber. By significantly boosting the *pump* power, an NEC of 0.176 ppm was achieved. This result is 2.6 times better than that previously reported for a non-intracavity configuration [45].

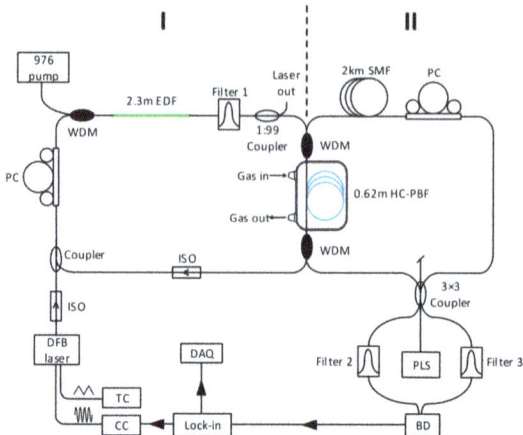

Figure 11. Experimental setup of the intracavity PT gas sensor. TC—temperature controller, CC—current controller, DAQ—data acquisition, EDF—erbium-doped fiber, WDM—wavelength division multiplexer, ISO—optical isolator, PLS—probe light source, HC-PBF IC—hollow core photonic bandgap fiber gas cell, BD—balanced detector, PC—polarization controller. Reprinted with permission from [56] C The Optical Society.

4.3. Intracavity PT Detection in a Mode-Locked Laser Configuration

An interesting approach to intracavity PT detection was proposed in Reference [16]. In this work, the Authors investigated the possibility of translating the PT-induced RI modulation directly into frequency deviation of a mode-locked laser. The schematic of the sensor is presented in Figure 12.

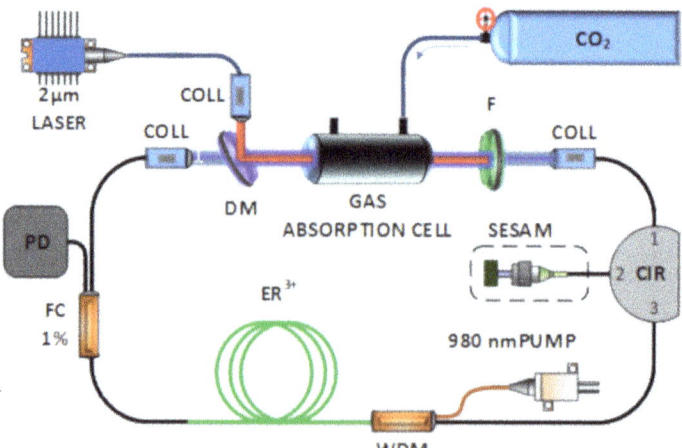

Figure 12. Experimental setup of the ML intracavity PT gas sensor. PD—photodiode, FC—fiber coupler with 1% out-coupling ratio, Er^{3+}—80-cm-long erbium doped fiber, WDM—wavelength division multiplexer, PUMP—980 nm singlemode pump laser, CIR—fiber circulator, SESAM—fiber pigtailed semiconductor saturable absorber mirror, COLL—fiber collimator, F—filter, DM—dichroic mirror. The gas absorption cell is 10 cm long. Reprinted with permission from [16] C The Optical Society.

The base of the sensor had standard configuration for a ring cavity Erbium-doped laser mode-locked by incorporating a semiconductor saturable absorber mirror (SESAM) [57]. The gain section was an 80-cm-long piece of erbium-doped fiber, which is optically pumped by a singlemode laser diode (λ = 980 nm) via a fiber wavelength division multiplexer (WDM). The fiber pigtailed SESAM was coupled to the cavity via a fiber circulator (CIR), which additionally ensured unidirectional operation of the laser. Parameters of the ML laser were monitored via a 1% fiber coupler with a photodiode. The gas sample was introduced into the ML laser via a 20-cm-long free-space open-path which was formed in the cavity with two fiber collimators. The gas under testing was carbon dioxide (CO_2) and was pumped into a 10-cm-long absorption cell. The gas sample was excited via an auxiliary 100 mW laser (DFB diode with an additional thulium doped fiber amplifier) targeting an isolated absorption line located at ~2003.5 nm. The *pump* beam was co-aligned with the *probe* beam of the ML laser using a dichroic mirror (DM). The ML laser had a pulse repetition frequency of f_{rep} ~ 21.226 MHz. Exciting the CO_2 molecules in the path-length of the *probe* beam of the ML laser resulted in modulation of the optical path-length, and hence a slight change in the repetition frequency. PT signal analysis was simplified by modulating the *pump* laser and registering the resulting repetition frequency deviation via an FM demodulator and an LIA. In this simplified PT gas sensor, the Authors were able to achieve an NEC of 311 ppmv for CO_2 at ambient pressure. It is worth noting that compared to other fiber-based sensors, no interferometric read-out of the signal is required and due to its configuration it is possible to use mid-IR wavelength *pump* sources to excite gas molecules.

5. Fabry-Perot-Based PTS Gas Sensors

Inspired by work done in the field of PT gas sensors, researchers investigated the possibility of designing unique configurations in which the induced RI modulation would be probed using a Fabry-Perot (FP) cavity. As described before, variations of the RI of a sample (not only gaseous) are primarily connected with changes in the density of the sample. A gas sample excited by the *pump* laser experiences a localized heat gradient, and thus the refractive index decreases. The idea of using an FP cavity as a PT effect readout is directly connected with the physics underlying the principle of their operation. FP interferometers are constructed using two partially reflecting parallel mirrors separated by a certain distance. The intensity of light transmitted I_t through such a cavity is described by [58]

$$I_t = I_0 \frac{1}{1 + \left(\frac{2F}{\pi}\right)^2 \sin^2\left(\frac{\Delta\varphi}{2}\right)} \qquad (2)$$

where I_0 is the intensity of the incident light, F is the cavity finesse, and $\Delta\varphi$ is the phase difference. Finesse of the cavity can be calculated as a function of the reflectivity of the mirrors R, $F = \pi \sqrt{R}(1-R)^{-1}$. Whereas the phase difference is given by:

$$\Delta\varphi = \Delta\varphi = \frac{2\pi}{\lambda_0} 2nd \cos \Theta \qquad (3)$$

where, n is the refractive index, λ_0 is the wavelength of the incident light in vacuum, and $\cos\theta$ is the angle of incidence. Based on the formulas described above, one can clearly see that a change in the refractive index inside the cavity will have a direct influence on the intensity of light transmitted through the FP interferometer, which can be conveniently detected. This approach was adopted in several PT gas sensors with great success.

5.1. Fabry-Perot PT Gas Sensor with 2f Modulation

In Ref [59]., Waclawek et al. proposed an FP-based PT gas with the layout shown in Figure 13.

The FP interferometer was based on two dielectrically-coated silica mirrors with reflectivity R = 0.85, which resulted in a finesse of the cavity of 19.3. The *probe* laser in this experiment was a single-mode tunable fiber-pigtailed DFB laser diode with a wavelength of 1600 nm. The radiation was injected into the FP cavity after focusing by a lens, and the beam exiting the interferometer was collected with a second lens onto a gallium indium arsenide (GaInAs) photodiode. In this experiment, a sulfur dioxide (SO_2) absorption line located at 1379.78 cm^{-1} was targeted with a QCL laser source providing up to 173 mW of CW power. The beam from the QCL was coupled perpendicularly into the FP cavity, coinciding with the *probe* beam. The QCL emission was wavelength modulated with a sinewave function to employ WMS detection at second harmonic using an LIA. To ensure a stable operation and linear response to the PT-induced RI changes, the wavelength of the *probe* laser was locked to one of the steep slopes of the FP cavity resonance by a slow feedback loop, which monitored the DC output of the photodiode. The proposed FP-based PT gas detection setup is simple, inexpensive and requires minimum alignment. The calculated NEC of this sensor reached 1.1 ppmv of SO_2 and an excellent linearity of the senor response (R-square value of 0.9998). This work clearly illustrates the high potential for further development of FP-based PT sensors for sensitive and selective gas detection. Moreover, optimized dimensions of the sensor limit the volume of gas required for measurements (0.7 cm^3).

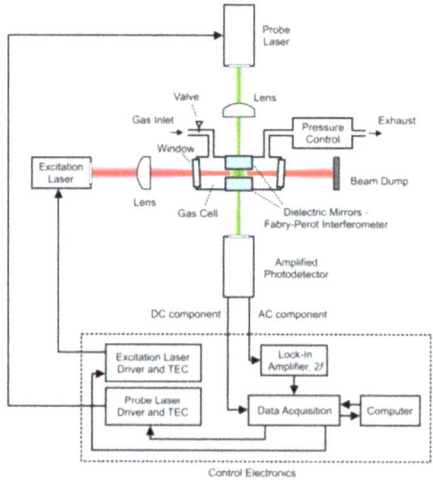

Figure 13. Schematic of the Fabry-Perot interferometer-based photothermal gas sensor. Reprinted with permission from [59] C The Optical Society.

5.2. Balanced Detection Fabry-Perot PT Gas Sensor

The work presented in Reference [59] was quickly followed by a second publication by the same Authors. In Reference [60], Waclawek et al. addressed the main issues of their previous sensor and proposed a similar configuration, but enriched with balanced detection of the PT-induced intensity modulation of the beam leaking from the FP cavity. The modified sensor layout is presented in Figure 14.

Compared to the previous configuration, in this improved version of the FP-based PT gas sensor, the *probe* beam is split into a reference beam and probing beam before entering the interferometer cavity (both beams travel through the same mirrors). The beam separation was 10 mm. Here, only one of the beams was intersected with the *pump* beam injected into the cavity perpendicularly to the *probe* beams. Both beams are detected by separate photodiodes after exiting the FP cavity, and the signals are subtracted using a differential amplifier. Both *probe* beams experience common noise and therefore can be easily cancelled out. In FP interferometers, the registered noise is enhanced proportionally to the finesse of the cavity. Balanced detection significantly limited the registered noise; therefore, this simple improvement allowed the authors to use higher reflectivity mirrors to achieve higher finesse of the cavity and increase the overall sensitivity of the sensor. In this experiment, mirrors with reflectivity R = 0.985 were used, forming a cavity with a finesse of ~89. The *pump* laser and the signal processing part of the sensor was identical to the previously reported, and was based on WMS detection technique via modulating the QCL laser with sinusoidal waveform in the vicinity of SO_2 absorption line (1390.93 cm^{-1}) and a LIA-based signal retrieval. This improved version of the sensor reached an NEC of 5 ppbv at an acquisition time of 1 s. This result is 240 times better than reported in the previous version of the sensor.

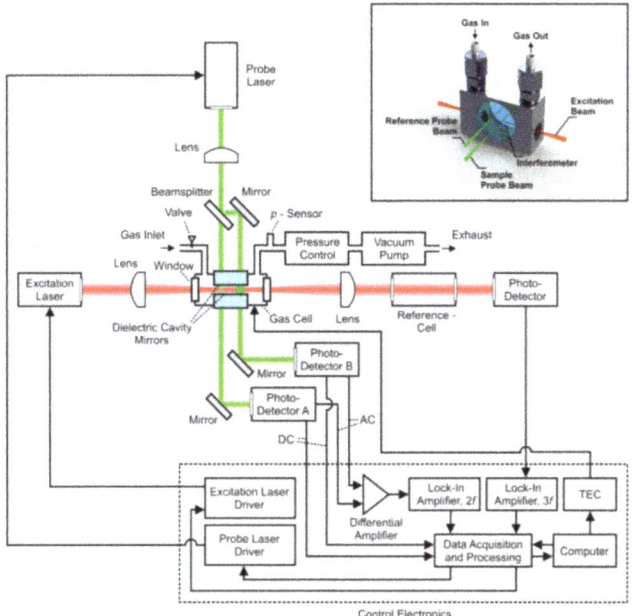

Figure 14. Balanced-detection Fabry-Perot interferometer PT gas sensor. Reprinted with permission from [60] C The Optical Society.

6. Conclusions

In this review, progresses in PT gas spectroscopy have been presented and discussed based on several sensor layouts and signal retrieval techniques. As presented in this review, PT gas detection can be accomplished in numerous dissimilar configurations, thus encouraging researchers to search for novel, yet unpublished, methods of utilizing the PT effect for precise and selective trace gas analysis. An appropriately designed PT gas sensor can reach sensitivities comparable with WMS or TDLAS techniques, without the necessity of using bulky and fragile multipass cells. Basic parameters of the sensors described in this review are summarized in Table 1.

Based on the summary provided in Table 1, the most promising PT gas detection techniques are the FP interferometer-based technique and the ones based on a hollow-core fiber. FP interferometric readout of the PT-induced RI modulation has proven to be very sensitive and foremost miniature. Moreover, with proper design, this configuration is capable of reaching strong absorption lines of molecules located in the mid-infrared wavelength region. Similar results, in terms of sensitivity, were achieved in HC-PCF-based PT sensors. This configuration is particularly interesting due to the fact that the sensors were constructed all in fiber, thus having limited susceptibility to environmental noise. The only limitation is the wavelength of operation. Currently available HC fibers have tremendous attenuation in the mid-IR wavelength region, therefore restricting the implementation of such PT sensors to gases having absorption lines in the near infrared. This configuration will surely be further investigated and improved based on novel antiresonant hollow-core fibers with limited losses in the 3–7 µm wavelength region. An interesting approach to PTS as proposed in Reference [16], where the induced RI modulation was directly translated to frequency variation of an ML laser repetition rate. This ensured non-complex signal processing and versatility of the constructed sensor. The most pronounced drawback of the sensor was the necessity of combining fiber-based components with bulk optics-based gas absorption cell. This issue could be easily mitigated by incorporating a section of a hollow-core fiber-based gas cell into the ML laser resonator. This would limit the influence of environmental noise and considerably improve the robustness and compactness of the sensor. Significant improvement in terms of the registered signal would also be possible in the least complicated, yet very sensitive FP-based PTS gas sensor. This would require designing the FP resonator to utilize an additional acoustic resonance inside the gas cell, similar to configurations used in quartz-enhanced photoacoustic spectroscopy technique (QEPAS) [17]. The acoustic resonance should increase the registered signal several times, making the sensor configuration sensitive, selective and miniaturized. Nevertheless, similar to PAS, in PTS, the sensor can be designed to target strong particle transitions in the mid-IR, but at the same time have signal readout based on inexpensive fiber components. Considering the fact that the PT-induced RI modulation can be probed in numerous different ways (e.g., as variations in laser intensity, frequency or in a interferometric configuration), there is there is still plenty of space for further development of these techniques, which will surely be perused by researchers working on laser-based gas sensing.

Table 1. Summarized sensitivities of PT gas sensors described in this review.

Configuration	Detected Gas	*Pump* Laser Type; Wavelength	*Pump* Laser Power	NEC	Reference
Bulk-optics based HeNe Mach-Zehnder interferometer	CH_3OH	CO_2 laser, 9.64 µm	20 W	3 ppbv	[18]
Unfolded Jamin HeNe interferometer	NH_3	CO_2 laser, 9.22 µm	7 W	250 pptv	[22]
Open-path fiber-based Mach-Zehnder interferometer	CO_2	Amplified DFB diode, 1573 nm	300 mW	400 ppmv	[24]
Fiber-based multipass-assisted Mach-Zehnder interferometer	CH_4	QCL, 7.2 µm	50 mW	10 ppmv	[32]
HC-PCF-based Mach-Zehnder interferometer	C_2H_2	Amplified DFB diode, 1530.37 nm	25 mW	2 ppbv	[41]
HC-PCF-based pulsed PT sensor with Sagnac loop interferometer	C_2H_2	Amplified DFB diode, 1530.37 nm	22.2 mW	3.3 ppmv	[45]
Intracavity PT sensor based on a HeNe laser	SF_6	CO_2, 10,560 nm	0.8 W	3 ppbv	[55]
PT sensor with intracavity *pump* signal enhancement	C_2H_2	DFB diode, 1530.37 nm	20 mW	0.176 ppm	[56]
PT gas sensor in a intracavity mode-locked fiber laser configuration	CO_2	Amplified DFB diode, 2003 nm	100 mW	311 ppmv	[16]
Fabry-Perot interferometer-based PT gas sensor	SO_2	QCL, 7246 nm	173 mW	1.1 ppmv	[59]
Balanced detection Fabry-Perot interferometer-based PT gas sensor	SO_2	QCL, 7246 nm	174 mW	5 ppbv	[60]

Funding: This work was financially supported by the statutory funds of the Chair of EM Field Theory, Electronic Circuits and Optoelectronics, Wroclaw University of Science and Technology (0401/0030/18).

Conflicts of Interest: The author declares no conflict of interest.

References

1. Boccara, A.C.; Fournier, D.; Jackson, W.; Amer, N.M. Sensitive photothermal deflection technique for measuring absorption in optically thin media. *Opt. Lett.* **1980**, *5*, 377–379. [CrossRef] [PubMed]
2. Cabrera, H.; Akbar, J.; Korte, D.; Ramírez-Miquet, E.E.; Marín, E.; Niemela, J.; Ebrahimpour, Z.; Mannatunga, K.; Franko, M. Trace detection and photothermal spectral characterization by a tuneable thermal lens spectrometer with white-light excitation. *Talanta* **2018**, *183*, 158–163. [CrossRef] [PubMed]
3. Bialkowski, S. *Photothermal Spectroscopy Methods for Chemical Analysis*; John Wiley & Sons: New York, NY, USA, 1996; Volume 134.
4. Kosterev, A.A.; Bakhirkin, Y.A.; Curl, R.F.; Tittel, F.K. Quartz-enhanced photoacoustic spectroscopy. *Opt. Lett.* **2002**, *27*, 1902–1904. [CrossRef] [PubMed]
5. Harren, F.J.M.; Mandon, J.; Cristescu, S.M. Photoacoustic spectroscopy in trace gas monitoring. Encyclopedia of Analytical Chemistry: Applications. *Theory Instrum.* **2006**. [CrossRef]
6. Sigrist, M.W. Trace gas monitoring by laser photoacoustic spectroscopy and related techniques (plenary). *Rev. Sci. Instrum.* **2003**, *74*, 486–490. [CrossRef]
7. Roller, C.; Namjou, K.; Jeffers, J.D.; Camp, M.; Mock, A.; McCann, P.J.; Grego, J. Nitric oxide breath testing by tunable-diode laser absorption spectroscopy: Application in monitoring respiratory inflammation. *Appl. Opt.* **2002**, *41*, 6018–6029. [CrossRef] [PubMed]
8. Werle, P.O.; Mücke, R.; Slemr, F. The limits of signal averaging in atmospheric trace-gas monitoring by tunable diode-laser absorption spectroscopy (TDLAS). *Appl. Phys. B* **1993**, *57*, 131–139. [CrossRef]
9. Cassidy, D.T.; Reid, J. Atmospheric pressure monitoring of trace gases using tunable diode lasers. *Appl. Opt.* **1982**, *21*, 1185–1190. [CrossRef]
10. Bialkowski, S.E.; Chartier, A. Diffraction effects in single-and two-laser photothermal lens spectroscopy. *Appl. Opt.* **1997**, *36*, 6711–6721. [CrossRef]
11. Bialkowski, S.E. Photothermal lens aberration effects in two laser thermal lens spectrophotometry. *Appl. Opt.* **1985**, *24*, 2792–2796. [CrossRef]
12. Korte, D.; Cabrera, H.; Toro, J.; Grima, P.; Leal, C.; Villabona, A.; Franko, M. Optimized frequency dependent photothermal beam deflection spectroscopy. *Laser Phys. Lett.* **2016**, *13*, 125701. [CrossRef]
13. Jackson, W.B.; Amer, N.M.; Boccara, A.C.; Fournier, D. Photothermal deflection spectroscopy and detection. *Appl. Opt.* **1981**, *20*, 1333–1344. [CrossRef] [PubMed]
14. Supplee, J.M.; Whittaker, E.A.; Lenth, W. Theoretical description of frequency modulation and wavelength modulation spectroscopy. *Appl. Opt.* **1994**, *33*, 6294–6302. [CrossRef] [PubMed]
15. Schilt, S.; Thevenaz, L.; Robert, P. Wavelength modulation spectroscopy: Combined frequency and intensity laser modulation. *Appl. Opt.* **2003**, *42*, 6728–6738. [CrossRef] [PubMed]
16. Krzempek, K.; Dudzik, G.; Abramski, K. Photothermal spectroscopy of CO_2 in an intracavity mode-locked fiber laser configuration. *Opt. Express* **2018**, *26*, 28861–28871. [CrossRef] [PubMed]
17. Ma, Y. Review of recent advances in QEPAS-based trace gas sensing. *Appl. Sci.* **2018**, *8*, 1822. [CrossRef]
18. Davis, C.C.; Petuchowski, S.J. Phase fluctuation optical heterodyne spectroscopy of gases. *Appl. Opt.* **1981**, *20*, 2539–2554. [CrossRef]
19. Wysocki, G.; Weidmann, D. Molecular dispersion spectroscopy for chemical sensing using chirped mid-infrared quantum cascade laser. *Opt. Express* **2010**, *18*, 26123–26140. [CrossRef]
20. Stone, J. Thermooptical technique for the measurement of absorption loss spectrum in liquids. *Appl. Opt.* **1973**, *12*, 1828–1830. [CrossRef]
21. Murty, M.V.R.K. Some modifications of the Jamin interferometer useful in optical testing. *Appl. Opt.* **1964**, *3*, 535–538. [CrossRef]
22. Owens, M.A.; Davis, C.C.; Dickerson, R.R. A photothermal interferometer for gas-phase ammonia detection. *Anal. Chem.* **1999**, *71*, 1391–1399. [CrossRef] [PubMed]
23. Altmann, J.; Baumgart, R.; Weitkamp, C. Two-mirror multipass absorption cell. *Appl. Opt.* **1981**, *20*, 995–999. [CrossRef] [PubMed]
24. Krzempek, K.; Dudzik, G.; Abramski, K.; Wysocki, G.; Jaworski, P.; Nikodem, M. Heterodyne interferometric signal retrieval in photoacoustic spectroscopy. *Opt. Express* **2018**, *26*, 1125–1132. [CrossRef] [PubMed]
25. Giles, C.R.; Desurvire, E. Modeling erbium-doped fiber amplifiers. *J. Lightwave Technol.* **1991**, *9*, 271–283. [CrossRef]

26. Gibilisco, S. *Teach Yourself Electricity and Electronics*; McGraw-Hill Professional: New York, NY, USA, 2002; p. 477. ISBN 978-0-07-137730-0.
27. Murvay, P.S.; Silea, I. A survey on gas leak detection and localization techniques. *J. Loss Prev. Process Ind.* **2012**, *25*, 966–973. [CrossRef]
28. Chraim, F.; Erol, Y.B.; Pister, K. Wireless gas leak detection and localization. *IEEE Trans. Ind. Inf.* **2015**, *12*, 768–779. [CrossRef]
29. Sivathanu, Y. *Natural Gas Leak Detection in Pipelines*; Technology Status Report; En'Urga Inc.: West Lafayette, IN, USA, 2003.
30. Detto, M.; Verfaillie, J.; Anderson, F.; Xu, L.; Baldocchi, D. Comparing laser-based open-and closed-path gas analyzers to measure methane fluxes using the eddy covariance method. *Agric. For. Meteorol.* **2011**, *151*, 1312–1324. [CrossRef]
31. Grant, W.B.; Kagann, R.H.; McClenny, W.A. Optical remote measurement of toxic gases. *J. Air Waste Manag. Assoc.* **1992**, *42*, 18–30. [CrossRef]
32. Krzempek, K.; Hudzikowski, A.; Głuszek, A.; Dudzik, G.; Abramski, K.; Wysocki, G.; Nikodem, M. Multi-pass cell-assisted photoacoustic/photothermal spectroscopy of gases using quantum cascade laser excitation and heterodyne interferometric signal detection. *Appl. Phys. B* **2018**, *124*, 74. [CrossRef]
33. Faist, J.; Capasso, F.; Sivco, D.L.; Sirtori, C.; Hutchinson, A.L.; Cho, A.Y. Quantum cascade laser. *Science* **1994**, *264*, 553–556. [CrossRef]
34. Parry, J.P.; Griffiths, B.C.; Gayraud, N.; McNaghten, E.D.; Parkes, A.M.; MacPherson, W.N.; Hand, D.P. Towards practical gas sensing with micro-structured fibres. *Meas. Sci. Technol.* **2009**, *20*, 075301. [CrossRef]
35. Smolka, S.; Barth, M.; Benson, O. Highly efficient fluorescence sensing with hollow core photonic crystal fibers. *Opt. Express* **2007**, *15*, 12783–12791. [CrossRef] [PubMed]
36. Cubillas, A.M.; Silva-Lopez, M.; Lazaro, J.M.; Conde, O.M.; Petrovich, M.N.; Lopez-Higuera, J.M. Methane detection at 1670-nm band using a hollow-core photonic bandgap fiber and a multiline algorithm. *Opt. Express* **2007**, *15*, 17570–17576. [CrossRef] [PubMed]
37. Charlton, C.; Temelkuran, B.; Dellemann, G.; Mizaikoff, B. Midinfrared sensors meet nanotechnology: Trace gas sensing with quantum cascade lasers inside photonic band-gap hollow waveguides. *Appl. Phys. Lett.* **2005**, *86*, 194102. [CrossRef]
38. Wynne, R.M.; Barabadi, B.; Creedon, K.J.; Ortega, A. Sub-minute response time of a hollow-core photonic bandgap fiber gas sensor. *J. Lightwave Technol.* **2009**, *27*, 1590–1596. [CrossRef]
39. Yang, F.; Jin, W.; Cao, Y.; Ho, H.L.; Wang, Y. Towards high sensitivity gas detection with hollow-core photonic bandgap fibers. *Opt. Express* **2014**, *22*, 24894–24907. [CrossRef] [PubMed]
40. Nikodem, M.; Krzempek, K.; Dudzik, G.; Abramski, K. Hollow core fiber-assisted absorption spectroscopy of methane at 3.4 μm. *Opt. Express* **2018**, *26*, 21843–21848. [CrossRef] [PubMed]
41. Jin, W.; Cao, Y.; Yang, F.; Ho, H.L. Ultra-sensitive all-fibre photothermal spectroscopy with large dynamic range. *Nat. Commun.* **2015**, *6*, 6767. [CrossRef]
42. Rosenthal, A.; Kellnberger, S.; Sergiadis, G.; Ntziachristos, V. Wideband fiber-interferometer stabilization with variable phase. *IEEE Photonics Technol. Lett.* **2012**, *24*, 1499–1501. [CrossRef]
43. Smith, C.M.; Venkataraman, N.; Gallagher, M.T.; Müller, D.; West, J.A.; Borrelli, N.F.; Koch, K.W. Low-loss hollow-core silica/air photonic bandgap fibre. *Nature* **2003**, *424*, 657. [CrossRef]
44. Roberts, P.J.; Couny, F.; Sabert, H.; Mangan, B.J.; Williams, D.P.; Farr, L.; Russell, P.S.J. Ultimate low loss of hollow-core photonic crystal fibres. *Opt. Express* **2005**, *13*, 236–244. [CrossRef] [PubMed]
45. Lin, Y.; Jin, W.; Yang, F.; Ma, J.; Wang, C.; Ho, H.L.; Liu, Y. Pulsed photothermal interferometry for spectroscopic gas detection with hollow-core optical fibre. *Sci. Rep.* **2016**, *6*, 39410. [CrossRef] [PubMed]
46. Goldberg, L.; Mehuys, D. High power superluminescent diode source. *Electron. Lett.* **1994**, *30*, 1682–1684. [CrossRef]
47. Bristow, A.D.; Rotenberg, N.; Van Driel, H.M. Two-photon absorption and Kerr coefficients of silicon for 850–2200 nm. *Appl. Phys. Lett.* **2007**, *90*, 191104. [CrossRef]
48. Nikodem, M.; Gomółka, G.; Klimczak, M.; Pysz, D.; Buczyński, R. Laser absorption spectroscopy at 2 μm inside revolver-type anti-resonant hollow core fiber. *Opt. Express* **2019**, *27*, 14998–15006. [CrossRef] [PubMed]
49. Belardi, W.; Knight, J.C. Hollow antiresonant fibers with low bending loss. *Opt. Express* **2014**, *22*, 10091–10096. [CrossRef] [PubMed]

50. Belardi, W.; Knight, J.C. Hollow antiresonant fibers with reduced attenuation. *Opt. Lett.* **2014**, *39*, 1853–1856. [CrossRef] [PubMed]
51. Belardi, W. Design and properties of hollow antiresonant fibers for the visible and near infrared spectral range. *J. Lightwave Technol.* **2015**, *33*, 4497–4503. [CrossRef]
52. Hasan, M.I.; Akhmediev, N.; Chang, W. Mid-infrared supercontinuum generation in supercritical xenon-filled hollow-core negative curvature fibers. *Opt. Lett.* **2016**, *41*, 5122–5125. [CrossRef]
53. Wang, Z.; Belardi, W.; Yu, F.; Wadsworth, W.J.; Knight, J.C. Efficient diode-pumped mid-infrared emission from acetylene-filled hollow-core fiber. *Opt. Express* **2014**, *22*, 21872–21878. [CrossRef]
54. Sollapur, R.; Kartashov, D.; Zürch, M.; Hoffmann, A.; Grigorova, T.; Sauer, G.; Chemnitz, M. Resonance-enhanced multi-octave supercontinuum generation in antiresonant hollow-core fibers. *Light: Sci. Appl.* **2017**, *6*, e17124. [CrossRef] [PubMed]
55. Fung, K.H.; Lin, H.B. Trace gas detection by laser intracavity photothermal spectroscopy. *Appl. Opt.* **1986**, *25*, 749–752. [CrossRef] [PubMed]
56. Zhao, Y.; Jin, W.; Lin, Y.; Yang, F.; Ho, H.L. All-fiber gas sensor with intracavity photothermal spectroscopy. *Opt. Lett.* **2018**, *43*, 1566–1569. [CrossRef] [PubMed]
57. Keller, U.; Weingarten, K.J.; Kartner, F.X.; Kopf, D.; Braun, B.; Jung, I.D.; Der Au, J.A. Semiconductor saturable absorber mirrors (SESAM's) for femtosecond to nanosecond pulse generation in solid-state lasers. *IEEE J. Sel. Top. Quantum Electron.* **1996**, *2*, 435–453. [CrossRef]
58. Reider, G.A. *Photonics*; Springer: New York, NY, USA, 2016.
59. Waclawek, J.P.; Bauer, V.C.; Moser, H.; Lendl, B. 2f-wavelength modulation Fabry-Perot photothermal interferometry. *Opt. Express* **2016**, *24*, 28958–28967. [CrossRef] [PubMed]
60. Waclawek, J.P.; Kristament, C.; Moser, H.; Lendl, B. Balanced-detection interferometric cavity-assisted photothermal spectroscopy. *Opt. Express* **2019**, *27*, 12183–12195. [CrossRef]

© 2019 by the author. Licensee MDPI, Basel, Switzerland. This article is an open access article distributed under the terms and conditions of the Creative Commons Attribution (CC BY) license (http://creativecommons.org/licenses/by/4.0/).

Review

A Review of Femtosecond Laser-Induced Emission Techniques for Combustion and Flow Field Diagnostics

Bo Li [1], Dayuan Zhang [1], Jixu Liu [1], Yifu Tian [1], Qiang Gao [1,*] and Zhongshan Li [1,2]

[1] State Key Laboratory of Engines, Tianjin University, Tianjin 300072, China; boli@tju.edu.cn (B.L.); 2015201040@tju.edu.cn (D.Z.); jixuliu@tju.edu.cn (J.L.); tianyifu@tju.edu.cn (Y.T.); zhongshan.li@forbrf.lth.se (Z.L.)
[2] Division of Combustion Physics, Lund University, S-22100 Lund, Sweden
* Correspondence: qiang.gao@tju.edu.cn; Tel.: +86-022-2740-7623-8022

Received: 27 March 2019; Accepted: 6 May 2019; Published: 9 May 2019

Abstract: The applications of femtosecond lasers to the diagnostics of combustion and flow field have recently attracted increasing interest. Many novel spectroscopic methods have been developed in obtaining non-intrusive measurements of temperature, velocity, and species concentrations with unprecedented possibilities. In this paper, several applications of femtosecond-laser-based incoherent techniques in the field of combustion diagnostics were reviewed, including two-photon femtosecond laser-induced fluorescence (fs-TPLIF), femtosecond laser-induced breakdown spectroscopy (fs-LIBS), filament-induced nonlinear spectroscopy (FINS), femtosecond laser-induced plasma spectroscopy (FLIPS), femtosecond laser electronic excitation tagging velocimetry (FLEET), femtosecond laser-induced cyano chemiluminescence (FLICC), and filamentary anemometry using femtosecond laser-extended electric discharge (FALED). Furthermore, prospects of the femtosecond-laser-based combustion diagnostic techniques in the future were analyzed and discussed to provide a reference for the relevant researchers.

Keywords: combustion diagnostic; femtosecond laser; two-photon femtosecond laser-induced fluorescence; femtosecond laser-induced breakdown spectroscopy; femtosecond laser electronic excitation tagging; filament-induced nonlinear spectroscopy; femtosecond laser-induced plasma spectroscopy

1. Introduction

Development of advanced combustion technologies with high efficiency and low emission requires an understanding of the complex processes involving multi-scale kinetic reactions and turbulent flows. Numerous key parameters and information rely on accurate diagnostics techniques. Conventional diagnostic techniques include pressure measurement with sensors [1], velocity measurement with pitot tubes [2], and temperature measurement with thermocouples [3]. These methods, however, are all intrusive and have profound interference to combustion and flow fields, and the information extracted might be invalid, especially when high turbulence gets involved, so instantaneous measurements are needed. The advent of laser-based techniques dramatically promotes the development of diagnostic tools in these fields [4,5], which provides more reliable and new information to support further research. Compared with intrusive techniques, laser-based techniques can prominently reduce or even avoid the interference. In addition, high sensitivity, high temporal, and spatial resolution can be obtained in real time, and simultaneous multi-parameter measurements can also be achieved. Aldén [4] mentioned that "laser diagnostic techniques have for more than 30 years added very valuable input for a deepened understanding of combustion processes". Kohse-Höinghaus [5] further stated that laser-based diagnostic technique "is the only way to 'spy' on the fundamentals of combustion inside". The light source is the core component of laser diagnostic techniques, a typical example of which is nanosecond (ns) lasers. Over recent decades, ns-laser diagnostics have been widely adopted in characterizing many aspects of combustion and flow fields, e.g., flow and flame structure visualization [6], species measurement [7], temperature measurement [8], velocity measurement [9], and soot characteristics analysis [10]. Various ns-laser diagnostics, however, all have their own limitations. For example, in two-photon laser-induced fluorescence (TPLIF), it is hard to find the balance between high laser fluence, which is needed to compensate for the reduced absorption cross-section, and the laser-induced photolysis, which is proportional to laser fluence and is something we would like to avoid. Other challenges faced by ns-laser techniques include measuring in the near-wall region and in the sooty environment, and so on. Hence, it is necessary to develop new techniques to solve the problems mentioned above.

The advent of femtosecond (fs) lasers [11] has significantly enriched the laser-based diagnostic techniques. At present, there are many commercial fs lasers, and the average power of the fs laser pulse could reach the order of tens of watts, and the repetition rate can reach the order of GHz. Among them, the most commonly used fs laser for diagnostics in combustion and flow field is the Ti:sapphire laser system, whose wavelength centered at ~800 nm with a pulse duration of ~45 fs, a pulse energy of 8 mJ, and a repetition rate of 1 kHz. Compared with ns lasers, fs lasers have the characteristics of a narrower pulse duration, a higher peak power, and a broader line width. Hence, when fs lasers interact with gases, different behaviors and phenomena will be observed, such as fs laser filamentation [12,13]. To understand these behaviors and phenomena, a direct and convenient way is to monitor the emission that arises from the volume of the fs laser-gas interaction. By doing so, we can extract information from the gas field of interest, which will not be stimulated by ns lasers. Hopefully, this will shed light on solving the difficulties faced by ns-laser diagnostic techniques. For instance, femtosecond multiphoton laser-induced fluorescence (fs-MPLIF) has the advantages of high multiphoton excitation efficiency and, at the same time, low photolysis interference. Hence, the primary purpose of this paper is to discuss in detail these advantages along with the emerging diagnostic techniques based on them and to introduce the applications of fs laser-induced emission techniques for combustion and flow field diagnostics.

In this paper, we put these incoherent techniques based on fs laser-induced emission into four categories. In Section 2, we focus on a resonant excitation technique: Femtosecond two-photon laser-induced fluorescence (fs-TPLIF). In Section 3, we introduce a non-resonant excitation technique: Femtosecond laser-induced breakdown spectroscopy (fs-LIBS). In Section 4, we describe techniques that rely on fs laser-induced filamentation: Filament-induced nonlinear spectroscopy (FINS), femtosecond laser-induced plasma spectroscopy (FLIPS), femtosecond laser electronic excitation tagging (FLEET) and femtosecond laser-induced cyano chemiluminescence (FLICC). In Section 5, we outline a technique based on fs laser-guiding electric discharge: Filamentary anemometry using femtosecond laser-extended electric discharge (FALED). The paper is closed with a summary in Section 6.

2. Femtosecond Two-Photon Laser-Induced Fluorescence (fs-TPLIF)

Laser-induced fluorescence (LIF) is one of the most common techniques for combustion diagnostics [7], which uses a resonant laser beam to excite a selected transition from a special molecule (or atom) to a higher excited state and detects the induced fluorescence. Various combustion intermediate species can be measured by LIF [14,15]. Based on fluorescence imaging, two-dimensional visualization can also be achieved by planner LIF [16].

For LIF measurements, one-photon excitation is preferred, since it has a linear response and high absorption cross-section. However, many key combustion intermediates, such as atomic hydrogen (H) [17], atomic oxygen (O) [18] and carbon dioxide (CO) [19], have a large energy gap between the ground and the first excited electronic states, which makes one-photon LIF impossible as the wavelength falls into the region of vacuum-ultraviolet (VUV). To address this complication, researchers developed a two-photon excitation scheme. Two-photon laser-induced fluorescence (TPLIF) was first demonstrated for H and deuterium measurements by Bokor et al. [20], and then it was applied to detect many other species, including O [21–23] and N [22]. A limitation of two-photon excitation is that high laser fluence is required in order to overcome the issue of small absorption cross-section. The high laser fluence could cause significant photo-dissociation. For example, in TPLIF detection of H, a substantial amount of additional H can be produced via photodissociation [24] of some hydrogen-containing radicals, such as H_2O_2 [25] and CH_3 [26], which makes it difficult to distinguish the natively generated H in the flame and the photolytic H. Even if the photolytic products are different from the species of interest, they may also cause interference. For example, when performing CO-TPLIF measurements in a sooty flame, the ns-laser at 230 nm, which is used to excite CO molecules, will generate photolytic C_2 radicals. Moreover, the so-called Swan-band emissions from C_2 will interfere with the CO detection, since they partially overlap with the CO fluorescence.

To circumvent the above-mentioned issues, researchers use fs lasers in place of ns lasers. By assuming that the molecular transitions have a linewidth much narrower than the excitation laser energy distribution [27], the efficiency of two-photon excitation can be expressed as

$$S_{\text{LIF}} \propto \sigma_0 \cdot I_0^2 \frac{1}{c_\omega \cdot c_t}, \tag{1}$$

where S_{LIF} is the fluorescence signal, σ_0 is the two-photon rate coefficient, I_0 is the laser pulse energy of the excitation laser, c_ω is the laser spectral linewidth of the laser, and c_t is the laser pulse duration, so the relative TPLIF efficiency can be estimated by the product of $c_t \cdot c_\omega$ of the excitation laser pulse. For a Fourier transform limited (FTL) laser pulse, $c_t \cdot c_\omega$ is a constant and has the smallest value, which is the most favorable for the two-photon excitation. Most fs laser pulses are close to FTL, while for most ns or picosecond (ps) laser pulses, the $c_t \cdot c_\omega$ values are generally larger by a factor of around 100. For detection of molecular species, the possibility of a simultaneous excitation of the whole vibration band (or even including some hot bands) forms another extra advantage of fs-TPLIF in contrast to the ns and ps cases, where only a couple of rotational lines can be excited. This effect will be more evident in flames, where much more rotational lines and even hot vibrational bands can be populated.

Therefore, even if the pulse energy of an fs laser is two orders of magnitude lower than that of a typical ns laser, the fluorescence intensity is similar to that in ns or ps TPLIF measurements [28,29]. Considering that the photolytic interferences are often linearly proportional to the laser pulse energy, the photodissociation problem will be mitigated to a large extent for fs-TPLIF. Besides the advantage of photolytic interference suppression, the effective utilization of the whole broadband profile of the fs laser pulse due to the existence of multi photon-pairs in matching the two-photon resonant excitation has also been mentioned by Kulatilaka et al. [30].

So far, fs-TPLIF has been applied to measure various combustion intermediates and products, including molecules (e.g., CO), atoms (e.g., H and O), and radicals (e.g., OH). Inert gases (e.g., Kr and Xe) are also of interest because crucial parameters, like mixing fraction, could be extracted by measuring them. The excitation and detection strategies of the abovementioned species are listed in Table 1.

Table 1. Excitation and detection strategies of various species by multi-photo femtosecond laser induced fluorescence.

Species	Excitation		Detection		Ref.
	Wavelength	Transition	Wavelength	Transition	
OH	620 nm*2	$A^2\Sigma^+ \leftarrow X^2\Pi$	~310 nm	$A^2\Sigma^+ \to X^2\Pi$	[31,32]
CO	230 nm*2	$B^1\Sigma^+ \leftarrow X^1\Sigma^+$	400–600 nm	$B^1\Sigma^+ \to A^1\Pi^+$	[27,33–38]
	230 nm*2	$B^1\Sigma^+ \leftarrow X^1\Sigma^+$	280–380 nm	$b^3\Sigma^+ \to a^3\Pi$	[34]
H	205 nm*2	$n=3 \leftarrow n=1$	656 nm	$n=3 \to n=2$	[30,39–50]
	307 nm*3	$n=3 \leftarrow n=1$	656 nm	$n=3 \to n=2$	[40,47]
	243 nm*2 + 486 nm	$n=4 \leftarrow n=1$	656 nm	$n=3 \to n=2$	[51]
O	226 nm*2	$3p^3P \leftarrow 2p^3P$	845 nm	$3p^3P \to 3s^3S$	[49,52,53]
Kr	204.1 nm*2	$5p'[3/2]_2 \leftarrow 4p^6[^1S_0]$	826 nm	$5p'[3/2]_2 \to 5s'[1/2]_1$	[49,50,54]
	204.1 nm*2	$5p'[3/2]_2 \leftarrow 4p^6[^1S_0]$	750–840 nm		[55,56]
	212.6 nm*2	$4s^24p^55p/4s^24p^55p' \leftarrow 4s^24p^6$	759 nm		[57]
Xe	225 nm*2	$6p'[3/2]_2 \leftarrow 5p^{61}S_0$	835 nm	$6p'[3/2]_2 \leftarrow 6s'[1/2]_1$	[49,53]

Note: Refs. [50–52,55] are measured in plasma.

The measurement of CO by fs-TPLIF was first performed by Richardson et al. [37] in a steady flame. In their research, excitation wavelength at 230.1 nm was used to induce two-photon transition ($B^1\Sigma^+ \leftarrow X^1\Sigma^+$) of CO, and fluorescence in the range of 362–516 nm was captured. Later on, this excitation scheme has been extended into high-pressure flames [34], sooty flames [34], and piloted liquid-spray flames [33]. Our group also employed this scheme in a premixed laminar jet flame [27]. A hot band (1,n) together with the conventional band (0,n) of $B^1\Sigma^+ \to A^1\Pi^+$ transitions were observed in the burned zone of the flame. As shown in Figure 1b, the $B^1\Sigma^+ \leftarrow X^1\Sigma^+(1,1)$ band was excited due to the broadband nature of the fs laser pulse. We have seen that the two vibrational bands are sensitive to temperature and can potentially be used for accurate flame temperature measurements. In addition, the CO fs-TPLIF signal recorded across the focal point of the excitation beam shows a relatively flat intensity distribution despite the steep laser intensity variation, which is beneficial for CO imaging in contrast to ns and ps TPLIF.

For atomic intermediates in combustion, fs-TPLIF has been applied to H and O measurements. Atomic hydrogen is a critical species in the combustion of hydrogen or hydrocarbon fuels [58], which gets involved in flame ignition, propagation, and extinction. The H fs-TPLIF technique was first demonstrated in combustion by Kulatilaka et al. [30,46]. In their scheme, H was excited by a 205 nm fs laser, and the subsequent H_α fluorescence at 656 nm from the transition $n=3 \to n=2$ was detected, as shown by the red line in Figure 2. This excitation scheme has also been adopted for quantitative measurements [44] and 2D imaging [45]. Another extinction scheme, shown as the blue line in Figure 2, was proposed by our group [51]. In the scheme, H is electronically excited through a two-photon process by a 243 nm laser from the ground state to its first excited state ($n=3 \leftarrow n=1$), where the excited H is instantaneously re-excited through another one-photon process by a relay laser (486 nm) to excited state of $n=3$. In order to avoid the stray light interference from the 486 nm laser, the secondary fluorescence at 656 nm ($n=3 \to n=2$) was collected. Consequently, interference-free H imaging was achieved in a laminar jet flame. Furthermore, a three-photon (307.7 nm) excitation scheme ($n=3 \leftarrow n=1$) with detection fluorescence at 656 nm was demonstrated by Jain et al. [40], as shown by the green line in Figure 2.

Appl. Sci. **2019**, *9*, 1906

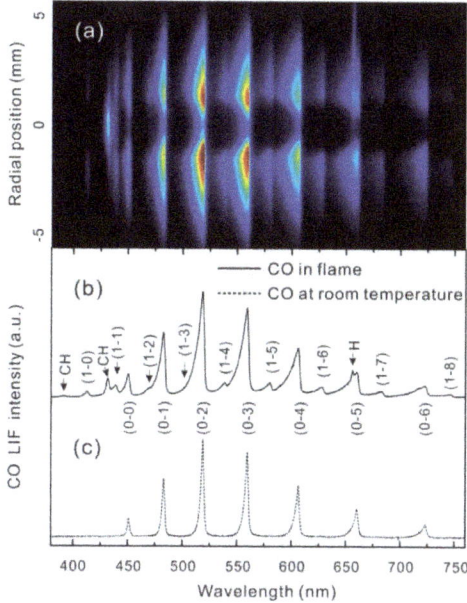

Figure 1. Spectra of CO femtosecond two-photon laser-induced fluorescence (fs-TPLIF): (**a**) Spatially resolved imaging spectrum recorded in a laminar premixed CH$_4$/air jet flame (Φ = 1.5); (**b**) integrated spectral curve of the flame; (**c**) spectral curve of a gas mixture of CO/N$_2$ at room temperature [27]. (© 2017 Optical Society of America).

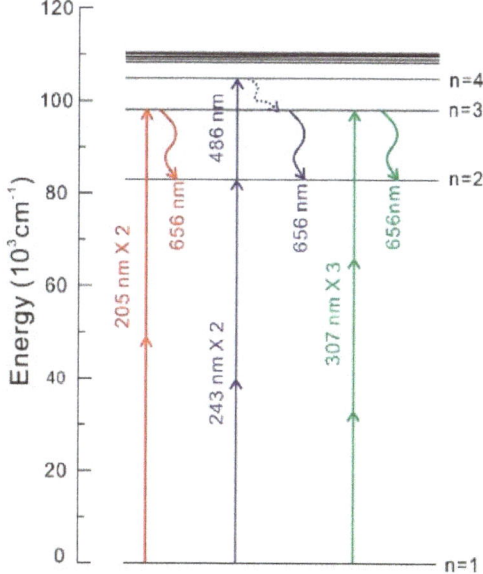

Figure 2. Excitation and detection strategy of atomic hydrogen using the multiphoton femtosecond laser-induced fluorescence (MP-fs-LIF).

Kulatilaka et al. [52] also performed an fs-TPLIF diagnosis for O in premixed laminar flame. An fs laser at 226 nm was used for two-photon excitation of O to its excited state ($3p^3P \leftarrow 2p^3P$), and the subsequent O fluorescence at 845 nm from the transition $3p^3P \rightarrow 3s^3S$ was detected. The advantage of fs lasers excitation over traditional ns lasers excitation in O measurements is that the photolysis of vibrationally excited CO_2, known as the main interference that produces additional O atoms [59], can be eliminated.

Combustion radicals can also be measured by fs-TPLIF. Hydroxyl (OH) is one of the essential combustion radicals, and OH-PLIF is widely used to visualize the reaction zone and product zone of a flame [60–63]. Different from CO, H, and O, OH can be excited with single-photon strategy. A typical single-photon OH-LIF uses an ns laser at ~283 nm to excite OH through a transition line in $A^2\Sigma^+ \leftarrow X^2\Pi$ (0, 1) band, which allows observation of the fluorescence emission from the (1, 1) and (0, 0) bands at ~310 nm [64]. There are many researches of OH-LIF, including ns-laser excitation [62,63] and ps-laser excitation [65]. Stauffer et al. [31,32] performed the fs-TPLIF of OH measurements with two-photon excitation at 620 nm. However, they did not directly observe any obvious advantages of two-photon fs-laser excitation relative to single-photon ns- or ps-laser excitation.

Femtosecond-TPLIF has also been used to measure inert gases such as Kr [54–57] and Xe [49]. Although these studies do not directly measure the combustion species, they are still relevant to combustion research. For example, fs-TPLIF imaging of Kr was achieved to measure the mixture fraction in gaseous flows [57], which is a crucial parameter in the combustion flow field. Femtosecond-TPLIF measurement of Xe was demonstrated in the plasma induced by discharges [49], which might be useful for studying the ignition behavior of flammable gases.

In summary, fs-TPLIF has shown great potential for combustion species measurements, owing to its advantages of high multi-photon efficiency and little photolysis interference. More studies are yet to be performed. For example, fs-TPLIF can be further used to measure other combustion intermediates such as atomic C [66,67] and N [22]. In addition, the technique has not been applied to measure any polyatomic molecule, such as NH_3 [68], which is also of great interest to the combustion community. Furthermore, two-color fs-TPLIF is another strategy that is worth trying. For single-color fs-TPLIF, only one tunable UV laser is needed, but the problem is that its pulse energy is quite limited, while for the two-color strategy, we could use two lasers at two wavelengths. Hence, we have the flexibility to choose at least one laser with strong pulse energy, e.g., from the fourth-harmonic output (200 nm) of an fs Ti:sapphire laser, and then select a tunable UV laser to match the strong laser. In doing so, the excitation efficiency might be improved.

3. Femtosecond Laser-Induced Breakdown Spectroscopy (fs-LIBS)

Laser-induced breakdown spectroscopy (LIBS) is broadly applied to combustion diagnostics, largely encouraged by its simplicity in terms of both experimental setup and data evaluation, which have been applied to many fields such as industry [69], chemistry [70], biology [71,72], nuclear [73], earth sciences [74–76], and cultural heritage [77]. Different from LIF, LIBS technique involves non-resonant excitation, and an ns laser with a fixed wavelength meets the demand. For the current development of ns-LIBS, readers can refer to [78,79]. In the field of combustion, LIBS is mainly used to measure the equivalence ratio. When a pulsed laser is focused at a premixed fuel/air mixture, a breakdown will be generated, which acts as a virtual measurement probe. When the assumptions of local thermal equilibrium (LTE) and optically thin plasma are satisfied [80,81], the intensities of the spectral emission lines of the plasma can be used to determine the equivalence ratio quantitatively. In most of the previous studies, the ratios of certain spectral lines of LIBS could be used to correlate and quantify mixture fraction including H (656 nm)/O (777 nm) [82–84], H (656 nm)/N (568 nm) [85], C (833 nm)/N (744 nm) [86], C (833 nm)/O (777 nm) [86], C (711 nm)/(N (745 nm)+O (777 nm)) [87], and C_2 (516.5 nm)/CN (388 nm) [88].

With the development of ultrafast lasers, LIBS with fs lasers as a light source (fs-LIBS) was introduced [89]. In recent years, fs-LIBS has been applied to combustion diagnostics, and it was also mainly used to measure equivalence ratio. Femtosecond-LIBS is similar to ns-LIBS. The fs laser is focused by a spherical lens with short focal length (e.g., 65 mm focal length [90]) to generate a breakdown (or plasma) in the flow field. By measuring and analyzing the emission spectrum of the plasma, the parameters, such as equivalence ratio, can be obtained. Femtosecond-LIBS has the potential to solve some technical defects faced by traditional ns-LIBS. For example, in ns-LIBS measurements, the intrinsic unstable character of avalanche ionization will lead to volatile signal and hence, result in the instability of the equivalence ratio measurements. The situation will be even worse under high pressure or with the presence of a high level of soot [91]. Femtosecond-LIBS, on the other hand, can avoid the problem to some extent [92].

For the first time, Kotzagianni et al. [90,93] performed fs-LIBS technique in a premixed CH_4/air flame. They used the spectral intensity of CN at 388 nm to calibrate equivalence ratio, as shown in Figure 3, and in situ equivalent ratio measurements were realized. In addition, fs-LIBS was also applied by Couris et al. [94]. They used the intensity ratio of H (656 nm)/O (777 nm) and C_2 (516.5 nm)/CN (388 nm) to calibrate the fuel concentration in flames. Patnaik et al. [95] developed a multivariate analysis to understand and mitigate the measurement instability by varying the pulse duration of the light source from ns to fs. The results indicated that LIBS measurements with fs pulse excitation could reduce the instability in equivalence ratio measurements under high pressure.

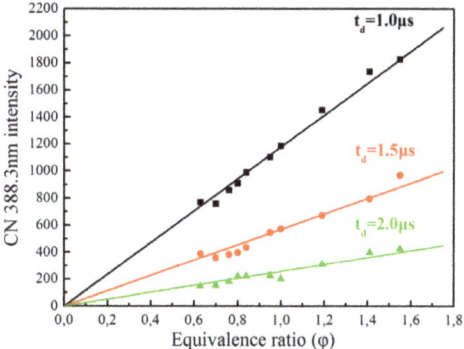

Figure 3. Intensity variation of the CN (388.3 nm) band as a function of the equivalence ratio at different delay times in a CH_4/air flame [93]. (Reproduced from Kotzagianni M., and Couris S., Femtosecond laser induced breakdown for combustion diagnostics. *Appl. Phys. Lett.* 2012, *100*, 264104, with the permission of AIP Publishing).

Although its feasibility in equivalence ratio measurements has been confirmed, more works are yet to be done to fully exploit the advantages of fs-LIBS, e.g., its capability for one-dimensional or high spatial resolution measurements. In addition, similar to ns-LIBS, drastic breakdown happens in fs-LIBS based on tight focusing of the fs laser. Therefore, the spectrum of fs-LIBS is dominated by atomic emissions, which might miss some crucial information related to molecules. Furthermore, fs-LIBS has continuum background interference [90,93], which may disturb the measurements. If using a spherical lens with long focal length, this issue will be reduced. Hence, many techniques depending on femtosecond laser-induced emission have been developed.

4. Femtosecond Laser-Induced Filament Emission

Different from the strong focusing in fs-LIBS, when an fs laser is focused with a longer focal length lens, the focused femtosecond laser can form a stable plasma channel in the optical medium. This phenomenon is also known as filamentation, and the plasma channel is called a filament. Femtosecond laser-induced filamentation is a unique phenomenon that appears during the propagation of a high-intensity ultrashort laser in a transparent medium. This phenomenon was first observed by Braun et al. [96] in 1995. The appearance of fs filament is due to the dynamic balance between the optical Kerr effect-induced self-focusing and the defocusing effect of the self-generated plasma. Chin et al. [13] suggested that filament could be defined as the propagation zone where there is intensity clamping, and its length is at least several times longer than Rayleigh range.

The clamping intensity inside the plasma channel induced by fs laser is up to 10^{13}~10^{14} W/cm^2 [13], which is intense enough to excite atoms/molecules into higher excited states through photolysis, ionization, dissociation, and collision. Then, the excited atoms/molecules release characteristic fluorescence. Therefore, the emission spectra of the plasma channel are rich in information, and the component information of the measured flow field can be obtained qualitatively or quantitatively through spectral analysis. Such non-resonant spectroscopy techniques include filament-induced nonlinear spectroscopy (FINS) and femtosecond laser-induced plasma spectroscopy (FLIPS). In addition, the plasma channel can also excite molecules into higher excited states through laser-induced photochemical reactions, which can extend the fluorescence "lifetime". These long-life signals can be used to measure the velocity of the flow field. Such tagging velocimetry includes femtosecond laser electronic excitation tagging (FLEET) and femtosecond laser-induced cyano chemiluminescence (FLICC).

4.1. Spectroscopy Techniques Based on Femtosecond Laser-Induced Filament Emission

Filament-induced nonlinear spectroscopy (FINS) is a non-resonant spectroscopy technique, which utilizes the short-lifetime signal of the fs laser-induced filament. FINS technique has already demonstrated its potential application prospect in many fields [97,98], such as remote sensing of greenhouse gases [99], pollutants measurements in atmosphere [100], humidity monitoring [101], and heavy-metal contaminants detection in water [102]. The applications of FINS in combustion diagnostics have emerged in recent years.

In 2013, Li et. al. [103] demonstrated the proof-of-concept for FINS in combustion diagnostics, and simultaneous monitoring of multiple combustion intermediates was realized in ethanol/air flame. They compared the emission spectra obtained by FINS, ns-LIBS and spontaneous emission of ethanol/air flame [104], as shown in Figure 4. They found that the signal of FINS in flame mainly results from the interaction between fs laser pulses and the combustion intermediates, such as OH, CH, and C_2. They expanded the application of FINS into different alkanol/air flames [105], i.e., methanol (CH_3OH), ethanol (CH_3CH_2OH), n-propanol ($CH_3(CH_2)_2OH$), n-butanol ($CH_3(CH_2)_3OH$), and n-pentanol ($CH_3(CH_2)_4OH$). Li et al. [106] also demonstrated that FINS could be utilized to sense the combustion intermediates by measuring the backward fluorescence spectra of filament emission in the flame.

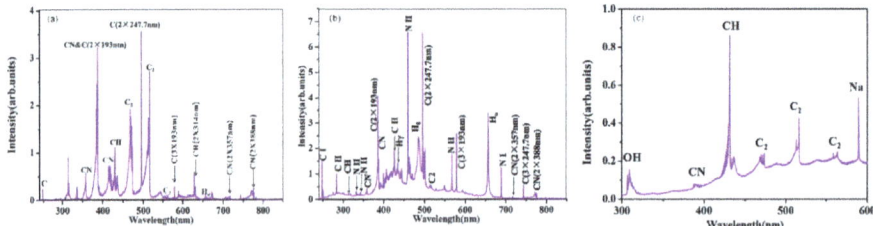

Figure 4. A comparison of emission spectra obtained by (**a**) femtosecond filament excitation, (**b**) nanosecond laser-induced breakdown excitation, and (**c**) without any laser excitation [104]. (Reprinted from Sensors and Actuators: B Chemical, 203, Li H., Wei X., Xu H., Chin S., Yamanouchi K., and Sun H., femtosecond laser filamentation for sensing combustion intermediates: A comparative study, 887, 2014, with permission from Elsevier; Reprinted from The Lancet, 203, Li H., Wei X., Xu H., Chin S., Yamanouchi K., and Sun H., femtosecond laser filamentation for sensing combustion intermediates: A comparative study, 887, 2014, with permission from Elsevier).

Li et al. [107] also obtained critical power and clamping intensity of filament in the combustion flow field. The results indicate that the critical power in the flame is 4–5 times smaller than the usually quoted one in air, and the clamping intensity inside the filament is roughly half of that in the air.

They also found the lasing action [108] and third-harmonic generation [109] in flame. These results and observations can provide insights into the understanding of FINS for practical applications in combustion diagnostics.

There are some advantages of FINS in the application of combustion diagnostics. First of all, FINS can measure multiple combustion intermediates simultaneously with a simple optical setup, which is based on the unique nonlinear optical phenomenon of the filament. Compared with the ns-LIBS technique, the spectra of FINS consist of not only atoms but also molecules, which provide more valuable information. Consequently, FINS can reflect the composition of the measured flow field more realistically. However, the current exploitation of FINS is not complete. For example, previous works have focused on FINS in an alcohol burner array, as shown in Figure 5, which lacks standard data such as temperature or component concentration for comparison. It is thus more meaningful to perform FINS in a standard burner, such as a McKenna burner [110].

Figure 5. The photo of the flame on the alcohol burner array together with the femtosecond laser-induced filament [108]. (Reproduced from Chu W., Li H., Ni J., Zeng B., Yao J., Zhang H., Li G., Jing C., Xie H., Xu H.; Yamanouchi K, and Cheng Y, Lasing action induced by femtosecond laser filamentation in ethanol flame for combustion diagnosis, with the permission of AIP Publishing).

Besides FINS, another technique called femtosecond laser-induced plasma spectroscopy (FLIPS) should also fall in this category. FLIPS can acquire component information based on quantitative spectral analysis of the plasma emission. As mentioned above, different from fs-LIBS, FLIPS generally uses an fs pulse focused lens with a longer focal length, and the clamping intensity of the generated filament is about 10^{13}~10^{14} W/cm^2 [12,13]. By contrast, the breakdown threshold of fs-LIBS is about 2~4×10^{14} W/cm^2 in air at λ = 800 nm [89,111]. As a consequence, FLIPS does not suffer from interference of continuous background and contains substantial molecular bands, which can provide more information for spectral analysis.

Hsu et al. [112] used FLIPS to achieve a stable and reliable gas sensing at elevated pressures. The results showed that the signal level of FLIPS increases with the increase of pressure while maintaining the stability in the pressure range of 1–40 bar.

Our group [113] developed FLIPS for instantaneous one-dimensional equivalence ratio measurement in a free jet with non-reacting premixed CH$_4$/air. By measuring the spatially resolved spectra of FLIPS, we found that the spectral peak area ratios of CH (431 nm)/N$_2$ (337 nm), CH (431 nm)/N$_2$ (357 nm), and CH (431 nm)/O (777 nm) can be utilized to achieve one-dimensional local equivalence ratio measurement. Among them, the CH peak at ~431 nm and the O peak at ~777 nm is strong enough to be used to achieve single-shot measurements, and a spatial resolution of 150 µm was achieved, which is important for turbulent flow fields. Figure 6 indicates the results of FLIPS technique used in both laminar and turbulent flow fields for instantaneous one-dimensional equivalence ratio measurement.

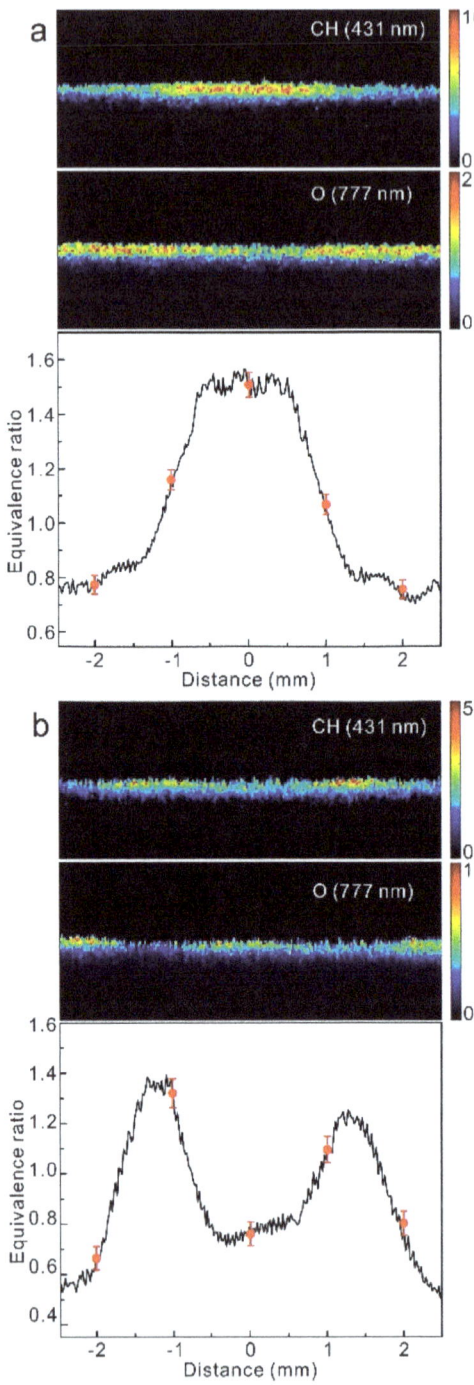

Figure 6. Single-shot images of femtosecond laser-induced plasma spectroscopy (FLIPS) in laminar (**a**) and turbulent (**b**) flow field with two intensified charge-coupled device (ICCD) cameras and two filters [113]. (© 2019 Optical Society of America).

The main advantages of FLIPS are as follows. Firstly, FLIPS can achieve instantaneous one-dimensional measurement with a simple experimental setup. Secondly, the spatial resolution of FLIPS is on the order of 100 microns, which is at least one order of magnitude lower than that of ns-LIBS. Thirdly, compared with traditional ns-LIBS, FLIPS technique is free from broadband bremsstrahlung interference. Fourthly, FLIPS technique also has feasibility under high-pressure conditions.

Femtosecond laser-induced emissions can last for a while, and the filament offers a one-dimensional spatial resolution. Combining these two features, researches established another application to achieve velocity measurements in the flow field.

4.2. Tagging Velocimetry Based on Femtosecond Laser-Induced Emission

Velocity measurement in the flow field is another important application of fs laser-induced filament emission. Velocity is one of the most significant parameters in the combustion flow field. Techniques that can be used for velocity measurements include invasive techniques, such as pitot tube [2] and hot wire anemometer [114], and non-invasive techniques, such as particle imaging velocimetry (PIV) [115,116], laser Doppler velocimetry (LDV) [117], and molecular tagging velocimetry (MTV) [118]. As a cutting-edge technique, MTV has developed rapidly in recent decades, and it has been extensively studied in the combustion flow field. MTV measures flow field velocity by time-of-flight analysis. The molecules in the flow field are tagged through a "write" process, which is followed by a "read" process in order to probe the tagged molecules at a known delay time. Consequently, the movement of the tagged molecules gives the velocity. MTV techniques can be categorized by whether the atomic/molecular tracer is "seeded" or "unseeded". In "seeded" MTV methods, the flow is seeded with molecules (or atoms) such as ketone [119,120], ester [121], nitric oxide [122,123], and metal atoms [124,125]. Different atomic/molecular tracers are seeded in terms of different flow fields. Therefore, these methods have a wide scope of applications and high accuracy. However, seeding of these atomic/molecular tracers is often undesirable due to expense, toxicity, corrosivity, and so on. Furthermore, when these methods are applied in the combustion flow field, it is necessary to consider whether the additional tracers have effects on the combustion reactions. Various "unseeded" MTV methods were developed as well, e.g., Raman excitation plus laser-induced electronic fluorescence (RELIEF) with O_2 as tracer [126,127], air photolysis and recombination tracking (APART) with NO as tracer [128], ozone tagging velocimetry (OTV) with O_3 as tracer [129,130], and hydroxyl tagging velocimetry (HTV) with OH as tracer [131–133]. For detailed information about MTV, readers can refer to a review [118].

Traditional MTV techniques use ns lasers as the light source. With the evolution of fs lasers and their gradual applications to the field of combustion, some velocity measurement techniques based on fs lasers have been developed, including femtosecond laser electronic excitation tagging (FLEET) and femtosecond laser-induced cyano chemiluminescence (FLICC).

FLEET is a recently developed fs-laser-based tagging velocimetry in nitrogen-containing flow field, which was proposed by Michael et al. [134]. It is the first attempt of an fs laser in gaseous flow field velocity measurement. The FLEET nitrogen tagging mechanism is shown in Figure 7. When an fs laser with a wavelength of 800 nm was focused at the nitrogen-containing flow field, the laser pulse ionized and dissociated molecular nitrogen into atomic nitrogen, which produced long-lived fluorescence as the nitrogen atoms recombined into excited electronic states of molecular nitrogen. Through imaging N_2 fluorescence of first positive band ($B^3\Pi_g \rightarrow A^3\Sigma_u^+$) at different delays and analyzing, the flow field velocity can be obtained. However, FLEET is a non-resonant process, as shown by the red transition of Figure 8. FLEET process requires absorption of at least eight photons to overcome the 9.8 eV [135] dissociation energy of N_2. As a result, in order to induce multi-photon dissociation, the femtosecond laser with an energy of several mJ is needed, which can photo-dissociate many other species, heat the probing volume, or alter local chemistry. One way to avoid these problems is to use low laser pulse energy. Therefore, selective two-photon absorptive resonance FLEET (STARFLEET) approach was developed by Jiang et al. [135–138], which significantly reduces the per-pulse energy to 30 µJ. As shown by the blue transition in Figure 8, STARFLEET is designed to exploit the resonant excitation of the N_2 $a''^1\Sigma_g^+ \leftarrow X1\Sigma_g^+$ transition via two-photon absorption at ~202.25 nm. Then the pre-dissociated N_2 at $a''^1\Sigma_g^+$ absorbs the third photon, and dissociation of N_2 molecule happens. STARFLEET approach has the same capabilities as FLEET, but is more efficient than FLEET.

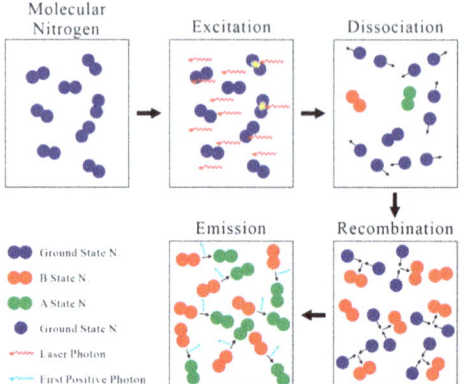

Figure 7. Diagram of the femtosecond laser electronic excitation tagging velocimetry (FLEET) nitrogen tagging mechanism. The incident 100 fs laser dissociates the nitrogen molecules into atoms, which subsequently recombine into the excited B state of molecular nitrogen, which emits in the red and near infrared through the first positive transition to the A state.

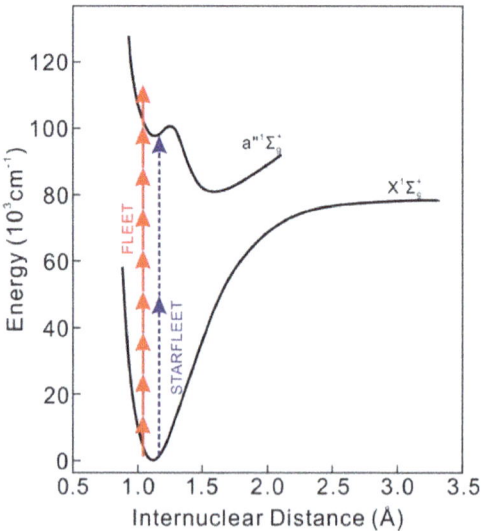

Figure 8. The nitrogen excitation strategy of FLEET and selective two-photon absorptive resonance FLEET (STARFLEET).

FLEET is used in a wide range of applications, including subsonic [134,139–141], transonic [136, 142–144], and hypersonic [145–148] flow fields. At present, FLEET has been successfully applied in nitrogen [134,140,141,149–152], argon [139,153–155], air [141,147,156,157], 1,1,1,2-Tetrafluoroethane [158], helium [159], carbon dioxide [159], oxygen [159], and combustion [141,148,160] flow fields. Since the advent of FLEET, its measurement potential has been fully explored. FLEET can be used not only for velocity measurement but also for measurements of other parameters in the flow field. For example, by measuring the FLEET spectra, based on two-line thermometry, the flow field temperature can be obtained [150,157]. By analyzing the FLEET images, the boundary layer [146,161], shear layer [147,162], vorticity [163], and mixture fraction [156] can be obtained. By analyzing the FLEET signal intensity and lifetime, the pressure information of the flow field can be inverted [164].

The FLEET technique has the following advantages. Firstly, FLEET can achieve velocity measurement with only one fs laser, whose experimental system is simple. Secondly, FLEET uses fs laser as a light source, which can induce efficient multi-photon photolysis, therefore, it can be used to realize multi-dimensional measurement [140,153,165]. Thirdly, FLEET uses molecular nitrogen as a tracer, so the interference on the reactions of the flow field can be minimized. Fourthly, FLEET enables the measurement of other parameters in the flow field. However, there are still some technical defects in the current FLEET technique, such as poor signal-to-noise ratio (SNR). Many researchers have tried to enhance the signal intensity by using different wavelengths or adding inert gases [139]. At present, the most effective way to enhance signal intensity is using laser-induced chemiluminescence of the radicals that have strong emission intensity.

Another fs-laser-based tagging velocimetry is femtosecond laser-induced cyano chemiluminescence (FLICC) [166], which is proposed by our group. When an fs laser focused at methane-seeded nitrogen gas flows, the high-intensity emission originating from CN ($B^2\Sigma^+ \rightarrow X^2\Sigma^+$) fluorescence was observed. The emission is strong and can last for hundreds of microseconds with a proper methane concentration. Therefore, velocity measurement can be achieved through CN fluorescence imaging. As shown in Figure 9, FLICC is an ideal tool for near-wall velocity measurements, which can be used to study, e.g., the boundary layer structure in supersonic flows. Although the FLICC technique requires hydrocarbons to be seeded in the measured flow field, it has a natural advantage when it comes to combustion flow fields.

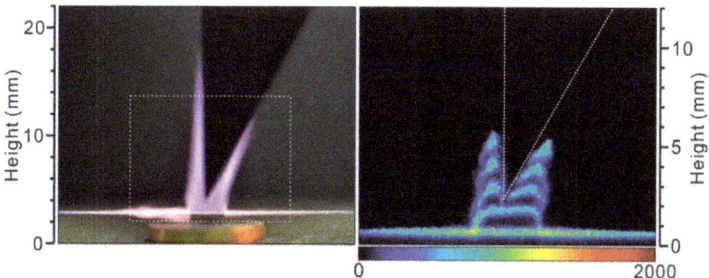

Figure 9. Following the movement of femtosecond laser-induced cyano chemiluminescence (FLICC) signal using a single lens reflex camera (**left**) and an ICCD camera in the on-chip multi-exposure mode (**right**).

5. Femtosecond Lasers Guided High Voltage Electric Discharges

In addition to the above-mentioned fs-laser-based techniques, fs lasers can also be used to guide high voltage electric discharges, which can be used for flow field measurements and plasma applications. Plasma is widely used in the fields of industry [167], environment [168], and medical science [169,170] due to its conductive and germicidal properties. High voltage discharges are the most commonly used methods to generate plasma, such as corona discharge [171,172], spark discharge [173], and dielectric barrier discharge [174,175]. Under atmospheric or high pressures, electric discharge in the air can form a long thin filamentary volume of ionized plasma [176]. However, no matter the kind of discharges, they are all based on a strong electrical field, which suffers the problem of random discharges in both time and space. Controlling the time and space of discharges simultaneously and accurately is so hard that it brings many limitations to its practical applications.

In the 1970s–1990s, researches on controlling high voltage discharges using pulse lasers have been reported [177]. The high voltage discharge can be well controlled in the air through nanosecond laser guiding [178]. Due to the limitation of the Rayleigh range of the focused laser beam, discharge guidance over the range of about 1 m was achieved. The fs laser-induced filament appears to be ideally suitable for guiding a long-range atmospheric discharge.

Femtosecond laser-induced filament is the result of a dynamical competition between the Kerr effect and plasma defocus effect. The length of the filament depends on the femtosecond laser pulse energy and can be extended to several kilometers [13] when the laser energy is high enough. The long-distance transmission of the filament offers many potential applications. Wolf et al. [179] reported a free space laser telecommunication through the fog. They use a filament-induced shock wave to radially expel the droplets out of the beam from the air it sweeps, then providing a clean channel for telecom laser transmission, which has the potential to be applied to Earth-satellite free-space optical communications and secure ground-based optical communications.

The fs laser-induced filament is a weakly ionized plasma channel with a diameter of ~100 μm and an electron density of ~10^{16} cm^{-3} [180]. This plasma channel is weakly conductive and can be used for guiding high voltage discharges. The most practical application of fs laser-induced filament is laser-triggered switching. Many applications rely on the initiation of a well-controlled discharge between two charged electrodes, especially for high current equipment, where a gas-filled trigger is necessary. Since the first laser-triggered switching experiment in the early 1960s, a wide variety of spark gap geometries and laser types have been investigated. Using ns lasers to trigger discharges can achieve an excellent temporal stability of switched pulse with a jitter time of one nanosecond [181]. Better temporal stability, within picoseconds, was achieved by using an fs laser to trigger discharges [182,183], and the discharge path is strictly defined along the filament. Femtosecond laser-induced filament guiding discharge provides a safe, remote, and highly accurate controllable switch for high current equipment [184], which perfectly solves the problem of random discharges in both time and space.

Since the time and space of discharges can be accurately controlled, many applications based on laser-induced filament were developed [185,186]. Polynkin et al. [187] reported a multi-pulse scheme for laser-guided electrical breakdown of air, which is shown to be suitable for guiding discharges propagating in either direction or along curved paths. An example of curved discharges guided by timed sequences of three laser filaments in the dome and zigzag formations is shown in Figure 10, and a scheme for the guidance of natural lightning based on the application of multiple chirped femtosecond laser pulses was proposed.

Our group developed an fs laser-induced filament guiding discharges tagging velocimetry [188], named filamentary anemometry using femtosecond laser-extended electric discharge (FALED). FALED technique uses the filament to ignite a pulsed electric discharge between two electrodes. The laser-guided thin filamentary discharge plasma column was blown up perpendicularly by an air jet placed beneath in-between the two electrodes. The conductivity of the plasma channel was observed to sustain much longer so that a sequence of discharge filaments was generated as the plasma channel was blown up by the jet flow. The sequential bright thin discharge filaments can be photographed using a household camera to calculate the flow velocity distribution of the jet flow, as shown in Figure 11. The velocity measured by FALED agrees well with that measured by FLEET [188], and FALED has a better signal to noise ratio and a thinner tagged line. Compared with FLEET and FLICC, FALED largely extends the application scenario and does not suffer from quenching. Furthermore, it should be emphasized that the discharge current is supposed to be extremely low in order to reduce intrusiveness.

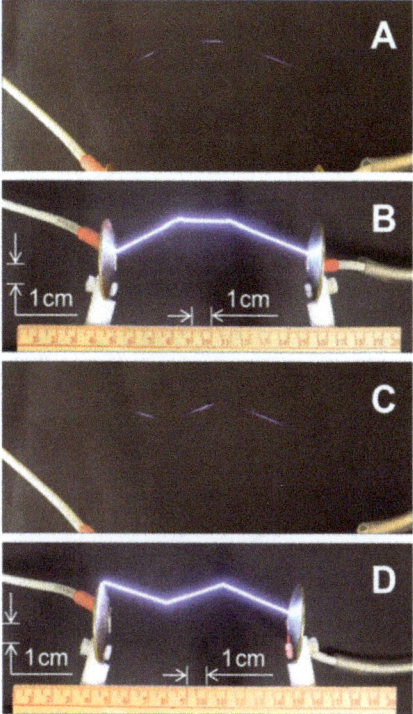

Figure 10. Examples of curved discharges guided by timed sequences of three laser filaments in the dome and zigzag formations. (**A**), (**C**) Photographs of fluorescence by the filament plasmas, averaged over 100 laser shots. (**B**), (**D**) Single-shot images of the corresponding guided discharges [187]. (Reproduced from Polynkin P, Multi-pulse scheme for laser-guided electrical breakdown of air. *Appl. Phys. Lett.* 2017, *111*, 161102, with the permission of AIP Publishing).

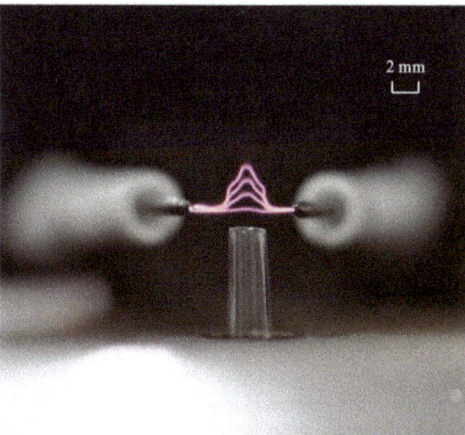

Figure 11. Filamentary anemometry using femtosecond laser-extended electric discharge (FALED) photo taken by the single lens reflex (SLR) camera of four sequential discharges with a separation of 40 µs and a gas flow speed of 35 m/s.

In summary, fs laser guiding discharges can have a lot of applications, such as laser-triggered switching, guiding natural lightning, and velocity measurements. There are many exciting phenomena that are worthy of examination. Shown in Figure 12 are spatially resolved plasma spectra of fs laser-induced filament guiding discharges. The plasma spectra indicate that the species in the discharged filament are totally different when the discharge current is changed by varying current limiting resistors. This phenomenon may be used to develop a method for species measurement with the potential to achieve one-dimensional measurement. The accurate controlling of discharges in both time and space also provides a possible way to study the spatiotemporal evolution processes of discharges plasma.

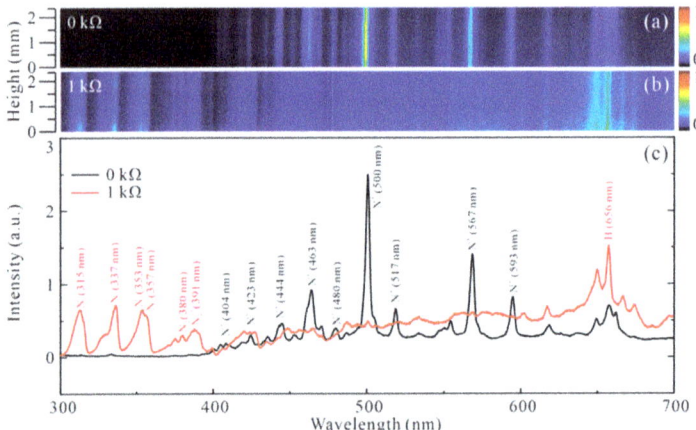

Figure 12. Emission spectra from the filamentary column of different plasmas collected with spectrometer slit, vertically orientated perpendicular to the thin plasma columns. (**a**) Imaging spectra of FALED with a resistance of 0 kΩ; (**b**) Imaging spectra of FALED along the jet flow with a resistance of 1 kΩ; (**c**) Spectral curve of emission from the filamentary plasma with the resistances of 0 kΩ (black line) and 1 kΩ (red line).

6. Conclusions

In summary, the development and application of a number of fs laser-based techniques for combustion and flow field diagnostics are reviewed. These novel techniques include fs two-photon laser-induced fluorescence (fs-TPLIF), fs laser-induced breakdown spectroscopy (fs-LIBS), fs laser electronic excitation tagging (FLEET), filament-induced nonlinear spectroscopy (FINS) and fs laser-induced plasma spectroscopy (FLIPS). In spite of their relatively short history, significant impacts of these techniques have been demonstrated in measuring velocity, temperature and intermediates in combustion and flow field.

However, there is still plenty of space for further development of these techniques, as all the techniques mentioned above have the temporal-resolved measurement capability, and the time resolution is in the range of nanosecond to sub-microsecond level. The fundamental mechanisms governing the chemical and physical processes introduced by the ultra-short laser pulse and ultra-high peak power are still to be investigated to form a knowledge-based understanding of the interaction between the fs laser and the combustion flow field and the mechanisms of filamentary electric discharge guided by fs lasers. It is predicted that the fs laser-based technologies will play an important role in combustion and flow field diagnostics in the near future.

Author Contributions: Conceptualization, Z.L.; Resources, D.Z., J.L. and Y.T.; Data curation, D.Z.; Writing-original draft preparation, B.L., D.Z. and Z.L.; Writing-review and editing, Z.L., B.L. and Q.G.; Supervision, Q.G. and Z.L.

Funding: This work was financially supported by the National Natural Science Foundation of China (NSFC) (91741205, 51776137).

Conflicts of Interest: The authors declare no conflict of interest.

References

1. Mock, R.; Meixner, H. A miniaturized high-temperature pressure sensor for the combustion chamber of a spark ignition engine. *Sens. Actuators A* **1990**, *25*, 103–106. [CrossRef]
2. Klopfenstein, R., Jr. Air velocity and flow measurement using a Pitot tube. *ISA Trans.* **1998**, *37*, 257–263. [CrossRef]
3. Heitor, M.V.; Moreira, A.L.N. Thermocouples and sample probes for combustion studies. *Prog. Energy Combust. Sci.* **1993**, *19*, 259–278. [CrossRef]
4. Aldén, M.; Bood, J.; Li, Z.; Richter, M. Visualization and understanding of combustion processes using spatially and temporally resolved laser diagnostic techniques. *Proc. Combust. Inst.* **2011**, *33*, 69–97. [CrossRef]
5. Kohse-Höinghaus, K.; Barlow, R.S.; Aldén, M.; Wolfrum, J. Combustion at the focus: Laser diagnostics and control. *Proc. Combust. Inst.* **2005**, *30*, 89–123. [CrossRef]
6. Hanson, R.K.; Seitzman, J.M.; Paul, P.H. Planar laser-fluorescence imaging of combustion gases. *Appl. Phys. B* **1990**, *50*, 441–454. [CrossRef]
7. Crosley, D.R.; Smith, G.P. Laser-induced fluorescence spectroscopy for combustion diagnostics. *Opt. Eng.* **1983**, *22*, 545–553. [CrossRef]
8. Aldén, M.; Omrane, A.; Richter, M.; Särner, G. Thermographic phosphors for thermometry: A survey of combustion applications. *Prog. Energy Combust. Sci.* **2011**, *37*, 422–461. [CrossRef]
9. Chan, V.S.S.; Turner, J.T. Velocity measurement inside a motored internal combustion engine using three-component laser Doppler anemometry. *Opt. Laser Technol.* **2000**, *32*, 557–566. [CrossRef]
10. Shaddix, C.R.; Smyth, K.C. Laser-induced incandescence measurements of soot production in steady and flickering methane, propane, and ethylene diffusion flames. *Combust. Flame* **1996**, *107*, 418–452. [CrossRef]
11. Spence, D.E.; Kean, P.N.; Sibbett, W. 60-fsec pulse generation from a self-mode-locked Ti: Sapphire laser. *Opt. Lett.* **1991**, *16*, 42–44. [CrossRef] [PubMed]
12. Chin, S.L.; Xu, H. Tunnel ionization, population trapping, filamentation and applications. *J. Phys. B At. Mol. Opt. Phys.* **2016**, *49*, 222003. [CrossRef]
13. Chin, S.L. *Femtosecond Laser Filamentation*; Springer: New York, NY, USA, 2010; Volume 55, p. 137.
14. Eckbreth, A.C. Recent advances in laser diagnostics for temperature and species concentration in combustion. *Symp. Combust.* **1981**, *18*, 1471–1488. [CrossRef]
15. Hassel, E.P.; Linow, S. Laser diagnostics for studies of turbulent combustion. *Meas. Sci. Technol.* **2000**, *11*, R37. [CrossRef]
16. Hanson, R.K. Combustion diagnostics: Planar imaging techniques. *Symp. Combust.* **1986**, *21*, 1677–1691. [CrossRef]
17. Lucht, R.P.; Salmon, J.T.; King, G.B.; Sweeney, D.W.; Laurendeau, N.M. Two-photon-excited fluorescence measurement of hydrogen atoms in flames. *Opt. Lett.* **1983**, *8*, 365–367. [CrossRef] [PubMed]
18. Gathen, V.S.V.D.; Niemi, K. Absolute atomic oxygen density measurements by two-photon absorption laser-induced fluorescence spectroscopy in an RF-excited atmospheric pressure plasma jet. *Plasma Sources Sci. Technol.* **2005**, *14*, 375.
19. Rosell, J.; Sjöholm, J.; Richter, M.; Aldén, M. Comparison of three schemes of two-photon laser-induced fluorescence for CO detection in flames. *Appl. Spectrosc.* **2013**, *67*, 314–320. [CrossRef]
20. Bokor, J.; Freeman, R.R.; White, J.C.; Storz, R.H. Two-photon excitation of the level in H and D atoms. *Phys. Rev. A At. Mol. Opt. Phys.* **1981**, *24*, 612–614. [CrossRef]
21. Aldén, M.; Edner, H.; Grafström, P.; Svanberg, S. Two-photon excitation of atomic oxygen in a flame. *Opt. Commun.* **1982**, *42*, 244–246. [CrossRef]
22. Bischel, W.K.; Perry, B.E.; Crosley, D.R. Two-photon laser-induced fluorescence in oxygen and nitrogen atoms. *Chem. Phys. Lett.* **1981**, *82*, 85–88. [CrossRef]
23. DiMauro, L.F.; Gottscho, R.A.; Miller, T.A. Two-photon laser-induced fluorescence monitoring of O atoms in a plasma etching environment. *J. Appl. Phys.* **1984**, *56*, 2007–2011. [CrossRef]

24. Kulatilaka, W.D.; Frank, J.H.; Patterson, B.D.; Settersten, T.B. Analysis of 205-nm photolytic production of atomic hydrogen in methane flames. *Appl. Phys. B* **2009**, *97*, 227–242. [CrossRef]
25. Li, B.; Jonsson, M.; Algotsson, M.; Bood, J.; Li, Z.S.; Johansson, O.; Aldén, M.; Tunér, M.; Johansson, B. Quantitative detection of hydrogen peroxide in an HCCI engine using photofragmentation laser-induced fluorescence. *Proc. Combust. Inst.* **2013**, *34*, 3573–3581. [CrossRef]
26. Li, B.; Zhang, D.; Yao, M.; Li, Z. Strategy for single-shot CH_3 imaging in premixed methane/air flames using photofragmentation laser-induced fluorescence. *Proc. Combust. Inst.* **2017**, *36*, 4487–4495. [CrossRef]
27. Li, B.; Li, X.; Zhang, D.; Gao, Q.; Yao, M.; Li, Z. Comprehensive CO detection in flames using femtosecond two-photon laser-induced fluorescence. *Opt. Express* **2017**, *25*, 25809–25818. [CrossRef]
28. Settersten, T.B.; Dreizler, A.; Farrow, R.L. Temperature- and species-dependent quenching of CO B probed by two-photon laser-induced fluorescence using a picosecond laser. *J. Chem. Phys.* **2002**, *117*, 3173–3179. [CrossRef]
29. Frank, J.H.; Settersten, T.B. Two-photon LIF imaging of atomic oxygen in flames with picosecond excitation. *Proc. Combust. Inst.* **2005**, *30*, 1527–1534. [CrossRef]
30. Kulatilaka, W.D.; Gord, J.R.; Katta, V.R.; Roy, S. Photolytic-interference-free, femtosecond two-photon fluorescence imaging of atomic hydrogen. *Opt. Lett.* **2012**, *37*, 3051–3053. [CrossRef]
31. Stauffer, H.U.; Kulatilaka, W.D.; Gord, J.R.; Roy, S. Laser-induced fluorescence detection of hydroxyl (OH) radical by femtosecond excitation. *Opt. Lett.* **2011**, *36*, 1776–1778. [CrossRef]
32. Stauffer, H.U.; Gord, J.R.; Roy, S.; Kulatilaka, W.D. Detailed calculation of hydroxyl (OH) radical two-photon absorption via broadband ultrafast excitation. *J. Opt. Soc. Am. B Opt. Phys.* **2011**, *29*, 40–52. [CrossRef]
33. Wang, Y.; Jain, A.; Kulatilaka, W. CO imaging in piloted liquid-spray flames using femtosecond two-photon LIF. *Proc. Combust. Inst.* **2019**, *37*, 1305–1312. [CrossRef]
34. Wang, Y.; Kulatilaka, W.D. Detection of carbon monoxide (CO) in sooting hydrocarbon flames using femtosecond two-photon laser-induced fluorescence (fs-TPLIF). *Appl. Phys. B* **2018**, *124*. [CrossRef]
35. Rahman, K.A.; Patel, K.S.; Slipchenko, M.N.; Meyer, T.R.; Zhang, Z.; Wu, Y.; Gord, J.R.; Roy, S. Femtosecond, two-photon, laser-induced fluorescence (TP-LIF) measurement of CO in high-pressure flames. *Appl. Opt.* **2018**, *57*, 5666–5671. [CrossRef] [PubMed]
36. Wang, Y.; Kulatilak, W.D. Investigation of Femtosecond two-photon LIF of CO at elevated pressures. In *Laser Applications to Chemical, Security and Environmental Analysis*; Optical Society of America: Orlando, FL, USA, 2018.
37. Richardson, D.R.; Roy, S.; Gord, J.R. Femtosecond, two-photon, planar laser-induced fluorescence of carbon monoxide in flames. *Opt. Lett.* **2017**, *42*, 875–878. [CrossRef]
38. Richardson, D.R.; Roy, S.; Gord, J.R. Carbon monoxide femtosecond TALIF in turbulent flames. In Proceedings of the AIAA Aerospace Sciences Meeting, Kissimmee, FL, USA, 8–12 January 2018.
39. Jain, A.; Wang, Y.; Kulatilaka, W.D. Effect of H-atom concentration on soot formation in premixed ethylene/air flames. *Proc. Combust. Inst.* **2019**, *37*, 1289–1296. [CrossRef]
40. Jain, A.; Kulatilaka, W.D. Investigation of multi-photon excitation schemes for detecting atomic hydrogen in flames. In Proceedings of the 2018 Conference on Lasers and Electro-Optics (CLEO), San Jose, CA, USA, 13–18 May 2018; pp. 1–2.
41. Ding, P.; Ruchkina, M.; Liu, Y.; Alden, M.; Bood, J.; Department, O.P.; Lund, U.; Lunds, U.; Fysiska, I. Femtosecond two-photon-excited backward lasing of atomic hydrogen in a flame. *Opt. Lett.* **2018**, *43*, 1183–1186. [CrossRef]
42. Ding, P.; Ruchkina, M.; Liu, Y.; Ehn, A.; Alden, M.; Bood, J. Picosecond backward-propagating lasing of atomic hydrogen via femtosecond 2-photon-excitation in a flame. In Proceedings of the 2018 Conference on Lasers and Electro-Optics (CLEO), San Jose, CA, USA, 13–18 May 2018; pp. 1–2.
43. Ruchkina, M.; Ding, P.; Ehn, A.; Alden, M.; Bood, J. Spatially-resolved hydrogen atom detection in flames using backward lasing. In Proceedings of the 2018 Conference on Lasers and Electro-Optics (CLEO), San Jose, CA, USA, 13–18 May 2018; pp. 1–2.
44. Hall, C.A.; Kulatilaka, W.D.; Gord, J.R.; Pitz, R.W. Quantitative atomic hydrogen measurements in premixed hydrogen tubular flames. *Combust. Flame* **2014**, *161*, 2924–2932. [CrossRef]
45. Kulatilaka, W.D.; Gord, J.R.; Roy, S. Femtosecond two-photon LIF imaging of atomic species using a frequency-quadrupled Ti:sapphire laser. *Appl. Phys. B* **2014**, *116*, 7–13. [CrossRef]

46. Kulatilaka, W.; Roy, S.; Gord, J. Multi-photon fluorescence imaging of flame species using femtosecond excitation. In Proceedings of the Aerodynamic Measurement Technology, Ground Testing, and Flight Testing Conference, New Orleans, LA, USA, 25–28 June 2012.
47. Jain, A.; Kulatilaka, W.D. Two- and three-photon LIF detection of atomic hydrogen using femtosecond laser Pulses. In Proceedings of the Laser Applications to Chemical, Security and Environmental Analysis, Orlando, FL, USA, 25–28 June 2018.
48. Jain, A.; Wang, Y.; Kulatilaka, W.D. Spatially resolved atomic hydrogen concentration measurements in sooting hydrocarbon flames using femtosecond two-photon LIF. In Proceedings of the 33rd AIAA Aerodynamic Measurement Technology and Ground Testing Conference, Denver, CO, USA, 5–9 June 2017.
49. Schmidt, J.B.; Roy, S.; Kulatilaka, W.D.; Shkurenkov, I.; Adamovich, I.V.; Lempert, W.R.; Gord, J.R. Femtosecond, two-photon-absorption, laser-induced-fluorescence (fs-TALIF) imaging of atomic hydrogen and oxygen in non-equilibrium plasmas. *J. Phys. D Appl. Phys.* **2017**, *50*, 015204. [CrossRef]
50. Schmidt, J.B.; Kulatilaka, W.D.; Roy, S.; Frederickson, K.; Lempert, W.R.; Gord, J.R. Femtosecond TALIF imaging of atomic hydrogen in pulsed, non-equilibrium plasmas. In Proceedings of the 52nd Aerospace Sciences Meeting, National Harbor, MD, USA, 13–17 January 2015.
51. Li, B.; Zhang, D.; Li, X.; Gao, Q.; Yao, M.; Li, Z. Strategy of interference-free atomic hydrogen detection in flames using femtosecond multi-photon laser-induced fluorescence. *Int. J. Hydrog. Energy* **2017**, *42*, 3876–3880. [CrossRef]
52. Kulatilaka, W.D.; Roy, S.; Jiang, N.; Gord, J.R. Photolytic-interference-free, femtosecond, two-photon laser-induced fluorescence imaging of atomic oxygen in flames. *Appl. Phys. B* **2016**, *122*. [CrossRef]
53. Schmidt, J.B.; Sands, B.L.; Kulatilaka, W.D.; Roy, S.; Scofield, J.; Gord, J.R. Femtosecond, two-photon laser-induced-fluorescence imaging of atomic oxygen in an atmospheric-pressure plasma jet. *Plasma Sources Sci. Technol.* **2015**, *24*, 032004. [CrossRef]
54. Wang, Y.; Capps, C.; Kulatilaka, W.D. Mixture fraction imaging using femtosecond TPLIF of Krypton. In Proceedings of the AIAA Aerospace Sciences Meeting, Grapevine, TX, USA, 9–13 January 2017.
55. Wang, Y.; Kulatilaka, W.D. Two-photon laser induced fluorescence of krypton using femtosecond pulses. In Proceedings of the Conference on Lasers and Electro-Optics (CLEO), San Jose, CA, USA, 14–19 May 2017; pp. 1–2.
56. Wang, Y.; Capps, C.; Kulatilaka, W.D. Femtosecond two-photon laser-induced fluorescence of krypton for high-speed flow imaging. *Opt. Lett.* **2017**, *42*, 711–714. [CrossRef] [PubMed]
57. Richardson, D.R.; Jiang, N.; Stauffer, H.U.; Kearney, S.P.; Roy, S.; Gord, J.R. Mixture-fraction imaging at 1 kHz using femtosecond laser-induced fluorescence of krypton. *Opt. Lett.* **2017**, *42*, 3498–3501. [CrossRef] [PubMed]
58. Burluka, A. *Combustion Physics*; Wiley-VCH Verlag GmbH & Co. KGaA: Weinheim, Germany, 2010.
59. Settersten, T.B.; Dreizler, A.; Patterson, B.D.; Schrader, P.E.; Farrow, R.L. Photolytic interference affecting two-photon laser-induced fluorescence detection of atomic oxygen in hydrocarbon flames. *Appl. Phys. B* **2003**, *76*, 479–482. [CrossRef]
60. Singh, S.; Musculus, M.P.B.; Reitz, R.D. Mixing and flame structures inferred from OH-PLIF for conventional and low-temperature diesel engine combustion. *Combust. Flame* **2009**, *156*, 1898–1908. [CrossRef]
61. Donbar, J.M.; Driscoll, J.F.; Carter, C.D. Reaction zone structure in turbulent nonpremixed jet flames-from CH-OH PLIF images. *Combust. Flame* **2000**, *122*, 1–19. [CrossRef]
62. Li, Z.S.; Li, B.; Sun, Z.W.; Bai, X.S.; Aldén, M. Turbulence and combustion interaction: High resolution local flame front structure visualization using simultaneous single-shot PLIF imaging of CH, OH, and CH_2O in a piloted premixed jet flame. *Combust. Flame* **2010**, *157*, 1087–1096. [CrossRef]
63. Duwig, C.; Li, B.; Li, Z.S.; Aldén, M. High resolution imaging of flameless and distributed turbulent combustion. *Combust. Flame* **2012**, *159*, 306–316. [CrossRef]
64. Laurendeau, N.M.; Goldsmith, J.E.M. Comparison of hydroxyl concentration profiles using five laser-induced fluorescence methods in a lean subatmospheric-pressure $H_2/O_2/Ar$ flame. *Combust. Sci. Technol.* **1989**, *63*, 139–152. [CrossRef]
65. Renfro, M.W.; Guttenfelder, W.A.; King, G.B.; Laurendeau, N.M. Scalar time-series measurements in turbulent $CH_4/H_2/N_2$ nonpremixed flames: OH. *Combust. Flame* **2000**, *123*, 389–401. [CrossRef]
66. Westblom, U.; Bengtsson, P.E.; Aldén, M. Carbon atom fluorescence and C_2 emission detected in fuel-rich flames using a UV laser. *Appl. Phys. B* **1991**, *52*, 371–375. [CrossRef]

67. Das, P.; Ondrey, G.S.; van Veen, N.; Bersohn, R. Two photon laser induced fluorescence of carbon atoms. *J. Chem. Phys.* **1983**, *79*, 724–726. [CrossRef]
68. Zhang, D.Y.; Gao, Q.; Li, B.; Li, Z.S. Ammonia measurements with femtosecond laser-induced plasma spectroscopy. *Appl. Opt. B* **2019**, *58*, 1210–1214. [CrossRef]
69. Bol'Shakov, A.A.; Yoo, J.H.; Liu, C.; Plumer, J.R.; Russo, R.E. Laser-induced breakdown spectroscopy in industrial and security applications. *Appl. Opt.* **2010**, *49*, C132–C142. [CrossRef]
70. Choi, I.; Chan, G.C.Y.; Mao, X.; Perry, D.L.; Russo, R.E. Line selection and parameter optimization for trace analysis of uranium in glass matrices by laser-induced breakdown spectroscopy (LIBS). *Appl. Spectrosc.* **2013**, *67*, 1275–1284. [CrossRef]
71. Bonta, M.; Gonzalez, J.J.; Quarles, C.D., Jr.; Russo, R.E.; Hegedus, B.; Limbeck, A. Elemental mapping of biological samples by the combined use of LIBS and LA-ICP-MS. *J. Anal. At. Spectrom.* **2016**, *1*, 252–258. [CrossRef]
72. Rehse, S.J.; Salimnia, H.; Miziolek, A.W. Laser-induced breakdown spectroscopy (LIBS): An overview of recent progress and future potential for biomedical applications. *J. Med. Eng. Technol.* **2012**, *36*, 77–89. [CrossRef] [PubMed]
73. Singh, M.; Karki, V.; Mishra, R.K.; Kumar, A.; Kaushik, C.P.; Mao, X.; Russo, R.E.; Sarkar, A. Analytical spectral dependent partial least squares regression: A study of nuclear waste glass from thorium based fuel using LIBS. *J. Anal. At. Spectrom.* **2015**, *3*, 2507–2515. [CrossRef]
74. Quarles, C.D.; Gonzalez, J.J.; East, L.J.; Yoo, J.H.; Morey, M.; Russo, R.E. Fluorine analysis using Laser Induced Breakdown Spectroscopy (LIBS). *J. Anal. At. Spectrom.* **2014**, *29*, 1238–1242. [CrossRef]
75. Colao, F.; Fantoni, R.; Lazic, V.; Paolini, A.; Fabbri, F.; Ori, G.G.; Marinangeli, L.; Baliva, A. Investigation of LIBS feasibility for in situ planetary exploration: An analysis on Martian rock analogues. *Planet. Space Sci.* **2004**, *52*, 117–123. [CrossRef]
76. Harmon, R.S.; Remus, J.; McMillan, N.J.; McManus, C.; Collins, L.; Gottfried, L.J.L., Jr.; DeLucia, F.C.; Miziolek, A.W. LIBS analysis of geomaterials: Geochemical fingerprinting for the rapid analysis and discrimination of minerals. *Appl. Geochem.* **2009**, *24*, 1125–1141. [CrossRef]
77. Gaudiuso, R.; Dell'Aglio, M.; De Pascale, O.; Senesi, G.S.; De Giacomo, A. Laser induced breakdown spectroscopy for elemental analysis in environmental, cultural heritage and space applications: A review of methods and results. *Sensors* **2010**, *10*, 7434–7468. [CrossRef] [PubMed]
78. Haddad, J.E.; Canioni, L.; Bousquet, B. Good practices in LIBS analysis: Review and advices. *Spectrochim. Acta B* **2014**, *101*, 171–182. [CrossRef]
79. Michel, A.P.M. Review: Applications of single-shot laser-induced breakdown spectroscopy. *Spectrochim. Acta B* **2010**, *65*, 185–191. [CrossRef]
80. Rhodes, W.T. Laser-induced breakdown spectroscopy. In *Springer Series in Optical Sciences*; Springer: Berlin, Germany, 2014.
81. Cremers, D.A.; Multari, R.A.; Knight, A.K. Laser-induced breakdown spectroscopy. In *Encyclopedia of Analytical Chemistry*; 2016; Available online: https://onlinelibrary.wiley.com/doi/10.1002/9780470027318.a5110t.pub3/figures (accessed on 25 April 2019).
82. Tian, Z.; Dong, M.; Li, S.; Lu, J. Spatially resolved laser-induced breakdown spectroscopy in laminar premixed methane-air flames. *Spectrochim. Acta B* **2017**, *136*, 8–15. [CrossRef]
83. Kiefer, J.; Zhou, B.; Li, Z.; Aldén, M. Impact of plasma dynamics on equivalence ratio measurements by laser-induced breakdown spectroscopy. *Appl. Opt.* **2015**, *54*, 4221–4226. [CrossRef]
84. Michalakou, A.; Stavropoulos, P.; Couris, S. Laser-induced breakdown spectroscopy in reactive flows of hydrocarbon-air mixtures. *Appl. Phys. Lett.* **2008**, *92*, 081501. [CrossRef]
85. Do, H.; Carter, C.D.; Liu, Q.; Ombrello, T.M.; Hammack, S.; Lee, T.; Hsu, K.Y. Simultaneous gas density and fuel concentration measurements in a supersonic combustor using laser induced breakdown. *Proc. Combust. Inst.* **2015**, *35*, 2155–2162. [CrossRef]
86. Zimmer, L.; Yoshida, S. Feasibility of laser-induced plasma spectroscopy for measurements of equivalence ratio in high-pressure conditions. *Exp. Fluids* **2012**, *52*, 891–904. [CrossRef]
87. Ferioli, F.; Buckley, S.G. Measurements of hydrocarbons using laser-induced breakdown spectroscopy. *Combust. Flame* **2006**, *144*, 435–447. [CrossRef]
88. Kotzagianni, M.; Yuan, R.; Mastorakos, E.; Couris, S. Laser-induced breakdown spectroscopy measurements of mean mixture fraction in turbulent methane flames with a novel calibration scheme. *Combust. Flame* **2016**, *167*, 72–85. [CrossRef]

89. Labutin, T.A.; Lednev, V.N.; Ilyincd, A.A.; Popov, A.M. Femtosecond laser-induced breakdown spectroscopy. *J. Anal. At. Spectrom.* **2016**, *31*, 90–118. [CrossRef]
90. Kotzagianni, M.; Couris, S. Femtosecond laser induced breakdown spectroscopy of air-methane mixtures. *Chem. Phys. Lett.* **2013**, *561–562*, 36–41. [CrossRef]
91. Patnaik, A.K.; Wu, Y.; Hsu, P.S.; Gragston, M.; Zhang, Z.; Gord, J.R.; Roy, S. Simultaneous LIBS signal and plasma density measurement for quantitative insight into signal instability at elevated pressure. *Opt. Express* **2018**, *26*, 25750. [CrossRef] [PubMed]
92. Patnaik, A.K.; Hsu, P.S.; Stolt, A.J.; Estevadeordal, J.; Gord, J.R.; Roy, S. Advantages of ultrafast LIBS for high-pressure diagnostics. *Imaging Appl. Opt.* **2018**. [CrossRef]
93. Kotzagianni, M.; Couris, S. Femtosecond laser induced breakdown for combustion diagnostics. *Appl. Phys. Lett.* **2012**, *100*, 264104. [CrossRef]
94. Couris, S.; Kotzagianni, M.; Baskevicius, A.; Bartulevicius, T.; Sirutkaitis, V. Combustion diagnostics with femtosecond laser radiation. *J. Phys.* **2014**, *548*, 12056. [CrossRef]
95. Patnaik, A.K.; Hsu, P.S.; Roy, S.; Wu, Y.; Gragston, M.; Zhang, Z.; Gord, J.R. Ultrafast laser-induced-breakdown spectroscopy (LIBS) for F/A-ratio measurement of hydrocarbon flames. In Proceedings of the AIAA Aerospace Sciences Meeting, Kissimmee, FL, USA, 8–12 January 2018.
96. Braun, A.; Korn, G.; Liu, X.; Du, D.; Squier, J.; Mourou, G. Self-channeling of high-peak-power femtosecond laser pulses in air. *Opt. Lett.* **1995**, *20*, 73–75. [CrossRef]
97. Kasparian, J.; Wolf, J.P. Physics and applications of atmospheric nonlinear optics and filamentation. *Opt. Express* **2008**, *16*, 466–493. [CrossRef]
98. Couairon, A.; Mysyrowicz, A. Femtosecond filamentation in transparent media. *Phys. Rep.* **2007**, *441*, 47–189. [CrossRef]
99. Xu, H.L.; Daigle, J.F.; Luo, Q.; Chin, S.L. Femtosecond laser-induced nonlinear spectroscopy for remote sensing of methane. *Appl. Phys. B* **2006**, *82*, 655–658. [CrossRef]
100. Xu, H.; Cheng, Y.; Chin, S.; Sun, H. Femtosecond laser ionization and fragmentation of molecules for environmental sensing. *Laser Photonics Rev.* **2015**, *9*, 275–293. [CrossRef]
101. Li, H.; Jiang, Y.; Li, S.; Chen, A.; Li, S.; Jin, M. Research on the fluorescence emission from water vapor induced by femtosecond filamentation in air. *Chem. Phys. Lett.* **2016**, *662*, 188–191. [CrossRef]
102. Li, H.; Zang, H.; Xu, H.; Sun, H.; Baltuška, A.; Polynkin, P. Robust Remote sensing of trace-level heavy-metal contaminants in water using laser filaments. *Glob. Chall.* **2018**, *3*, 1800070. [CrossRef]
103. Li, H.L.; Xu, H.L.; Yang, B.S.; Chen, Q.D.; Zhang, T.; Sun, H.B. Sensing combustion intermediates by femtosecond filament excitation. *Opt. Lett.* **2013**, *38*, 1250–1252. [CrossRef] [PubMed]
104. Li, H.; Wei, X.; Xu, H.; Chin, S.; Yamanouchi, K.; Sun, H. Femtosecond laser filamentation for sensing combustion intermediates: A comparative study. *Sens. Actuators B Chem.* **2014**, *203*, 887–890. [CrossRef]
105. Li, H.; Chu, W.; Xu, H.; Cheng, Y.; Chin, S.; Yamanouchi, K.; Sun, H. Simultaneous identification of multi-combustion-intermediates of alkanol-air flames by femtosecond filament excitation for combustion sensing. *Sci. Rep.* **2016**, *6*, 27340. [CrossRef]
106. Li, S.; Li, Y.; Shi, Z.; Sui, L.; Li, H.; Li, Q.; Chen, A.; Jiang, Y.; Jin, M. Fluorescence emission induced by the femtosecond filament transmitting through the butane/air flame. *Spectrochim. Acta A* **2018**, *189*, 32–36. [CrossRef] [PubMed]
107. Li, H.; Chu, W.; Zang, H.; Xu, H.; Cheng, Y.; Chin, S.L. Critical power and clamping intensity inside a filament in a flame. *Opt. Express* **2016**, *24*, 3424. [CrossRef] [PubMed]
108. Chu, W.; Li, H.; Ni, J.; Zeng, B.; Yao, J.; Zhang, H.; Li, G.; Jing, C.; Xie, H.; Xu, H.; et al. Lasing action induced by femtosecond laser filamentation in ethanol flame for combustion diagnosis. *Appl. Phys. Lett.* **2014**, *104*, 91106. [CrossRef]
109. Zang, H.; Li, H.; Su, Y.; Fu, Y.; Hou, M.; Baltuška, A.; Yamanouchi, K.; Xu, H. Third-harmonic generation and scattering in combustion flames using a femtosecond laser filament. *Opt. Lett.* **2018**, *43*, 615–618. [CrossRef] [PubMed]
110. Hartung, G.; Hult, J.; Kaminski, C.F. A flat flame burner for the calibration of laser thermometry techniques. *Meas. Sci. Technol.* **2006**, *17*, 2485–2493. [CrossRef]
111. Liu, X.; Lu, X.; Liu, X.; Xi, T.; Liu, F.; Ma, J.; Zhang, J. Tightly focused femtosecond laser pulse in air: From filamentation to breakdown. *Opt. Express* **2010**, *18*, 26007–26017. [CrossRef] [PubMed]

112. Hsu, P.S.; Patnaik, A.K.; Stolt, A.J.; Estevadeordal, J.; Roy, S.; Gord, J.R. Femtosecond-laser-induced plasma spectroscopy for high-pressure gas sensing: Enhanced stability of spectroscopic signal. *Appl. Phys. Lett.* **2018**, *113*, 214103. [CrossRef]
113. Zhang, D.; Gao, Q.; Li, B.; Zhu, Z.; Li, Z. Instantaneous one-dimensional equivalence ratio measurements in methane/air mixtures using femtosecond laser-induced plasma spectroscopy. *Opt. Express* **2019**, *27*, 2159–2169. [CrossRef] [PubMed]
114. Bruun, H.H. Hot-wire anemometry: Principles and Signal Analysis. *Meas. Sci. Technol.* **1996**, 7. [CrossRef]
115. Westerweel, J.; Elsinga, G.E.; Adrian, R.J. Particle image velocimetry for complex and turbulent flows. *Annu. Rev. Fluid Mech.* **2013**, *45*, 409–436. [CrossRef]
116. Wereley, S.T.; Meinhart, C.D. Recent advances in micro-particle image velocimetry. *Annu. Rev. Fluid Mech.* **2010**, *42*, 557–576. [CrossRef]
117. Tropea, C. Laser Doppler anemometry: Recent developments and future challenges. *Meas. Sci. Technol* **1995**, *6*, 605–619. [CrossRef]
118. Chen, F.; Li, H.; Hu, H. Molecular tagging techniques and their applications to the study of complex thermal flow phenomenal, 211. *Acta. Mech. Sin.* **2015**, *31*, 425–445. [CrossRef]
119. Lempert, W.R.; Jiang, N.; Sethuram, S.; Mo, S. Molecular tagging velocimetry measurements in supersonic microjets. *AIAA J.* **2015**, *40*, 1065–1070. [CrossRef]
120. Mittal, M.; Sadr, R.; Schock, H.J.; Fedewa, A.; Naqwi, A. In-cylinder engine flow measurement using stereoscopic molecular tagging velocimetry (SMTV). *Exp. Fluids* **2009**, *46*, 277–284. [CrossRef]
121. Krüger, S.; Grünefeld, G. Stereoscopic flow-tagging velocimetry. *Appl. Phys. B* **1999**, *69*, 509–512. [CrossRef]
122. Pan, F.; Sanchez-Gonzalez, R.; McIlvoy, M.H.; Bowersox, R.D.; North, S.W. Simultaneous three-dimensional velocimetry and thermometry in gaseous flows using the stereoscopic vibrationally excited nitric oxide monitoring technique. *Opt. Lett.* **2016**, *41*, 1376–1379. [CrossRef] [PubMed]
123. Sanchez-Gonzalez, R.; Bowersox, R.D.; North, S.W. Simultaneous velocity and temperature measurements in gaseous flowfields using the vibrationally excited nitric oxide monitoring technique: A comprehensive study. *Appl. Opt.* **2012**, *51*, 1216–1228. [CrossRef]
124. Rubinsztein-Dunlop, H.; Littleton, B.; Barker, P.; Ljungberg, P.; Malmsten, Y. Ionic strontium fluorescence as a method for flow tagging velocimetry. *Exp. Fluids* **2001**, *30*, 36–42. [CrossRef]
125. Barker, P.; Thomas, A.; Rubinsztein-Dunlop, H.; Ljungberg, P. Velocity measurements by flow tagging employing laser enhanced ionisation and laser induced fluorescence. *Spectrochim. Acta B* **1995**, *50*, 1301–1310. [CrossRef]
126. Miles, R.B.; Grinstead, J.; Kohl, R.H.; Diskin, G. The RELIEF flow tagging technique and its application in engine testing facilities and for helium-air mixing studies. *Meas. Sci. Technol.* **2000**, *11*, 1272–1281. [CrossRef]
127. Miles, R.; Cohen, C.; Connors, J.; Howard, P.; Huang, S. Velocity measurements by vibrational tagging and fluorescent probing of oxygen. *Opt. Lett.* **1987**, *11*, 861–863. [CrossRef]
128. Dam, N.; Kleindouwel, R.J.; Sijtsema, N.M.; Meulen, J.J. Nitric oxide flow tagging in unseeded air. *Opt. Lett.* **2001**, *26*, 36–38. [CrossRef] [PubMed]
129. Pitz, R.W.; Ribarov, L.A.; Wehrmeyer, J.A.; Batliwala, F.; Debarber, P.A. Ozone tagging velocimetry using narrowband excimer lasers. *AIAA J.* **1999**, *37*, 708–714.
130. Pitz, R.W.; Brown, T.M.; Nandula, S.P.; Skaggs, P.A.; Debarber, P.A.; Brown, M.S.; Segall, J. Unseeded velocity measurement by ozone tagging velocimetry. *Opt. Lett.* **1996**, *21*, 755–757. [CrossRef] [PubMed]
131. Lahr, M.D.; Pitz, R.W.; Douglas, Z.W.; Carter, C.D. Hydroxyl-tagging-velocimetry measurements of a supersonic flow over a cavity. *J. Propuls. Power* **2010**, *26*, 790–797. [CrossRef]
132. Ribarov, L.A.; Wehrmeyer, J.A.; Pitz, R.W.; Yetter, R.A. Hydroxyl tagging velocimetry (HTV) in experimental air flows. *Appl. Phys. B* **2002**, *74*, 175–183. [CrossRef]
133. Boedeker, L.R. Velocity measurement by H_2O photolysis and laser-induced fluorescence of OH. *Opt. Lett.* **1989**, *14*, 473–475. [CrossRef] [PubMed]
134. Michael, J.B.; Edwards, M.R.; Dogariu, A.; Miles, R.B. Femtosecond laser electronic excitation tagging for quantitative velocity imaging in air. *Appl. Opt.* **2011**, *26*, 5158–5162. [CrossRef]
135. Jiang, N.; Halls, B.R.; Stauffer, H.U.; Danehy, P.M.; Gord, J.R.; Roy, S. Selective two-photon absorptive resonance femtosecond-laser electronic-excitation tagging velocimetry. *Opt. Lett.* **2016**, *41*, 2225–2228. [CrossRef]

136. Reese, D.; Danehy, P. Application of STARFLEET velocimetry in the NASA langley 0.3-meter transonic cryogenic tunnel. In Proceedings of the Aerodynamic Measurement Technology and Ground Testing Conference, Atlanta, GA, USA, 25–29 June 2018.
137. Jiang, N.; Halls, B.R.; Stauffer, H.U.; Danehy, P.M.; Gord, J.R.; Roy, S. Selective two-photon absorptive resonance femtosecond-laser electronic-excitation tagging (STARFLEET) velocimetry in flow and combustion diagnostics. In Proceedings of the 32nd AIAA Aerodynamic Measurement Technology and Ground Testing Conference, Washington, DC, USA, 13–17 June 2016.
138. Jiang, N.; Stauffer, H.U.; Roy, S.; Danehy, P.M.; Halls, B.R.; Gord, J.R. Nitrogen molecular-tagging velocimetry techniques using ultrashort-pulse lasers. In Proceedings of the Laser Applications to Chemical, Security and Environmental Analysis, Heidelberg, Germany, 25–28 July 2016.
139. Zhang, Y.; Miles, R.B. Femtosecond laser tagging for velocimetry in argon and nitrogen gas mixtures. *Opt. Lett.* **2018**, *43*, 551–554. [CrossRef]
140. Pouya, S.; Van Rhijn, A.; Dantus, M.; Koochesfahani, M. Multi-photon molecular tagging velocimetry with femtosecond excitation (FemtoMTV). *Exp. Fluids* **2014**, *55*, 1791. [CrossRef]
141. Miles, R.B.; Edwards, M.R.; Michael, J.B.; Calvert, N.D.; Dogariu, A. Femtosecond laser electronic excitation tagging (FLEET) for imaging flow structure in unseeded hot or cold air or nitrogen. In Proceedings of the 51st AIAA Aerospace Sciences Meeting including the New Horizons Forum and Aerospace Exposition, Grapevine, TX, USA, 7–10 January 2013.
142. Burns, R.A.; Danehy, P.M. FLEET velocimetry measurements on a transonic airfoil. In Proceedings of the 55th AIAA Aerospace Sciences Meeting, Grapevine, TX, USA, 9–13 January 2017.
143. Burns, R.A.; Danehy, P.M.; Halls, B.R.; Jiang, N. Femtosecond laser electronic excitation tagging velocimetry in a transonic, cryogenic wind tunnel. *AIAA J.* **2016**, *55*, 680–685. [CrossRef]
144. Burns, R.; Danehy, P.M.; Jones, S.B.; Halls, B.R.; Jiang, N. Application of FLEET velocimetry in the NASA langley 0.3-meter transonic cryogenic tunnel. In Proceedings of the 31st AIAA Aerodynamic Measurement Technology and Ground Testing Conference, Dallas, TX, USA, 22–26 June 2015.
145. Dogariu, L.; Dogariu, A.; Smith, M.S.; Marineau, E.C.; Miles, R.B. Hypersonic flow velocity measurements using FLEET. In Proceedings of the 2018 Conference on Lasers and Electro-Optics (CLEO), San Jose, CA, USA, 13–18 May 2018.
146. Dogariu, L.E.; Dogariu, A.; Miles, R.B.; Smith, M.S.; Marineau, E.C. Non-intrusive hypersonic freestream and turbulent boundary-layer velocity measurements in AEDC tunnel 9 using FLEET. In Proceedings of the AIAA Aerospace Sciences Meeting, Kissimmee, FL, USA, 8–12 January 2018.
147. Zhang, Y.; Miles, R.B. Shear layer measurements along curved surfaces using the FLEET method. In Proceedings of the AIAA Aerospace Sciences Meeting, Kissimmee, FL, USA, 8–12 January 2018.
148. Deluca, N.J.; Miles, R.B.; Jiang, N.; Kulatilaka, W.D.; Patnaik, A.K.; Gord, J.R. FLEET velocimetry for combustion and flow diagnostics. *Appl. Opt.* **2017**, *56*, 8632–8638. [CrossRef]
149. Hsu, P.S.; Jiang, N.; Danehy, P.M.; Gord, J.R.; Roy, S. Fiber-coupled ultrashort-pulse-laser-based electronic-excitation tagging velocimetry. *Appl. Opt.* **2018**, *57*, 560–566. [CrossRef]
150. Edwards, M.R.; Dogariu, A.; Miles, R.B. Simultaneous temperature and velocity measurements in air with femtosecond laser tagging. *AIAA J.* **2015**, *53*, 2280–2288. [CrossRef]
151. Edwards, M.R.; Limbach, C.; Miles, R.B.; Tropina, A. Limitations on high-spatial resolution measurements of turbulence using femtosecond laser tagging. In Proceedings of the 53rd AIAA Aerospace Sciences Meeting, Kissimmee, FL, USA, 5–9 January 2015.
152. Burns, R.A.; Peters, C.J.; Danehy, P.M. Unseeded velocimetry in nitrogen for high-pressure, cryogenic wind tunnels: Part I. Femtosecond-laser tagging. *Meas. Sci. Technol.* **2018**, *29*, 115302. [CrossRef]
153. Zhang, Y.; Shneider, M.N.; Miles, R.B. Femtosecond laser excitation in argon-nitrogen mixtures. *AIAA J.* **2018**, *56*, 1060–1071. [CrossRef]
154. Zhang, Y.; Shneider, M.N.; Miles, R.B. Characterization of intermediate reactions following femtosecond laser excitation in argon-nitrogen mixtures. In Proceedings of the 55th AIAA Aerospace Sciences Meeting, Grapevine, TX, USA, 9–13 January 2017.
155. Zhang, Y.; Calvert, N.; Shneider, M.N.; Miles, R.B. Enhancement of FLEET in argon gas mixtures. In Proceedings of the 32nd AIAA Aerodynamic Measurement Technology and Ground Testing Conference, Washington, DC, USA, 13–17 June 2016.

156. Halls, B.R.; Jiang, N.; Gord, J.R.; Danehy, P.M.; Roy, S. Mixture-fraction measurements with femtosecond-laser electronic-excitation tagging. *Appl. Opt.* **2017**, *56*, E94–E98. [CrossRef] [PubMed]
157. Edwards, M.; Dogariu, A.; Miles, R. Simultaneous temperature and velocity measurement in unseeded air flows with FLEET. In Proceedings of the 51st AIAA Aerospace Sciences Meeting including the New Horizons Forum and Aerospace Exposition, Grapevine, TX, USA, 7–10 January 2013.
158. Zhang, Y.; Miles, R.B.; Danehy, P.M. Femtosecond laser tagging in 1,1,1,2-Tetrafluoroethane with trace quantities of air. In Proceedings of the AIAA Aerospace Sciences Meeting, Kissimmee, FL, USA, 8–12 January 2018.
159. Calvert, N.; Zhang, Y.; Miles, R.B. Characterizing FLEET for aerodynamic measurements in various gas mixtures and non-air environments. In Proceedings of the 32nd AIAA Aerodynamic Measurement Technology and Ground Testing Conference, Washington, DC, USA, 13–17 June 2016.
160. Zhang, D.; Li, B.; Gao, Q.; Li, Z. Applicability of femtosecond laser electronic excitation tagging in combustion flow field velocity measurements. *Appl. Spectrosc.* **2018**, *72*, 1807–1813. [CrossRef]
161. Calvert, N.; Dogariu, A.; Miles, R.B. FLEET boundary layer velocity profile measurements. In Proceedings of the 44th AIAA Plasmadynamics and Lasers Conference, San Diego, CA, USA, 24–27 June 2013.
162. Zhang, Y.; Calvert, N.; Dogariu, A.; Miles, R.B. Towards shear flow measurements using FLEET. In Proceedings of the 54th AIAA Aerospace Sciences Meeting, San Diego, CA, USA, 4–8 January 2016.
163. Calvert, N.D.; Dogariu, A.; Miles, R.B. 2-D velocity and vorticity measurements with FLEET. In Proceedings of the 30th AIAA Aerodynamic Measurement Technology and Ground Testing Conference, Atlanta, GA, USA, 16–20 June 2014.
164. Burns, R.A.; Peters, C.J.; Danehy, P.M. Multiparameter flowfield measurements in high-pressure, cryogenic environments using femtosecond lasers. In Proceedings of the 32nd AIAA Aerodynamic Measurement Technology and Ground Testing Conference, Washington, DC, USA, 13–17 June 2016.
165. Danehy, P.M.; Bathel, B.F.; Calvert, N.D.; Dogariu, A.; Miles, R.B. Three component velocity and acceleration measurement using FLEET. In Proceedings of the 30th AIAA Aerodynamic Measurement Technology and Ground Testing Conference, Atlanta, GA, USA, 16–20 June 2014.
166. Li, B.; Zhang, D.; Li, X.; Gao, Q.; Zhu, Z.; Li, Z. Femtosecond laser-induced cyano chemiluminescence in methane-seeded nitrogen gas flows for near-wall velocimetry. *J. Phys. D Appl. Phys.* **2018**, *51*, 295102. [CrossRef]
167. Li, X.; Shi, C.; Xu, Y.; Zhang, X.; Wang, K.; Zhu, A. Pulsed streamer discharge plasma over Ni/HZSM-5 catalysts for methane conversion to aromatics at atmospheric pressure. *Plasma Process. Polym.* **2007**, *4*, 15–18. [CrossRef]
168. Naseh, M.V.; Khodadadi, A.A.; Mortazavi, Y.; Pourfayaz, F.; Alizadeh, O.; Maghrebi, M. Fast and clean functionalization of carbon nanotubes by dielectric barrier discharge plasma in air compared to acid treatment. *Carbon* **2010**, *48*, 1369–1379. [CrossRef]
169. Mandolfino, C.; Lertora, E.; Gambaro, C. Influence of cold plasma treatment parameters on the mechanical properties of polyamide homogeneous bonded joints. *Surf. Coat. Technol.* **2017**, *313*, 222–229. [CrossRef]
170. Machala, Z.; Janda, M.; Hensel, K.; Jedlovský, I.; Leštinská, L.; Foltin, V.; Martišovitš, V.; Morvová, M. Emission spectroscopy of atmospheric pressure plasmas for bio-medical and environmental applications. *J. Mol. Spectrosc.* **2007**, *243*, 194–201. [CrossRef]
171. Wang, T.; Wei, Y.; Liu, Y.; Chen, N.; Liu, Y.; Ju, J.; Sun, H.; Wang, C.; Lu, H.; Liu, J.; et al. Direct observation of laser guided corona discharges. *Sci. Rep.* **2016**, *5*, 18681. [CrossRef]
172. Roth, G.J.; Gundersen, M.A. Laser-induced fluorescence images of NO distribution after needle pulsed negative corona discharge. *IEEE Trans. Plasma Sci.* **2002**, *27*, 28–29. [CrossRef]
173. Ballal, D.R.; Lefebvre, A.H. The influence of spark discharge characteristics on minimum ignition energy in flowing gases. *Combust. Flame* **1975**, *24*, 99–108. [CrossRef]
174. Horvatic, V.; Michels, A.; Ahlmann, N.; Jestel, G.; Veza, D.; Vadla, C.; Franzke, J. Time- and spatially resolved emission spectroscopy of the dielectric barrier discharge for soft ionization sustained by a quasi-sinusoidal high voltage. *Anal. Bioanal. Chem.* **2015**, *407*, 6689–6696. [CrossRef]
175. Ambrico, P.F.; Ambrico, M.; Šimek, M.; Colaianni, A.; Dilecce, G.; De Benedictis, S. Laser triggered single streamer in a pin-to-pin coplanar dielectric barrier discharge. *Appl. Phys. Lett.* **2009**, *94*, 231501. [CrossRef]
176. Zhu, J.; Gao, J.; Li, Z.; Ehn, A.; Aldén, A.; Larsson, A.; Kusano, Y. Sustained diffusive alternating current gliding arc discharge in atmospheric pressure air. *Appl. Phys. Lett.* **2014**, *105*, 234102. [CrossRef]

177. Woodworth, J.R.; Frost, C.A.; Green, T.A. UV laser triggering of high-voltage gas switches. *J. Appl. Phys.* **1982**, *53*, 4734–4739. [CrossRef]
178. Zhang, C.; Shao, T.; Yan, P.; Zhou, Y. Nanosecond-pulse gliding discharges between point-to-point electrodes in open air. *Plasma Sources Sci. Technol.* **2014**, *23*, 35004. [CrossRef]
179. Schimmel, G.; Produit, T.; Mongin, D.; Kasparian, J.; Wolf, J. Free space laser telecommunication through fog. *Optica* **2018**, *5*, 1338. [CrossRef]
180. Théberge, F.; Liu, W.; Simard, P.T.; Becker, A.; Chin, S.L. Plasma density inside a femtosecond laser filament in air: Strong dependence on external focusing. *Phys. Rev. E* **2006**, *74*, 36406. [CrossRef]
181. Lechien, K.R.; Savage, M.E.; Anaya, V.; Bliss, D.E.; Clark, W.T.; Corley, J.P.; Feltz, G.; Garrity, J.E.; Guthrie, D.W.; Hodge, K.C. Development of a 5.4 MV laser triggered gas switch for multimodule, multimegampere pulsed power drivers. *Rev. Mod. Phys.* **2008**, *11*, 060402. [CrossRef]
182. Arantchouk, L.; Houard, A.; Brelet, Y.; Carbonnel, J.; Larour, J.; André, Y.B.; Mysyrowicz, A. A simple high-voltage high current spark gap with subnanosecond jitter triggered by femtosecond laser filamentation. *Appl. Phys. Lett.* **2013**, *102*, 163502. [CrossRef]
183. Luther, B.M.; Furfaro, L.; Klix, A.; Rocca, J.J. Femtosecond laser triggering of a sub-100 picosecond jitter high-voltage spark gap. *Appl. Phys. Lett.* **2001**, *79*, 3248–3250. [CrossRef]
184. Arantchouk, L.; Point, G.; Brelet, Y.; Larour, J.; Carbonnel, J.; André, Y.B.; Mysyrowicz, A.; Houard, A. Compact 180-kV Marx generator triggered in atmospheric air by femtosecond laser filaments. *Appl. Phys. Lett.* **2014**, *104*, 103506. [CrossRef]
185. Point, G.; Arantchouk, L.; Thouin, E.; Carbonnel, J.; Mysyrowicz, A.; Houard, A. Long-lived laser-induced arc discharges for energy channeling applications. *Sci. Rep.* **2017**, *7*, 13801. [CrossRef] [PubMed]
186. Clerici, M.; Hu, Y.; Lassonde, P.; Milián, C.; Couairon, A.; Christodoulides, D.N.; Chen, Z.; Razzari, L.; Vidal, F.; Légaré, F.; et al. Laser-assisted guiding of electric discharges around objects. *Sci. Adv.* **2015**, *1*, e1400111. [CrossRef] [PubMed]
187. Polynkin, P. Multi-pulse scheme for laser-guided electrical breakdown of air. *Appl. Phys. Lett.* **2017**, *111*, 161102. [CrossRef]
188. Li, B.; Tian, Y.; Gao, Q.; Zhang, D.; Li, X.; Zhu, Z.; Li, Z. Filamentary anemometry using femtosecond laser-extended electric discharge—FALED. *Opt. Express* **2018**, *26*, 21132. [CrossRef] [PubMed]

© 2019 by the authors. Licensee MDPI, Basel, Switzerland. This article is an open access article distributed under the terms and conditions of the Creative Commons Attribution (CC BY) license (http://creativecommons.org/licenses/by/4.0/).

Review

Mid-Infrared Tunable Laser-Based Broadband Fingerprint Absorption Spectroscopy for Trace Gas Sensing: A Review

Zhenhui Du [1,2,*], Shuai Zhang [1], Jinyi Li [3], Nan Gao [2,4] and Kebin Tong [1]

1. State Key Laboratory of Precision Measuring Technology and Instruments, Tianjin University, Tianjin 300072, China; Shuai_zhang@tju.edu.cn (S.Z.); tongkebin@tju.edu.cn (K.T.)
2. Key Laboratory of Micro Opto-electro Mechanical System Technology, Tianjin University, Ministry of Education, Tianjin 300072, China; ngao@hebut.edu.cn
3. Key Laboratory of Advanced Electrical Engineering and Energy Technology, Tianjin Polytechnic University, Tianjin 300387, China; lijinyi@tjpu.edu.cn
4. School of Mechanical Engineering, Hebei University of Technology, Tianjin 300130, China
* Correspondence: duzhenhui@tju.edu.cn; Tel.: +86-138-205-95185

Received: 30 November 2018; Accepted: 11 January 2019; Published: 18 January 2019

Abstract: The vast majority of gaseous chemical substances exhibit fundamental rovibrational absorption bands in the mid-infrared spectral region (2.5–25 µm), and the absorption of light by these fundamental bands provides a nearly universal means for their detection. A main feature of optical techniques is the non-intrusive in situ detection of trace gases. We reviewed primarily mid-infrared tunable laser-based broadband absorption spectroscopy for trace gas detection, focusing on 2008–2018. The scope of this paper is to discuss recent developments of system configuration, tunable lasers, detectors, broadband spectroscopic techniques, and their applications for sensitive, selective, and quantitative trace gas detection.

Keywords: tunable laser absorption spectroscopy; mid-infrared fingerprint spectrum; broadband spectrum; trace gas detection; wavelength modulation spectroscopy; quantum cascade lasers; interband cascade lasers

---

1. Introduction

Laser-based trace gas sensing is becoming more popular in a wide variety of areas including urban and industrial emission measurement [1,2], environmental and pollution monitoring [3,4], chemical analysis and industrial process control [5,6], medical diagnostics [7,8], homeland security [9], and scientific research [10,11]. With the increase in global environmental, ecological, and energy issues, laser-based trace gas detection technology has attracted unprecedented attention. Laser gas sensing is based on the analysis of characteristic spectra of molecules, mostly known as tunable diode laser absorption spectroscopy (TDLAS) or tunable laser absorption spectroscopy (TLAS). Laser absorption spectrometers (LAS), also known laser gas analyzers (LGA), enjoy the merits of non-contact, fast response time, high sensitivity and selectivity, the potential to be calibration-free, low maintenance requirements, and a long life cycle. LAS is particularly suitable for in situ, online analysis, and real-time detection.

Laser gas sensing has undergone tremendous progress along with the advancement in tunable semiconductor lasers in the last few decades. Direct absorption spectroscopy (DAS) is the most common technique for simple optical configuration, signal processing, and potential absolute measurement. DAS often suffers from low sensitivity (absorbance $\sim 10^{-3}$) due to the interference from 1/f noise in the system and laser power fluctuation. There are basically two ways to improve the sensitivity in this situation: (1) reduce the noise in the signal; or (2) increase the absorption. The former can be achieved by using modulation technique, e.g., wavelength modulation spectroscopy (WMS) and frequency modulation spectroscopy (FMS), with a typical sensitivity of absorbance ~10^{-5}. The latter can be obtained by placing the gas inside a cavity in which the light passes through multiple times to increase the interaction length, e.g., multiple-pass or long path absorption cells, and cavity enhanced absorption spectroscopy (CEAS) [12,13]. Both ways of reducing noise and increasing absorption can be applied to a same system, e.g., cavity enhanced wavelength modulation spectrometry [13] and noise-immune cavity-enhanced optical heterodyne molecular spectroscopy (NICE-OHMS) [14,15].

FMS is a method of optical heterodyne spectroscopy capable of rapid measurement of the absorption or dispersion associated with narrow spectral features. The absorption or dispersion is measured by detecting the heterodyne beat signal that occurs when the FMS optical spectrum of the probe wave is distorted by the spectral feature of interest. Recently, dispersion spectroscopy, namely chirped laser dispersion spectroscopy (CLaDS) [16] or heterodyne phase sensitive dispersion spectroscopy (HPSDS) [17], has attracted attention for its immunity to optical intensity changes and superb linearity in the measurement of concentration.

CEAS and its new versions, e.g., cavity ring-down spectroscopy (CRDS) [18], broadband cavity ring-down spectroscopy [19], phase-shift cavity ring-down spectroscopy [20], integrated cavity output spectroscopy (ICOS) [21], and continuous wave cavity enhanced absorption spectrometry (cw-CEAS) [22], provide much larger pathlength enhancement by using a resonant cavity, and thus have highly sensitive absorbance ~10^{-7}–10^{-9}.

Practically, the simplest and most promising method to enhance the signal of trace gas detection is to perform the detection at wavelengths where the transitions have larger line strengths, e.g., using fundamental rovibrational bands or electronic transitions. The fundamental rovibrational bands of a vast majority of gaseous chemical substances, located at the mid-infrared spectral region (MIR, 2.5–25 μm), are due to the transitions of molecular rovibrational energy states. In general, these bands have stronger line strengths than the overtone and combination bands typically used in the visible and near-IR regions. The MIR spectrum depends on the physical properties of the molecule such as the number and type of atoms, the bond angles, and the bond strength. Thus, the MIR spectrum is uniquely characterized by highly specific spectroscopic features and is considered the molecular signature, which allows both the identification and quantification of the molecular species, especially suitable for larger molecules, e.g., volatile organic compounds (VOCs) [23].

VOCs are gaseous organic chemicals at the conditions of normal temperature and pressure (NTP, 293.15 K and 101.325 kPa). There are several hundred types of VOCs, some of which are dangerous to human health or cause harm to the environment. VOCs monitoring has attracted attention for long time. Commonly used spectroscopic techniques for VOCs detection are Fourier-transform infrared spectroscopy (FTIR) and differential optical absorption spectroscopy (DOAS) [24]; however, their sensitivity, selectivity, and fragile optical setup are not always sufficient for harsh applications.

Recently, newly commercialized MIR detectors and lasers, especially quantum cascade lasers (QCLs) [25] and interband cascade lasers (ICLs) [26], have stimulated the development of high-performance, compact, and rugged gas sensors. Traditionally, TLAS use a discrete narrow absorption lines of small molecules for gas sensing. For larger molecules, however, so many lines overlapping with each other results in the spectral features being broad and smooth, except for occasional spikes [23,27]. These spectral features are distinct from those of the discrete narrow absorption lines with a Lorentzian, Gaussian, or Voigt profile. Detection of trace gas with broadband absorption is much more difficult than with an isolated narrow spectral line. Extra effort should be made to cope with the challenges of the broadband of larger molecules.

In this paper, we primarily review tunable laser-based broadband absorption spectroscopy for trace gas detection in 2008–2018. After a brief overview of the principle (Section 2), we discuss the system configuration, including MIR tunable lasers, detectors, and optical configuration in Section 3. We discuss broadband spectroscopic techniques concerning derivative spectroscopy (Section 4), WMS

(Section 5), and optical frequency comb spectroscopy (Section 6). Section 7 is a collection of MIR gas sensing applications. Section 8 gives conclusions and prospects.

2. Principle

Quantitative spectral analysis is based on the Beer-Lambert law, which gives the relationship between the incident and the transmitted radiation through a gas cell filled with a molecular gas sample:

$$I(\nu) = I_0(\nu) \times exp\{-\sigma(\nu) \times L \times C\}, \quad (1)$$

where I_0 and I are the incident and transmitted radiant powers, respectively; σ is absorption cross section of the molecule in cm^2/molecule; L is absorption pathlength in cm; C is the density of the molecule in molecule/cm^3. Usually, the absorption cross section σ is also used to describe the absorption intensity. The line strength is retrieved by spectrally integrating the absorption line shape and applying the ideal gas law:

$$S(T) = \frac{K_B T A}{X_i L P \, r_{iso}}, \quad (2)$$

where K_B, T (K), and P (Pa) are the Boltzmann constant, gas temperature, and total pressure of the gas sample, respectively; X_i is the amount fraction of i species; A (cm^{-1}) is integral absorbance; r_{iso} is a correction factor for isotopic fractionation of the gas sample.

WMS is commonly used to improve the sensitivity of gas sensing. The WMS theory and signal model have been detailed previously [28], and so are only briefly reviewed here. A periodic sawtooth ramp ridden by a high-frequency sinusoidal is applied to the laser injection current, thus the laser wavenumber $\nu(t) = \nu_c + \nu_a \cos \omega t$ is scanned across the transition of gas to be detected, where ν_c and ν_a, are the laser center wavenumber and modulation depth, respectively; ω is the radian frequency. In case of ideal conditions, ignoring all kinds of interference, the modulated absorption signal is detected by a photodiode and then processed using a lock-in amplifier to demodulate the signal at the harmonics (1f, 2f, 3f, etc.). The second harmonic component (WMS-2f) is commonly used for calculating the concentration of target gas. In the case of optically thin ($\sigma(\nu) \cdot L \cdot C \leq 0.05$), the ideal 2$f$ signal is modeled as:

$$A_{ideal\ 2f} = \frac{2I_0 CL}{\pi} \int_0^{\pi} -\alpha(\nu_c + \nu_a \cos\theta)\cos 2\theta d\theta \propto I_0 CL, \quad (3)$$

where α is the absorption coefficient and $\theta = \omega t$ is the phase angle. When the incident laser intensity I_0 and optical path L are constant, the amplitude of WMS-2f signal is proportional to the gas concentration. Practically, apart from the 2f signal descript in Equation (3), the detected signal consists of random noise and the derivation of optical fringes [29,30]. The optical fringes appear as unpleasant spectral features that are usually mixed with the target absorption, and constitute one of the major obstacles in the gas detection. In a well-designed and -fabricated system, the optical fringes should be reduced, and only small residual fringes remain with sinusoidal waveforms, while random noise is seen as small time-varying wiggles superimposed on the true underlying signal, with small standard deviation. Thus, the detected signal could be described as:

$$A_{detected\ 2f} = e_n + \sum a_j(t) \times \cos(\omega_j(t) \times t) + A_{ideal\ 2f}, \quad (4)$$

where $\alpha_j(t)$ and $\omega_j(t)$ are the instantaneous amplitude and frequency of jth fringe component, respectively; A_{ideal} 2f is the WMS-2f signal modeled by Equation (3). The profiles of second harmonic of absorption, fringes, and noise will inherit the featuresof their origination. These profile differences among WMS-2f, harmonic of optical fringes, and noise will be a novel breakthrough point to distinguish and eliminate the interference from the signal (details in Section 5.3).

3. System Configuration

A typical LAS consists of a laser, a photodetector, and an optical configuration for light interaction with gas. For WMS-based LAS, there are, additionally, a laser modulator and a signal demodulator, the latter usually by a lock-in amplifier (LIA).

The laser is the LAS's key component; it usually needs to be continuously tunable mode-hop-free, reliable, single frequency with narrow linewidth (typically <1 MHz), and to have low noise intensity. Historically, lead-salt diode lasers have been developed in a MIR gas sensor. However, these lasers require cooling to liquid nitrogen temperatures and present problems for mode hops and multi-mode operation. Recently, great progress in laser technology has brought many types of excellent lasers, e.g., QCLs and ICLs.

High-sensitive and low-noise detectors are essential for trace gas detection. The most popular commercial infrared detector is a mercury-cadmium-telluride (MCT, or HgCdTe) photoconductive semiconductor-based detector. The MCT detector enjoys a very wide spectral response (2 to 25 μm) and higher speed of detection. Its main limitation is that it needs cooling to reduce noise due to the thermally excited current carriers. Alternatively, newly developed quantum heterostructure detectors could play a vital role in future infrared detection [31].

The optical configuration provides interaction between light and gas' the interaction length directly relates with the sensitivity of gas detection. Thus, a long interaction length is desired to achieve high sensitivity. Multiple-pass cells (MPCs) and open long path are commonly used in LAS to measure low-concentration components or to observe weak spectra in gas. The requirements of compactness, small sample volume, and fast response time stimulated the development of a new type of gas cell. Recently, a hollow waveguide (HWG)-based gas cell has been found to boast the advantages of small sample volume and fast response time [8,32], whereas substrate-integrated hollow waveguides (iHWG) are compact integrated sensors [33]. On the other hand, the need for non-fixed open-path gas detection, e.g., leak detection, aroused the development of standoff remote sensing without a retroreflector [34–36].

3.1. Mid-Infrared Tunable Lasers

MIR tunable laser technology has undergone tremendous development in the past decade, which involves QCL, ICL, difference frequency generator (DFG) [37], optical parametric oscillator (OPO) [38], fluoride fiber lasers [39], hollow-core fiber gas laser [40], VCSEL [41], and II–VI chalcogenides-based MIR lasers [42]. As there have been many excellent reviews of MIR light sources [25,43–45], here we present only a brief overview of the newly developed, highly reliable, and most widely used in spectroscopy, focusing on QCL, ICL, VCSEL, and optical frequency comb (OFC).

3.1.1. Quantum Cascade Lasers

QCL, first demonstrated by Faist et al. in 1994 [46], emits by intersubband transitions between energy levels inside superlattice quantum wells rather than the material bandgap energies in conventional lasers. The most attractive distributed feedback (DFB) QCLs were commercialized in 2004. Now, many commercial providers offer cw- and RT-operated QCLs in different configurations ranging from Fabry-Perot devices, to DFB resonators, to external cavities-based (EC-) QCLs, as well as high-power devices (with watts) in the infrared to terahertz spectral region. QCLs are attractive for infrared countermeasure, metrology, high-resolution spectroscopy, and chemical sensing applications [7,47–53].

QCLs enjoy a broad gain profile of hundreds of wavenumbers and a narrow linewidth about 0.5 MHz by providing monochromatic feedback, i.e., DFB and EC [54]. The natural linewidth can be as low as a few hundred hertz. The emission wavelength of DFB-QCLs can be tuned only a couple of wavenumbers, whereas EC-QCLs can provide hundreds of wavenumbers of coverage with drawbacks of slow mechanical speed, instability, alignment of multiple optical components, and high price. Broadband QCLs have been also built by an array of DFB QCLs with closely spaced emission wavelengths and fabricated monolithically with wavelength coverage of several micrometers [55]. QCL arrays enjoy broad tuning and high spectral resolution, which opens up a wide range of new possibilities for fast, compact, and mechanically robust solutions with high customizability [56].

Overall, QCLs have been greatly improved, but many challenges remain, such as intervalley scattering, heat removal from the core region, and interface scattering, which limit the performance of QCLs, especially at short wavelengths [57]. Although InP-based QCLs emitting at wavelengths of 3–4 µm have been demonstrated recently [58,59], they are still not commercially available.

3.1.2. Interband Cascade Lasers

ICL was presented by Yang in 1995 [60]. RT-cw operation of ICL was first demonstrated in 2008 [61]. Only a few years ago, DFB-ICLs became commercially available with an optical power of milliwatts and a spectral tuning range of a few wavenumbers by Nanoplus GmbH [26]. Like QCLs, ICLs employ the concept of bandstructure engineering to achieve an optimized laser design and reuse injected electrons to emit multiple photons. However, ICLs' photons are generated by interband transitions rather than the intersubband transitions used in QCLs, which allows ICLs to achieve lower input powers than is possible with QCLs.

DFB-ICLs provide single mode, narrow linewidth (~0.7 MHz) [62], low power consumption (hundreds of milliwatts), a compact system solution, and RT-cw emission in the 3–6 µm range, which fill the MIR gap perfectly. DFB-ICLs have been used for the ppb-level detection of many important gases, especially hydrocarbon species [63–69], which leads to many important applications in various areas, for example, clinical diagnostics [70,71], combustion probing [72], environmental monitoring [73,74], and remote sensing.

3.1.3. Mid-Infrared Vertical-Cavity Surface-Emitting Lasers

MIR VCSELs have been paid great attention in the past decade for their advantages of low power consumption, low beam divergence, narrow and single-fundamental-mode, high wavelength tunability, and on-wafer testing capability [45]. The buried tunnel junction (BTJ) concept yields high-performance VCSELs in the wavelength ranges of 1.3–2.6 µm [75] and 2.3–3.0 µm [76], respectively. An interband cascade VCSEL has achieved lasing to λ~3.4 µm in pulsed mode at temperatures up to 70 °C [77], and a 4 µm VCSEL has been developed by using a single-stage active region with eight type-II quantum wells combined with BTJ technology [78]. VCSELs operate with very low threshold currents of several mA and very low power consumption of milliwatts. They provide a single mode by distributed Bragg reflector with a moderate linewidth of tens of MHz. VCSELs are particularly suited for compact and battery-powered sensors.

3.1.4. Mid-Infrared Optical Frequency Comb

OFCs consist of a series of discrete, narrow, stable, equally spaced spectral lines that have a fixed phase relationship between them. These combs can span a broadband of frequency range that have found important applications in areas such as metrology, spectroscopy, and optical communications [79]. To date, OFC sources have extended from the ultraviolet, visible, infrared, and terahertz spectral regions [80]. We only focus on the MIR-OFC, including mode-locked, difference frequency generators (DFG), optical parametric oscillation (OPO), and direct modulation OFC.

Mode-Locked Laser. The most popular way of generating a frequency comb is with a mode-locked laser. Mode-locked lasers produce a series of optical pulses separated in time by the round-trip time of the laser cavity. The spectrum of such a pulse train approximates a series of Dirac delta functions separated by the repetition rate of the laser. In the past decade, there have been hundreds of demonstrations of generating OFC with various lasers including Ti: sapphire solid-state lasers [81], Er: fiber lasers [82], Kerr-lens mode-locked lasers [83], QCLs [84], and ICLs [85] with repetition rates typically between to MHz to 10 GHz. The issues with mode-locked OFC are lower output power, still experimental, and not commercially available.

Difference Frequency Generators. DFG is the most versatile method to generate mode-hop-free tunable broad laser by a nonlinear optical process. The two necessary conditions to achieve MIR OFC output are high-coherence pump lasers and a suitable order nonlinear crystal. There are more than a dozen demonstrated crystals, among which LiNbO3 and ZnGeP2 are the most popular. DFG-based OFC have more power and stability and do not require an oscillating cavity or a high threshold of the optical parametric process [86], whereas the drawbacks are high cost and a relatively low conversion efficiency [37,86,87].

Optical parametric oscillator. An OPO is another parametric nonlinear optical process used to provide a high output power and versatile sources of coherent radiation for spectral regions inaccessible to lasers. OPO has been shown providing high power OFCs based on Yb: fiber laser covering several micrometers [88,89]. Researchers have not only shown that degenerate synchronously OPOs are efficient tools to transfer near-infrared (NIR) frequency combs to the mid-infrared; also, they can utilize pump lasers in NIR and expand the spectrum to the mid-infrared [38,90]. OPO offer high output power with broad spectral coverage, but require a free-space resonator, external pumping sources, and many optical components [91]. This makes OPO sources bulky, vulnerable to any external disturbances, and very impractical in field applications.

Direct Frequency-Modulation Combs. Direct FM combs have been experimentally demonstrated in semiconductor lasers, such as QCLs [91–97], ICLs [85], quantum dot (QD) [98], and quantum dash lasers [99]. These lasers are passively mode-locked with cw or quasi-cw output. Direct FM combs enjoy broadband, wide repetition frequency from kHz to THz [100] with line width of kHz to MHz level. Direct FM comb offers the possibility of a portable, chip-scale device with low power consumption, which are desired in spectroscopic trace gas sensing.

3.2. Infrared Detector

There has been exciting progress in MCT junction technology, which could design and fabricate a high-performance MCT photovoltaic detector for operation in RT and in situ applications [101]. The commercialized detector benefits from the use of optimized material, device architecture, concentrators of radiation, enhanced absorption, and shields against thermal radiation, and can achieve directivity as high as 10^{11} cm· $\sqrt{Hz}$·W^{-1} in RT conditions [102]. Alternatively, more progress has been noted in quantum engineering-based detectors.

3.2.1. Quantum Heterostructure Detector

Quantum heterostructure-based infrared detectors, including quantum cascade (QC), quantum well (QW), quantum dot, and combinations of both QDs and QWs in a dot-in-a-well (DWELL) strategy detectors, have high detectivity and low dark currents [31,103,104]. QC detectors have low noise due to their low interference of background radiation. QW detectors are characterized by a narrow spectral band and are easily fabricated. However, the disadvantages of QW detectors include low operating temperatures and the requirement of a scattering filter to adjust the incident light angle [105]. QD detectors have higher operating temperatures and are capable of absorbing normally incident photons. However, they have much lower quantum efficiency due to the lower absorption and capture probabilities of incident light [104,106]. DWELL devices have shown promise in terms of detectivity and spectral range, though their operating temperatures remained fairly low.

Most quantum heterostructure detectors are still at the experimental stage, with different fabrication technologies and materials being used. Few commercial products have been reported, which leads to the restriction of such detectors' application in MIR detection. However, recent advances in this field such as new device-chip hybridization [107], antimonide-based membrane synthesis integration, and strain engineering [108] may be scalable methods for the fabrication of commercially available heterostructure detectors.

3.2.2. Infrared Avalanche Photodiodes

An avalanche photodiode (APD) provides an internal multiplication necessary to achieving high avalanche gain at low bias with low noise and high bandwidth [109]. Presently, MCT APD is the most promising for trace gases in standoff remote sensing with high sensitivity. The demands of night vision and LIDAR systems stimulate the development of MCT APD arrays [110]. Overall, commercialized high-sensitive and bandwidth MCT APD would play a vital role in many areas.

3.3. New Types of Gas Cells

Recently, new types of gas cells have been developed to achieve long optical path in a smaller volume, including modified MPCs [111–116], circular multi-reflection (CMR) cells [117–124], and HWG [32,63,69,125–128]. Guo and Sun reviewed the progress in modified MPCs and CMR cells and compared them in terms of optical pathlength (OPL), volume, and path-to-volume ratio (PVR) [116]. HWG-based gas cells have been discussed in detail in books and reviews [129–131]. We compared the main features among the modified MPCs, CMR cells and HWG, as shown in Table 1.

Table 1. Comparison of different types of gas cells.

	Type	Ref.	OPL/m	Volume/L	PVR/m/L	ADs	DADs
Modified MPCs	2 mirrors	[111]	22	0.55	40	(c) (d) (e)	②⑤
	6 mirrors	[112]	314	1.25	251.2	(b) (c) (e)	①②⑤
	3 mirrors	[113]	1.46	0.33	4.4	(c) (d)	①②⑤
	2 mirrors	[114]	57.6	0.225	256	(a) (b) (d) (e)	②⑤
	2 mirrors	[115]	26.4	0.28	94.3	(b) (c) (e)	②⑤
	6 mirrors	[116]	32.4	0.48	67.5	(c) (d) (e)	①②⑤
CMR cells	1 mirrors	[117]	1.05	0.078	13.5	(c) (d) (e)	③④
	6 mirrors	[118]	3.1	0.024	129.2	(a) (c) (d) (e)	③④
	1 mirrors	[119]	2.16	0.04	54	(a) (c) (d) (e)	③④
	1 mirrors	[120]	4.08	0.04	102	(a) (c) (d) (e)	③④
	6 mirrors	[121]	0.69	0.013	53.1	(a) (c) (d) (e)	③④
	1 mirrors	[122]	12.24	/	/	(a) (c) (d) (e)	③④
	1 mirrors	[123]	9.9	0.05	198	(a) (c) (d) (e)	③
	65 mirrors	[124]	10	0.14	71.4	(a) (c) (d) (e)	③④
HWGs	Ag/AgI-HWG	[32,69]	5	0.004	1250	(a) (d) (e)	④
	iHWG	[126]	0.25	0.001	250	(a) (d) (e)	④
	iHWG	[63]	0.075	0.0003	250	(a) (d) (e)	④
	PBF-HWG	[127]	1	5×10^{-6}	200,000	(a) (d) (e)	④⑤
	PBF-HWG	[128]	0.08	4.5×10^{-7}	177,778	(a) (d) (e)	④⑤

Advantages index	Disadvantage index
(a) easy configurations, fewer than three components (b) ultra-long pathlength (>50 m) (c) keeping focal properties, low-aberrations (d) compact structure (e) suitable for MIR	① multiple components (≥3 optical elements) and requiring adjustment ② large volume (>0.1 L) ③ hard to align, sensitive to vibrations or input conditions ④ spot diffusion introduced by aberrations ⑤ slow gas exchanging

There are three categories of HWG employed in spectroscopic gas sensing, namely Ag/AgI-coated HWG (Ag/AgI-HWG), photonic bandgaps HWG (PBG-HWG), and iHWG [131], and the comparison of them is as shown in Table 1. HWGs work as both an optical waveguide and gas transmission cell that provide an extended OPL yielding high sensitivity measurements [69]. It is worth noting that filling PBG-HWG with analyte gas for sensing is difficult owing to the considerable back-pressure building up in the hollow structure. HWGs are ideal candidates for gas cells due to their high PVR and ability to transmit light at a fairly wide wavelength range, which means building a MIR sensor platform is feasible.

3.4. Fully Integrated Sensors

Profiting from the development of MIR lasers, detectors, and gas cells, as described above, the laser spectrometer can be potentially integrated into a miniaturized and compact system for gas sensing, as shown in Figure 1, maintaining or even enhancing the achievable sensitivity [132]. In this subsection, we introduce the progress and applications of the fully integrated MIR laser spectrometer, focusing on QCL or ICL-coupled HWG gas sensors.

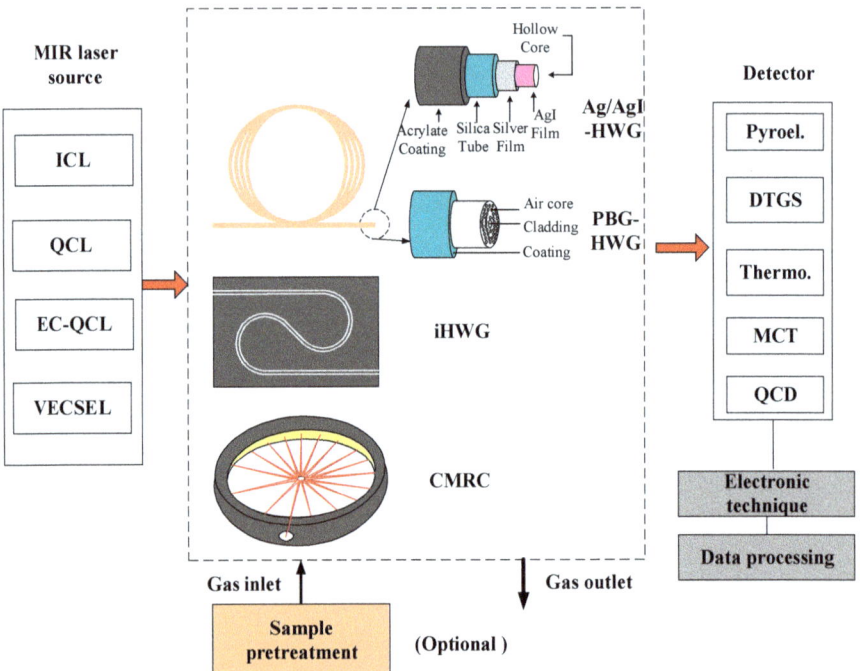

Figure 1. Overview of the recently emerging gas cell-based laser gas sensing principles. A gas cell with small volume and easy integration includes: Ag/AgI-HWG-Ag/AgI-coated hollow waveguides, PBG-HWG-photonic bandgaps hollow waveguides, iHWG-substrate-integrated hollow waveguides and CMRC-circular multi-reflection cell. The laser source includes: ICL—interband cascade laser, QCL—quantum cascade laser, EC-QCL—external cavity coupled QCL. The detector includes: Pyroel. —pyroelectric detector, DTGS—deuterated triglycine sulfate detector, Thermo. —thermopile detector, MCT—mercury cadmium telluride semiconductor detector, QCD—quantum cascade detector.

An integrated sensor consists of a compact MIR laser, a new gas cell, and a detector, as shown in Figure 1. The configuration provides an attractive solution for miniaturized and practical applications. The laser should be a QCL, ICL, EC-QCL, or VECSEL. The miniaturized gas cells, i.e., HWG or iHWG, cover the wavelength range of 3.0–11.0 μm, and provide a response time as fast as seconds [32] and a sample volume of sub-milliliter [131]. iHWG-based cells provide excellent modularity and mechanical stability, with effective OPL of hundreds of millimeters and larger losses per unit length than Ag/AgI-HWG-based cells [32], which could achieve an effective OPL of several meters. The sensor could operate in the DAS, WMS, or intrapulse modulation regime as well [33,133–136].

3.5. Open Path Detection without Retroreflectors

The requirements of non-fixed open-path gas detection, e.g., the atmospheric environmental monitoring, leak detection, security early warning, etc., promoted the standoff remote sensing. We focus on the open-path-averaged gas concentrations by the backscatter light from a remote hard target or topographic target. Other open-path systems that are deployed as point samplers or long-path with retroreflectors are beyond the scope of this review.

In order to realize standoff gas detection, a number of laser-based techniques are available, such as TDLAS [34,35,137,138], PAS [139], differential absorption lidar (DIAL) [140], CLaDS [141], and more recently active coherent laser spectrometers (ACLaS) [36]. The basic architectures of these techniques are shown in Figure 2. The OPL is often variable and unknown; therefore, performance is often quoted in a similar fashion to that of open path gas detectors, using the pathlength-integrated unit of ppm·m [142]. The detection limits for such systems commonly range from sub ppm·m to several hundred ppm·m [35,36,141], typically over distances of open path from several meters to hundreds of meters or even kilometers. The performance of this system is typically limited by the level of received laser power, which is dependent on the incident laser power, distance between receiver and scattering target, type of scattering materials, and size of the receiver aperture [142–144]. High-power and broadband tuning lasers, such as OPO and EC-QCLs, are desired to achieve a higher signal-to-noise ratio (SNR) [145–147]. However, high-sensitive and low-noise detectors are particularly important for applications in public areas where eye safety should be considered.

Recently, standoff trace detection has achieved tremendous progress and been applied in leaks [35], environmental monitoring [141], explosives [148,149], combustion [147], unmanned aerial vehicles (UAV) [150,151], and many other promising applications as well.

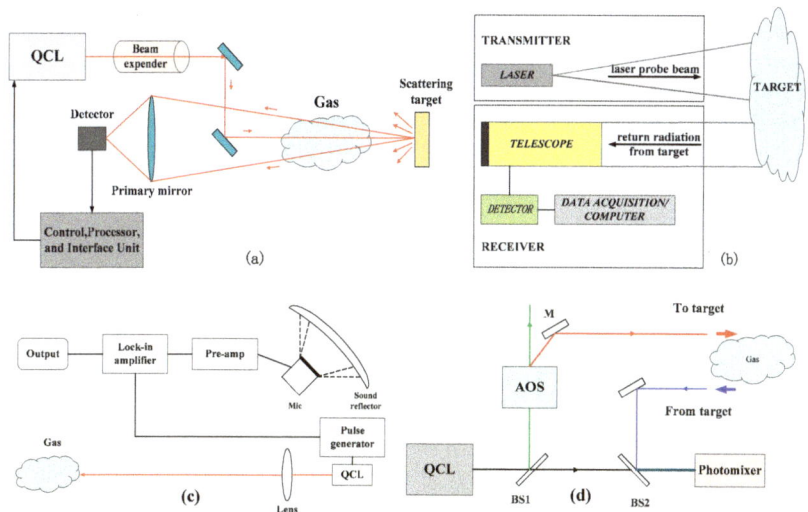

Figure 2. Diagrams of the basic architectures of standoff gas detection techniques with non-cooperators: (**a**) TDLAS [35], (**b**) DIAL [149], (**c**) PAS [139], (**d**) ACLaS [36].

4. Detection Methods: Derivative Spectroscopy

MIR spectroscopy is attractive for the strong fingerprint signature, and the commonly used DAS often suffers from low selectivity for the spectra overlapping of broadband absorption. Derivative spectroscopy uses first or higher derivatives of absorbance with respect to wavelength for qualitative analysis and for quantification with higher selectivity [152]. The derivatization of zero-order spectrum can lead to separation of overlapped signals or elimination of background caused by other compounds in a sample [153,154].

Derivative spectra could be obtained by the Savitzky-Golay (SG) smoothing/differentiation procedure, which is widely implemented in instrumental software or in packages for spectral data processing. The resolution enhancement in the second derivative spectrum depends on the data spacing in original spectra, absorption peak profile, parameters of SG (i.e., window size and polynomial order). To maximize the separation of the peaks in a second derivative spectrum, the original spectra should be recorded at high resolution and using appropriate parameters [155]. Other methods used to calculate the derivative spectra include numerical differentiation [156] and continuous wavelet transform [157]. The latter has been proven to be efficient in the analysis of overlapping spectra and is advantageous for providing higher SNR and flexibility in searching for absorption peaks.

Derivative spectroscopy can be used in various spectral region including UV [158], visible [159], NIR [160,161], and MIR. The method has a close dependence on instrumental parameters, like speed of scan, the linewidth, and SNR. The derivatization can amplify the noise signals in the resulting curves, which normally leads to a higher SNR. Another disadvantage is the non-robust character of the selected parameters of the elaborated methods. The selected parameters of this method are applicable only for the studied system and every change in composition requires re-optimization and the selection of new parameters of derivatization [153]. Without a homogeneous protocol of optimization, the parameters of the method vary, and most researchers did not describe the parameter selection in their published articles.

5. Detection Methods: Modulation Spectroscopy for Wideband Absorption

WMS has been demonstrated to have high sensitivity for sensing gas with an isolated spectral line. When the modulation index is small, WMS is approximately expressed as derivative spectroscopy. For the MIR spectral region, however, the fingerprint spectra are often broad, serried, crowded, and even overlap within the coverage of a tunable laser. To ensure the detection sensitivity and selectivity, the essential procedures include optimizing modulation index, varied modulation amplitude, removing fringes and noise interference, and multicomponent spectral fitting.

5.1. Optimizing the Modulation Index

Since the recorded harmonic signals used in WMS are heavily dependent on the spectral line profile and modulation index adopted in the WMS system, the situation for sensing a larger molecular gas with a broadband spectrum is quite different. Moreover, the similarity of signal waveform between a broadband spectrum and the intrinsic optical fringes interference will seriously complicate the signal processing [30] and deteriorate the sensing sensitivity and precision.

The modulation index always plays a pivotal role in WMS-based measurement. A modulation index of 2.2 is recognized as the optimum to achieve the maximum SNR with isolated Gaussian or Lorentzian line profile. For broadband spectrum, however, the reordered harmonic signal would broaden to overlap with the adjacent spectrum, interference, and optical fringes with the so-called modulation index optimum. The overlapping may deteriorate and even disable the WMS measurement. So the modulation index determination should balance the spectra discrimination and the SNR in WMS with broadband spectrum.

We investigated the determination of modulation index for the broadband absorption, taking the carbon disulfide (CS_2) spectrum around 2177.6 cm^{-1} as an example. To evaluate the spectra discrimination of the WMS-2f signal, we particularly defined a parameter, SD [27]:

$$SD = \frac{h_{CD}}{0.5 \times (h_A + h_B)}, \quad (5)$$

where A and B are the adjacent valleys of neighboring WMS-2f signals, C is the middle point of the line connecting A and B, point D at the signal curve is vertical to point C, h_A and h_B are the amplitude of absorption valley A and B, respectively. h_{CD} is the height difference between point C and D. The parameter SD is a constant from 0 to 1, which represents the WMS-2f completely overlapping and parting, respectively, as shown in the insert panel of Figure 3b.

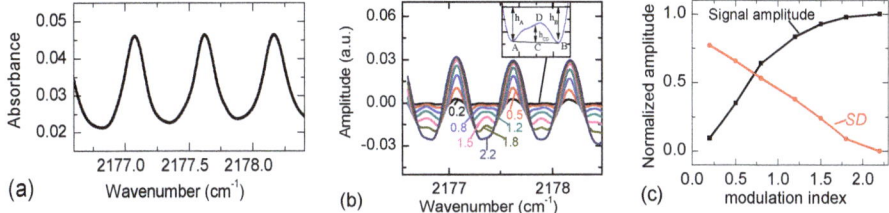

Figure 3. (a) broadband spectra of CS_2 around 2176 cm^{-1} with 30.5 ppm × 5 m; (b) simulated WMS-2f signals of Figure 1a with modulation index 0.2 to 2.2, the insert panel describing the definition of spectrum discrimination (SD); (c) the SD and normalized WMS-2f signals amplitude under various modulation index [27].

The parameter SD and normalized amplitude of WMS-2f signals under different modulation index was plotted in Figure 3c, which reverse with the modulation index. To balance the SD and normalized amplitude of WMS-2f signals, the modulation index around 1.0 should be a good compromise for optimizing WMS with the band spectrum of CS_2 detection.

5.2. Varied Modulation Amplitude

A modulation amplitude setting always confronts multi-spectrum with different widths, which creates a significant dilemma. Any single modulation amplitude cannot cover a large spectral width difference, e.g., a difference of the half width at half maximum is more than 50%. A practical way to achieve this is to use a varied modulation amplitude for multi-spectrum detection, namely WMS with varied modulation amplitude (WMS-VMA) [162].

The WMS-VMA was realized mainly by a homemade digital lock-in amplifier (DLIA), which performs the modulation, demodulation of WMS-1f and WMS-2f, and generation of reference signal by an integrated field-programmable gate array-based circuit. The DLIA generates an arbitrary waveform signal by direct digital frequency synthesis. The sine waveform with varied amplitude is prepared in advance and stored in the DLIA's random access memory, and then read out to control the sequence using the frequency controller integrated in the DLIA. The method has been verified by a multiparameter optical sensor with typical broadband spectroscopy [162].

5.3. Removing Fringes and Noise Interference

LAS always suffers interference from electric noise and optical fringes; the latter are caused by multiple reflections upon optical interface, i.e., the so-called "etalon effect." The interference causes low sensitivity and precision in the spectrometers. Since the optical fringes are small and sine-like in a well-designed and -fabricated system, the signal profiles of molecular absorption always exhibit distinct differences to those of optical fringes and electric noise. The WMS-2f signal profile would inherit their difference with an optimized modulation index. So, the molecular absorption can be distinguished from optical fringes and electric noise by the signal profile.

An empirical mode decomposition (EMD) algorithm has been successfully used in WMS to remove optical fringes [27] and noise [163] by characterization of the signal profile. The procedure for employing EMD to decompose and reconstruct the WMS-2f signal is described in Figure 4a. The detected 2f signal was decomposed into intrinsic mode functions (IMF), which have a specific physical source and are meaningful. The simplest criterion, whether IMF from molecular absorption or interference is the correlation coefficient (R) between an IMF and the reference 2f signal, is shown in Figure 4b,c.

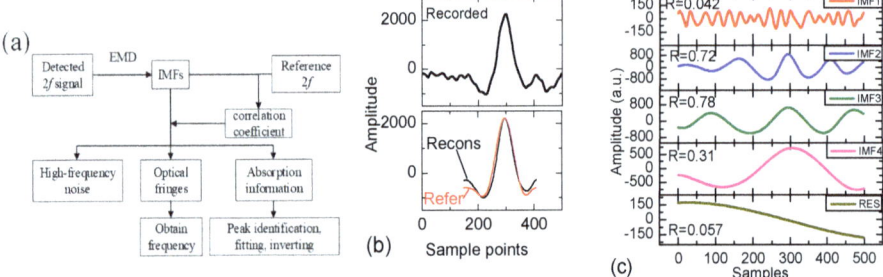

Figure 4. Schematic of removing optical fringes and noise with the empirical mode decomposition (EMD) algorithm. (**a**) flow chart of the algorithm; (**b**) WMS-2f signal: recorded in the upper panel, the reconstructed and simulated in lower panel; (**c**) the intrinsic mode functions (IMF) decomposed from the recorded WMS-2f [27].

The components with lower R most likely come from fringes or noise, while components with larger R with reference absorption must be the absorption. All these components with higher R can be added together to reconstruct a WMS-2f free from the interference of fringes and noise. Generally, this method can improve the sensor sensitivity by about 30% [164].

5.4. Multicomponent Spectral Fitting

Multicomponent spectral fitting is mainly used to eliminate spectral interference, which commonly occurs in MIR detection. Furthermore, benefitting from the use of multivariate regression and nonlinear least square fitting, it can calculate the multi-component concentration in a gas mixture, i.e., achieving multiple component detection simultaneously using a single DFB ICL [164,165]. The Levenberg-Marquardt (LM) algorithm, also known as the damped least-squares method, was applied for the data fitting. Reference WMS-2f signals of all the components were obtained beforehand. The concentrations of all components in the mixture comply with the constraint condition of non-negative parameters. In each iteration of the fitting routine, the WMS-2f signal of the mixture was simulated with the updated parameters. Once the routine converged, the best fitting parameters were determined as the concentrations of the components.

Multicomponent spectral detection could benefit from the redundancy of the multiple spectra, not only in the magnitude of the absorption but also in the line shape related to temperature and pressure broadening. To make full use of the information buried in the detected spectral lines, we presented an improved multicomponent spectral fitting algorithm for the sensor. We also applied the method of normalized WMS-2f by 1f for the sensor immunity of laser energy fluctuation [162].

6. Detection Methods: Optical Frequency Comb Spectroscopy

Frequency combs enjoy high spectral resolution and broad spectral coverage that make them a unique spectroscopic tool for precision spectroscopy and for multi-species detection [80,166]. Direct frequency comb spectroscopy (DFCS) employs optical frequency combs to probe spectral features in a parallel fashion [167]. We briefly review the DFCS including frequency comb-based Fourier transform spectroscopy (FC-FTS), cavity-enhanced direct frequency comb spectroscopy (CE-DFCS), and virtual imaging phased array spectroscopy (VIPAS).

6.1. Frequency Comb Fourier Transform Spectroscopy

FC-FTS is the measurement of the Fourier transform of the interferogram based on frequency combs source, which offers excellent spectral brightness and spatial coherence. There are two implementations of FC-FTS: Michelson interferometer-based Fourier transform spectroscopy and dual-comb spectroscopy (DCS) [168]. Each presents its own distinct advantages, but both rely on the same physical principle.

Michelson interferometer-based FC-FTS. In FC-FTS, OFC works as the light source for Fourier transform spectrometers (FTS). The high spectral brightness, together with the spatial and temporal coherence of the combs, enable acquisition times orders of magnitude shorter than in conventional FTIR spectroscopy. Thus, FC-FTS could be promising in standoff chemical sensing of transient, non-repeatable phenomenal like combustion, plasmas, and explosions [169]. Presently, the applications of FC-FTS is still limited by spectral width, lower comb intensities, mechanically scanned mirrors to record the interferogram, placing limits on the acquisition of spectra [170]. The interferometer records the interference pattern between two combs as slightly different because the light reflected by the moving mirror of the interferometer is Doppler shifted [171,172].

Dual comb spectroscopy. DCS uses two frequency combs of slightly differing line spacing, one for reference and the other for sample detection [168]. From each pair of optical lines, one from each comb, a radio frequency beat note is generated on a detector. In this way, optical frequencies are converted into radio frequencies such that the amplitude and phase changes caused by the interaction of one of the combs with a sample can be detected. DCS has an advantage similar to a Michelson interferometer but without moving parts and employing a single point detector, in which case a minimum detectable absorption ~1×10^{-8} cm^{-1} has been demonstrated [173].

Despite an intriguing potential for the measurement of molecular spectra spanning tens of nanometers within tens of microseconds at Doppler-limited resolution, the development of dual-comb spectroscopy is hindered by the demanding stability requirements of the laser combs [168,174,175]. For an ideal, the interference sampled waveform can be Fourier transformed to display the signal spectrum. In reality, the main difficulty comes from the time and phase fluctuations of the frequency comb. However, we experimentally demonstrate that the means of real-time dual-comb spectroscopy can be used to overcome this problem [176]. The true value of DCS for sensitive molecular detection lies in the MIR [177,178], and compact design will be obtained using the semiconductor laser comb technology in the future [179–181].

6.2. Cavity-Enhanced Direct Frequency Comb Spectroscopy

CE-DFCS combines broad spectral bandwidth, high spectral resolution, precise frequency calibration, and ultrahigh detection sensitivity all in one experimental platform based on an optical frequency comb interacting with a high-finesse optical cavity [166,182]. Michael et al. demonstrate a minimum detectable absorption of 8×10^{-10} cm^{-1}, a spectral resolution of 800 MHz, and 200 nm of spectral coverage [183]. Moreover, combined with VIPA technology, the minimum detectable concentration is 1.7×10^{11} cm^{-3} [184]. Whereas, a great deal of CE-DFCS applications are in the visible and NIR spectral region, MIR CE-DFCS is attracting a lot of attention [182,185,186]. It is noteworthy that the mode spacing between the cavity and comb should be well matched, or the operation will always too be complicated [186,187].

6.3. Virtual Imaging Phased Array Spectroscopy

VIPAS provides alternative approaches that circumvent this problem by directly measuring the power or phase of individual comb teeth that have interacted with the molecular gas [188]. In this case, a novel high-resolution crossed spectral disperser is employed to project the various frequency comb modes onto a two-dimensional digital camera. The distinguishing feature of the present approach is the use of a side-entrance etalon called a VIPA disperser in the visible spectral range. The multiple reflections within the VIPA etalon interfere such that the exiting beam has different frequencies emerging at different angles. Sensing with a broadband comb directly interrogates an absorbing sample, after which the spectrum is dispersed in two dimensions and sensed with detector arrays. Thus, the spectrometer transforms the one-dimensional comb into something more reminiscent of a two-dimensional 'brush'.

With the VIPA method, Diddams et al. [189] resolved 2200 comb modes covering a 6.5 THz span with resolution 1.2 GHz, while Gohle et al. [190] resolved 4000 modes covering a 4 THz span with resolution 1 GHz. However, due to the limitations in optical coating and array detector technology, VIPA is available in the visible and NIR ranges. Proof-of-principle demonstrations have been carried out in the MIR wavelength region spectral resolution 600 MHz (0.02 cm^{-1}) [191], and resolution 1 GHz (0.03 cm^{-1}) [192]. MID APD arrays may change this situation in the near future.

7. Summary of MIR Gas Sensing

MIR trace gas sensing in the molecular fingerprint region developed rapidly in the near decade mostly due to the commercialization of MIR tunable lasers, i.e., DFB-QCL and DFB-ICL, which could be proven by rough statistics on the number of studies published annually on TOPIC: (Mid-Infrared Lasers) AND TOPIC: (Sensing) in the Web of Science, as shown in Figure 5. Though many prominent works on MIR trace gas sensing were performed before 2008, NO was detected through the ν_1 band near 1875 cm^{-1} with limit of detection (LOD) of ppb level by DAS [193] and WMS [194], respectively, for industrial processes and vehicle emissions monitoring. We only summarize gas detection based on MIR absorption spectroscopy with tunable lasers in the near decade for the last 10 years; see Table 2 for single component detection and Table 3 for multi-component detection.

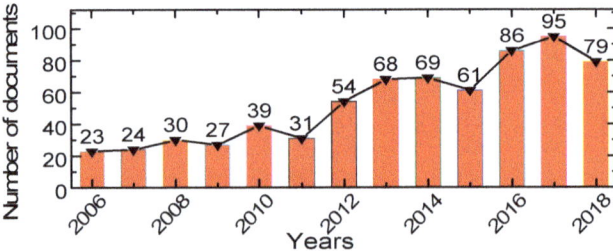

Figure 5. Census on mid-infrared lasers AND sensing (TOPIC) in the Web of Science.

Table 2. Summary of single component detection by MIR spectroscopy (2008–2018).

Species	Bands	Wavelength/nm	Laser Type	Techniques	LOD [1]/ppb	Refs	Applications
CH$_4$	ν_1	3392	ICL	WMS-2f	48@0.1 s	[8]	Exhaled breath analysis
		3451.9	DFG	HPSDS [2]	250	[195]	Technique research
	ν_3	3291	ICL	SA-DAS DAS	6.3@240 s; 2.25@2.5 s	[196, 197]	Atmospheric
				DAS	1.4@60 s	[198]	
				WMS	13.07@2 s	[199]	
		3300		DAS	15@60 s	[73]	
		3240		OF-CEAS [3]	3@2 s	[200]	Technique research
		3366		DAS	3.8×10^4	[63]	Industrial emission and process control
		3260	QW-DFB-DL [4]	PAS;WMS	1.5×10^4@12 s	[201]	Environmental
		3200	DFG	FCS	60@80 ms	[202]	Atmospheric
		3250	DROPO [5]	FCS	4@15 ms	[203]	
		3390	OPO	FCS	/	[204]	Technique research
		3270	OPO	NICE-OHMS [6]	0.09@20 s	[205]	Ultrasensitive detection research
			GaSb laser	WMS	13	[150]	Atmospheric
	ν_4	7791	QCL	CLaDS [7]	60 ppb@100 s	[141]	
C$_2$H$_6$	ν_{10}	3330	ICL	WMS	1.5@23 s	[206]	Technique research
		3340		WMS-2f; WMS-2f/1f DAS;	1.2@4 s; 1.0@4 s; 7.92@1 s	[65, 66]	Atmospheric
		3360	LD	WMS	0.13@1 s	[207]	Environmental
			LD	WMS-2f	0.24@1 s	[208]	Technique research
C$_2$H$_4$	ν_1	3266	ICL	WMS-2f	53@24 s	[64]	Industrial emission
C$_2$H$_2$	$\nu_4 + \nu_5$	7263	EC-QCL	WMS-2f/1f	3@110 s	[53]	Exhaled breath analysis
C$_3$H$_8$	ν_2	3370.4	ICL	WMS-2f	460@1 s	[209]	Leakage monitoring
C$_5$H$_8$	/	3333.3	OPO	FCS	7@30 s	[210]	Technique research
C$_6$H$_6$	ν_{14}	9640	QCL	DAS	12@200 s	[211]	Atmospheric
C$_{10}$H$_{22}$	/	3380	ICL	PAS	0.3	[212]	Industrial process
H$_2$CO	ν_4	3493	ICL	DAS	10^3	[213]	Workplace monitoring
		3356		WMS-2f	73@40 s	[214]	Combustion emission
				DAS; WMS	0.26@300 s; 0.069@90 s	[215]	Technique research
	ν_1	3599	ICL	WMS-2f	1.5@140 s	[216]	Atmospheric
				DF-RFM [8]	25@1 s	[67]	Technique research

111

Table 2. Cont.

Species	Bands	Wavelength/nm	Laser Type	Techniques	LOD [1]/ppb	Refs	Applications
CH_3OH	ν_9	3390	OPO	FCS	40@30 s	[210]	Technique research
C_2H_5OH	/	3367	OPO	FCS	40@30 s	[210]	Technique research
CH_3COCH_3	ν_9	3380	VECSEL [9]	DAS	13@300 s	[217]	Breath VOCs detection
		3389.8	OPO	FCS	9.1@30 s	[210]	Technique research
	ν_5	8000	ECQCL	WMS-2f	15@10 s	[52]	Spoilage monitoring of agricultural products
CO_2	ν_1	3300	OPO	DIAL [10]	1.2×10^5	[140]	Atmospheric
	ν_3	4330	QCL	I-QEPAS [11]	300 ppt@4 s	[218]	Technique research
		4200	ICL	DAS	5×10^4	[72]	Combustion diagnose
		4172			/	[219]	
CO	ν_1	4691.2	ICL	WMS	9@0.07 s	[71]	Exhaled breath analysis
		4600		DAS	500@14 s	[220]	Atmospheric
		4764	QCL	WMS	26@1 s	[221]	Indoor air
		4980		WMS	/	[222]	Combustion diagnose
NO	ν_1	5184	ICL	DAS	3×10^4@10 ms	[223]	Combustion emission
		5200	QCL	QEPAS [12]	120	[224]	Engine exhaust monitoring
		5250		WMS-2f	/	[225]	Technique research
		5030		DAS	/	[226]	Gas sensing in high temperature
		5263		I-QEPAS; WMS-2f	4.8@30 ms	[227]	Environmental
NO_2	ν_3	6250	QCL	WMS	360 (600 K); 760 (800 K)	[228]	Gas sensing in high temperature
	$\nu_1 + \nu_3$	3250–3550	OPO	PAS	14@170 s	[3]	Environmental pollution
N_2O	ν_3	7782	ECQCL	DAS	7.36×10^3	[229]	Toxic industrial chemical detection
	ν_2	8600	QCL/DFG	FCS	0.3	[230]	Atmospheric
	ν_1	4530	QCL	CLaDS	1.2×10^3	[231]	Atmospheric
NH_3	ν_2	10,400	ECQCL	PAS	1	[49]	Atmospheric
		10,340	QCL	QEPAS; 2f-WMS	6@1 s	[232]	Exhaled breath analysis
		9560		DAS	17.3@3 s	[233]	Atmospheric
		9060		WMS	0.3	[2]	Atmospheric
	ν_1	2958.5	OPO	FCS	25@30 s	[210]	Exhaled breath analysis
N_2H_4	ν_{12}	10,363	LD	DAS	400	[234]	Chemical analysis
O_3	ν_3	9697	QCL	DIAS [13]	300	[235]	Atmospheric

112

Table 2. Cont.

Species	Bands	Wavelength/nm	Laser Type	Techniques	LOD [1]/ppb	Refs	Applications
H_2O	ν_2	6700	QCL	OA-ICOS [14]	280	[21]	Chemical analysis
	ν_3	2666.7	OPO	FCS	5.3@30 s	[210]	Technique research
H_2O_2	ν_6	7730	QCL	QEPAS	12@100 s	[236]	Breath diagnosis
	ν_3	3760	OPO	FCS	8	[171]	Exhaled breath analysis
H_2S	ν_2	7900	QCL	QEPAS	450@3 s 330@30 s	[237]	Environmental pollution
SF_6	ν_3	10,540	QCL	QEPAS	0.05@1 s	[238, 239]	Technique research
CS_2	$\nu_1+\nu_3$	4590	QCL	QEPAS	28@1 s	[240]	Industrial process
				DOAS; WMS	10.5; 60@240 s	[27, 241]	Atmospheric
OCS	ν_3	4860	QCL	DAS	1.2@0.4 s	[242]	Exhaled breath analysis
CH_3SH	ν_2	3393	ICL	DAS	25@1.84 s	[68]	Industrial emission
		3392	ICL	WMS	7.1@295 s	[165, 243]	Atmospheric
CH_3SCH_3	ν_{18}	3370	ICL	WMS-2f/1f	2.8@125 s	[69]	Environmental
		3337	ICL	WMS	9.6@164 s	[164]	Atmospheric

[1] LOD: Limit of Detection; [2] HPSDS: Heterodyne phase sensitive dispersion spectroscopy; [3] OF-CEAS: Optical feedback cavity-enhanced absorption spectroscopy; [4] QW-DFB-DL: Quantum wells distributed feedback diode laser; [5] DROPO: Doubly resonant optical parametric oscillator; [6] NICE-OHMS: Noise-immune cavity-enhanced optical heterodyne molecular spectrometry; [7] CLaDS: Chirped laser dispersion spectroscopy; [8] DF-RFM: Dual-feedback RF modulation; [9] VECSEL: Vertical-external cavity surface-emitting laser; [10] DIAL: Differential absorption lidar; [11] I-QEPAS: Intracavity Quartz-Enhanced Photoacoustic Spectroscopy; [12] QEPAS: Quartz-enhanced photoacoustic spectroscopy; [13] DIAS: Differential absorption spectroscopy; [14] OA-ICOS: Off-axis integrated cavity output spectroscopy;

Table 3. Summary of information on multi-component simultaneous detection in the last 10 years.

Species	Bands	Wavelength/nm	Laser Type	Techniques	LOD/ppb	Refs	Applications
CH_4/C_2H_6	ν_3/ν_{10}	3291 3337	ICL	DAS	5@1 s 8@1 s	[244]	Atmospheric
CH_4/C_2H_6	ν_3/ν_{10}	3291 3337	ICL	DAS WMS-2f	2.7@1 s 2.6@3.4 s	[245]	Atmospheric
CH_4/C_2H_6	ν_3/ν_{10}	3337	ICL	WMS	17.4@4.6 s 2.4@4.6 s	[246,247]	Atmospheric
CH_4/C_2H_6	ν_3/ν_{10}	3404 3335.5	DFG	CLaDS	360@1 s; 60@100 s /	[248]	Technique research
$CH_4/C_2H_6/C_3H_8$	$\nu_3/\nu_{10}/\nu_1$	3345	ICL	QEPAS	90@1 s 7@1 s 3×10^3@1 s	[249]	Oil and gas industry monitoring
$CH_4/CO/H_2CO$	$\nu_4/\nu_1/\nu_2$	7880/4633 5683	QCL	WMS-2f	0.5 for H_2CO@2 s	[250]	Atmospheric
CH_4/N_2O	ν_4/ν_3	7700	QCL	DOAS	3×10^4 3.3×10^3	[35]	Remote gas leakage detection

Table 3. Cont.

Species	Bands	Wavelength/nm	Laser Type	Techniques	LOD/ppb	Refs	Applications
CH_4/N_2O	ν_4/ν_3	7800	QCL	WMS-2f	5.9@1 s 2.6@1 s	[114]	Atmospheric
$CH_4/H_2CO/C_2H_4/$ C_2H_2/CO	$\nu_3/\nu_1/\nu_{11}/\nu_3/\nu_1$	2500–5000	OPO	FCS	1.7 310 320 110 270	[251]	Technique research
CH_4/NO	ν_3/ν_1	3000–5400	DROPO	FCS	20 15	[172]	Technique research
$CH_4/N_2O/H_2O$	$\nu_4/\nu_3/\nu_2$	7710	QCL	DAS	23@1 s 6.5@1 s 6.2×10^4@1 s	[252]	Atmospheric
$CH_4/N_2O/H_2O$	$\nu_4/\nu_3/\nu_2$	8000	ECQCL	WMS-2f	4.8@1 s 0.9@1 s 3.1×10^4@1 s	[50]	Atmospheric
CH_3OH/C_2H_5OH	ν_8	10,100	ECQCL	DAS	130@1 s 1.2×10^3@1 s	[146]	Atmospheric VOCs
$C_2H_5OH/(C_2H_5)_2O/$ CH_3COCH_3	/	3800	QCL	CRDS	157 60 280	[18]	Atmospheric VOCs
CO_2/CO	ν_3/ν_1	4730	EC-QCL	WMS-2f	6.5×10^5 9	[7]	Exhaled breath analysis
CO_2/CO	ν_3/ν_1	4250 4860	QCL	WMS	10^6	[51]	Combustion diagnose
CO_2/CO	ν_3/ν_1	4193 4979	ICL QCL	DAS	/	[253]	Combustion diagnose
CO_2/N_2O	ν_3/ν_1	4466	QCL	DAS WMS	2.7; 0.2 4.3@1 s;/	[118]	Technique research
$H_2O/CO_2/CO$	$\nu_3/\nu_3/\nu_1$	2551 4176 4865	LD ICL QCL	WMS-2f/1f	1.4×10^7 6×10^6 4×10^6	[147]	Combustion diagnose
$NO/CO/$ N_2O	ν_1	5263 4566	QCL	DAS	0.5@1 s 0.8@1 s	[254]	Exhaled breath analysis
CO/N_2O	ν_1	4500	QCL	WMS	0.36 0.15	[255]	Atmospheric
CO/N_2O	ν_1	4610	QCL	QEPAS	0.09@5 s 0.05@5 s	[256]	Atmospheric
NO/NO_2	ν_1/ν_3	5263 6135	QCL	DAS	597.3@1 s 438.3@1 s	[257]	Atmospheric
NO/NO_2	ν_1/ν_3	5263 6134	QCL	FMS	4@1 s 9@1 s	[258]	Technique research
NO/NO_2	ν_1/ν_3	5263 6134	QCL	WMDM-2f [1]	0.75@100 s 0.9@200 s	[259]	Atmospheric
NO/NO_2	ν_1/ν_3	5263 6250	QCL	DAS	1.5@100 s 0.5@100 s	[260]	Vehicle exhaust emission
NO/NO_2	ν_1/ν_3	5250 6250	QCL	DAS	1.5@100 s 0.5@100 s	[261]	Environmental monitoring

Table 3. Cont.

Species	Bands	Wavelength/nm	Laser Type	Techniques	LOD/ppb	Refs	Applications
NO/NO$_2$/NH$_3$	$v_1/v_1/v_3$	5263	QCL	WMS	0.2@100 s; 0.96@1 s	[262,263]	Industrial emission
		6250			0.12@100 s; 0.94@1 s		
		9063			0.1@100 s; 0.86@1 s		
NH$_3$/C$_4$H$_{10}$	v_2/v_{24}	8500	QCL	MHS [2]	/	[48]	Technique research
NO$_2$/HONO	v_1/	6234	QCL	TILDAS [3]	0.03	[264]	Atmospheric pollution
		6024			0.3		
SO$_2$/SO$_3$	v_3	7500	QCL	DAS	$(1-2) \times 10^3$	[265]	Industrial emission
		7160					
C$_2$H$_5$Cl/CH$_2$Cl$_2$/CHCl$_3$	//v_4	7949	EC-QCL	DAS	4	[266]	Atmospheric VOCs
					7×10^3		
					11		

[1] WMDM-2f: Wavelength modulation-division multiplexing-2f; [2] MHS: Multiheterodyne spectroscopy; [3] TILDAS: Tunable infrared laser differential absorption spectroscopy.

8. Conclusions and Future Prospects

MIR spectral trace gas sensing is particularly attractive for its unique and strong fingerprint absorption. Higher sensitivity has been achieved by solving all the challenges of the spectral feature, i.e., broad, serried, crowding, and even overlapping in the MIR region. Benefiting from the methods mentioned above, multicomponent simultaneous detection is an expected achievement.

The invention and commercialization of high-performance MIR lasers, i.e., DFB-QCL and DFB-ICL, promoted the development and application of MIR tunable laser-based trace gas sensors, especially in miniaturized and portable applications. Though more than 100 types of gas have been detected by tunable laser-based sensors, there are huge demands for higher detection sensitivity, or, in extreme conditions or scientific exploration, more other types of gas need to be detected. Thus, we believe greater progress will be achieved in the next decade, which may include:

(1) Compact integrated gas sensors. Compact sensors with lower power consumption or battery power could benefit from VCSEL, DFB-ICL, and iHWG. Even a sensor system on a chip will become possible using integrated optics with the fabrication of miniaturized devices integrating the electronics and optics. These sensors could play a vital role in portable and wearable applications that could be applied for breath analysis, diagnostics, metabolomics, environmental safety detection, and related applications.

(2) Multicomponent gas sensors. Multicomponent sensors will achieve more progress, benefiting from the wider wavelength coverage by integrated laser arrays, OFC, or EC-QCL. More species could be detected simultaneously by a particularly devised broadband laser, which could expand their applications in scientific research, including combustion diagnosis, chemical reaction process dynamics, exhaled breath analysis, and metabolomics.

(3) Standoff remote sensing. The techniques of open-path standoff detection by backscattered MIR light provide a promising method of prompt and flexible assessment of atmospheric environmental, leaks, explosive, and security in handheld devices or UAV. The detection sensitivity could be substantially improved by newly developed high-performance MIR detectors and progress in high-power DFB-QCL.

(4) Ultra-sensitive sensing. With the development of a mid-infrared laser source and high-performance detector, combined with cavity enhancement technology and noise immunity technology, ultra-high detection sensitivity becomes possible.

Author Contributions: This review article was jointly written and proof-read by all authors. Z.D. proposed the idea, drafted the outline and structure, and contributed to the principle, detection methods as well as conclusion. S.Z. collected references, composed the whole manuscript, and contributed to the summary. J.L. contributed to the system configuration and detection methods. N.G. contributed to the system configuration. K.T. contributed to the detection methods.

Funding: This research was funded by the National Natural Science Foundation of China (61505142), the Tianjin Natural Science Foundation (16JCQNJC02100), the Science & Technology Development Fund of the Tianjin Education Commission for Higher Education (2017KJ085), and the Natural Science Foundation of Hebei Province (F2014202065).

Acknowledgments: This work was supported by the open project of Key Laboratory of Micro Opto-electro Mechanical System Technology, Tianjin University, Ministry of Education.

Conflicts of Interest: The authors declare no conflict of interest.

References

1. Jane, H.; Ralph, P.T. Optical gas sensing: A review. *Meas. Sci. Technol.* **2013**, *24*. [CrossRef]
2. Miller, D.J.; Sun, K.; Tao, L.; Khan, M.A.; Zondlo, M.A. Open-path, quantum cascade-laser-based sensor for high-resolution atmospheric ammonia measurements. *Atmos. Meas. Tech.* **2014**, *7*, 81–93. [CrossRef]
3. Lassen, M.; Lamard, L.; Balslev-Harder, D.; Peremans, A.; Petersen, J.C. Mid-infrared photoacoustic spectroscopy for atmospheric NO_2 measurements. In Proceedings of the SPIE OPTO, San Francisco, CA, USA, 30 January–1 February 2018; p. 7.
4. Huszár, H.; Pogány, A.; Bozóki, Z.; Mohácsi, Á.; Horváth, L.; Szabó, G. Ammonia monitoring at ppb level using photoacoustic spectroscopy for environmental application. *Sens. Actuator B-Chem.* **2008**, *134*, 1027–1033. [CrossRef]
5. Lackner, M. Tunable diode laser absorption spectroscopy (TDLAS) in the process industries—A review. *Rev. Chem. Eng.* **2007**, *23*, 65–147. [CrossRef]
6. Haas, J.; Mizaikoff, B. Advances in Mid-Infrared Spectroscopy for Chemical Analysis. *Annu. Rev. Anal. Chem.* **2016**, *9*, 45–68. [CrossRef] [PubMed]
7. Ghorbani, R.; Schmidt, F.M. Real-time breath gas analysis of CO and CO_2 using an EC-QCL. *Appl. Phys. B-Lasers Opt.* **2017**, *123*. [CrossRef]
8. Liu, L.; Xiong, B.; Yan, Y.; Li, J.; Du, Z. Hollow Waveguide-Enhanced Mid-Infrared Sensor for Real-Time Exhaled Methane Detection. *IEEE Photonics Technol. Lett.* **2016**, *28*, 1613–1616. [CrossRef]
9. Bauer, C.; Sharma, A.K.; Willer, U.; Burgmeier, J.; Braunschweig, B.; Schade, W.; Blaser, S.; Hvozdara, L.; Müller, A.; Holl, G. Potentials and limits of mid-infrared laser spectroscopy for the detection of explosives. *Appl. Phys. B-Lasers Opt.* **2008**, *92*, 327–333. [CrossRef]
10. Röpcke, J.; Welzel, S.; Lang, N.; Hempel, F.; Gatilova, L.; Guaitella, O.; Rousseau, A.; Davies, P.B. Diagnostic studies of molecular plasmas using mid-infrared semiconductor lasers. *Appl. Phys. B-Lasers Opt.* **2008**, *92*, 335–341. [CrossRef]
11. Bolshov, M.A.; Kuritsyn, Y.A.; Romanovskii, Y.V. Tunable diode laser spectroscopy as a technique for combustion diagnostics. *Spectroc. Acta Part B-Atom. Spectrosc.* **2015**, *106*, 45–66. [CrossRef]
12. O'Keefe, A.; Deacon, D.A.G. Cavity ring-down optical spectrometer for absorption measurements using pulsed laser sources. *Rev. Sci. Instrum.* **1988**, *59*, 2544–2551. [CrossRef]
13. Zybin, A.; Kuritsyn, Y.A.; Mironenko, V.R.; Niemax, K. Cavity enhanced wavelength modulation spectrometry for application in chemical analysis. *Appl. Phys. B-Lasers Opt.* **2004**, *78*, 103–109. [CrossRef]
14. Foltynowicz, A.; Schmidt, F.M.; Ma, W.; Axner, O. Noise-immune cavity-enhanced optical heterodyne molecular spectroscopy: Current status and future potential. *Appl. Phys. B-Lasers Opt.* **2008**, *92*, 313. [CrossRef]
15. Talicska, C.N.; Porambo, M.W.; Perry, A.J.; McCall, B.J. Mid-infrared concentration-modulated noise-immune cavity-enhanced optical heterodyne molecular spectroscopy of a continuous supersonic expansion discharge source. *Rev. Sci. Instrum.* **2016**, *87*, 063111. [CrossRef] [PubMed]
16. Nikodem, M.; Krzempek, K.; Karwat, R.; Dudzik, G.; Abramski, K.; Wysocki, G. Chirped laser dispersion spectroscopy with differential frequency generation source. *Opt. Lett.* **2014**, *39*, 4420–4423. [CrossRef] [PubMed]

17. Martín-Mateos, P.; Acedo, P. Heterodyne phase-sensitive detection for calibration-free molecular dispersion spectroscopy. *Opt. Express* **2014**, *22*, 15143–15153. [CrossRef] [PubMed]
18. Zhou, S.; Han, Y.; Li, B. Simultaneous detection of ethanol, ether and acetone by mid-infrared cavity ring-down spectroscopy at 3.8 μm. *Appl. Phys. B-Lasers Opt.* **2016**, *122*. [CrossRef]
19. Takehiro, H.; Takayuki, O.; Masafumi, I.; Norihiko, N.; Masaru, H. Optical-Fiber-Type Broadband Cavity Ring-Down Spectroscopy Using Wavelength-Tunable Ultrashort Pulsed Light. *Jpn. J. Appl. Phys* **2013**, *52*, 040201. [CrossRef]
20. Grilli, R.; Ciaffoni, L.; Orr-Ewing, A.J. Phase-shift cavity ring-down spectroscopy using mid-IR light from a difference frequency generation PPLN waveguide. *Opt. Lett.* **2010**, *35*, 1383–1385. [CrossRef]
21. Moyer, E.J.; Sayres, D.S.; Engel, G.S.; St. Clair, J.M.; Keutsch, F.N.; Allen, N.T.; Kroll, J.H.; Anderson, J.G. Design considerations in high-sensitivity off-axis integrated cavity output spectroscopy. *Appl. Phys. B-Lasers Opt.* **2008**, *92*, 467–474. [CrossRef]
22. Van Helden, J.H.; Lang, N.; Macherius, U.; Zimmermann, H.; Röpcke, J. Sensitive trace gas detection with cavity enhanced absorption spectroscopy using a continuous wave external-cavity quantum cascade laser. *Appl. Phys. Lett.* **2013**, *103*, 131114. [CrossRef]
23. Sigrist, M.W. Air monitoring by spectroscopic techniques. In *Proteomics*; John Wiley & Sons, Inc.: Hoboken, NJ, USA, 1994; pp. 335–338.
24. Du, Z.; Zhai, Y.; Li, J.; Hu, B. Techniques of On-Line Monitoring Volatile Organic Compounds in Ambient Air with Optical Spectroscopy. *Spectrosc. Spectr. Anal.* **2009**, *29*, 3199–3203. [CrossRef]
25. Wu, D.; Razeghi, M.; Lu, Q.Y.; Mcclintock, R.; Slivken, S.; Zhou, W. Recent progress of quantum cascade laser research from 3 to 12 μm at the Center for Quantum Devices. *Appl. Opt.* **2017**, *56*, H30–H44. [CrossRef]
26. Vurgaftman, I.; Weih, R.; Kamp, M.; Meyer, J.R.; Canedy, C.L.; Kim, C.S.; Kim, M.; Bewley, W.W.; Merritt, C.D.; Abell, J. Interband cascade lasers. *J. Phys. D-Appl. Phys.* **2015**, *48*, 123001. [CrossRef]
27. Du, Z.; Li, J.; Cao, X.; Gao, H.; Ma, Y. High-sensitive carbon disulfide sensor using wavelength modulation spectroscopy in the mid-infrared fingerprint region. *Sens. Actuator B-Chem.* **2017**, *247*, 384–391. [CrossRef]
28. Reid, J.; Labrie, D. Second-harmonic detection with tunable diode lasers—Comparison of experiment and theory. *Appl. Phys. B-Lasers Opt.* **1981**, *26*, 203–210. [CrossRef]
29. Werle, P. Accuracy and precision of laser spectrometers for trace gas sensing in the presence of optical fringes and atmospheric turbulence. *Appl. Phys. B-Lasers Opt.* **2011**, *102*, 313–329. [CrossRef]
30. Xiong, B.; Du, Z.; Li, J. Modulation index optimization for optical fringe suppression in wavelength modulation spectroscopy. *Rev. Sci. Instrum.* **2015**, *86*, 113104. [CrossRef]
31. Downs, C.; Vandervelde, E.T. Progress in Infrared Photodetectors Since 2000. *Sensors* **2013**, *13*, 5054–5098. [CrossRef]
32. Francis, D.; Hodgkinson, J.; Livingstone, B.; Black, P.; Tatam, R.P. Low-volume, fast response-time hollow silica waveguide gas cells for mid-IR spectroscopy. *Appl. Opt.* **2016**, *55*, 6797–6806. [CrossRef]
33. Wilk, A.; Carter, J.C.; Chrisp, M.; Manuel, A.M.; Mirkarimi, P.; Alameda, J.B.; Mizaikoff, B. Substrate-Integrated Hollow Waveguides: A New Level of Integration in Mid-Infrared Gas Sensing. *Anal. Chem.* **2013**, *85*, 11205–11210. [CrossRef] [PubMed]
34. Frish, M.B.; Wainner, R.T.; Laderer, M.C.; Green, B.D.; Allen, M.G. Standoff and Miniature Chemical Vapor Detectors Based on Tunable Diode Laser Absorption Spectroscopy. *IEEE Sens. J.* **2010**, *10*, 639–646. [CrossRef]
35. Diaz, A.; Thomas, B.; Castillo, P.; Gross, B.; Moshary, F. Active standoff detection of CH4 and N2O leaks using hard-target backscattered light using an open-path quantum cascade laser sensor. *Appl. Phys. B-Lasers Opt.* **2016**, *122*. [CrossRef]
36. Macleod, N.A.; Rose, R.; Weidmann, D. Middle infrared active coherent laser spectrometer for standoff detection of chemicals. *Opt. Lett.* **2013**, *38*, 3708–3711. [CrossRef] [PubMed]
37. Lee, K.F.; Hensley, C.J.; Schunemann, P.G.; Fermann, M.E. Midinfrared frequency comb by difference frequency of erbium and thulium fiber lasers in orientation-patterned gallium phosphide. *Opt. Express* **2017**, *25*, 17411–17416. [CrossRef] [PubMed]
38. Vodopyanov, K.L.; Sorokin, E.; Sorokina, I.T.; Schunemann, P.G. Mid-IR frequency comb source spanning 4.4–5.4 μm based on subharmonic GaAs optical parametric oscillator. *Opt. Lett.* **2011**, *36*, 2275–2277. [CrossRef] [PubMed]
39. Liu, J.; Wu, M.; Huang, B.; Tang, P.; Zhao, C.; Shen, D.; Fan, D.; Turitsyn, S.K. Widely Wavelength-Tunable Mid-Infrared Fluoride Fiber Lasers. *IEEE J. Sel. Top. Quantum Electron.* **2018**, *24*, 1–7. [CrossRef]

40. Xu, M.; Yu, F.; Knight, J. Mid-infrared 1 W hollow-core fiber gas laser source. *Opt. Lett.* **2017**, *42*, 4055–4058. [CrossRef]
41. Jayaraman, V.; Segal, S.; Lascola, K.; Burgner, C.; Towner, F.; Cazabat, A.; Cole, G.D.; Follman, D.; Heu, P.; Deutsch, C. Room temperature continuous wave mid-infrared VCSEL operating at 3.35um. In Proceedings of the SPIE OPTO, San Francisco, CA, USA, 19 February 2018; p. 7.
42. Mirov, S.B.; Fedorov, V.V.; Martyshkin, D.; Moskalev, I.S.; Mirov, M.; Vasilyev, S. Progress in Mid-IR Lasers Based on Cr and Fe-Doped II–VI Chalcogenides. *IEEE J. Sel. Top. Quantum Electron.* **2015**, *21*, 292–310. [CrossRef]
43. Schunemann, P.G.; Zawilski, K.T.; Pomeranz, L.A.; Creeden, D.J.; Budni, P.A. Advances in nonlinear optical crystals for mid-infrared coherent sources. *J. Opt. Soc. Am. B* **2016**, *33*, D36–D43. [CrossRef]
44. Jung, D.; Bank, S.; Lee, M.L.; Wasserman, D. Next-generation mid-infrared sources. *J. Opt.* **2017**, *19*, 123001. [CrossRef]
45. Tournié, E.; Baranov, A.N. Chapter 5—Mid-Infrared Semiconductor Lasers: A Review. In *Semiconductors and Semimetals*; Coleman, J.J., Bryce, A.C., Jagadish, C., Eds.; Elsevier: Amsterdam, The Netherlands, 2012; Volume 86, pp. 183–226.
46. Faist, J.; Capasso, F.; Sivco, D.L.; Sirtori, C.; Hutchinson, A.L.; Cho, A.Y. Quantum Cascade Laser. *Science* **1994**, *264*, 553. [CrossRef] [PubMed]
47. Curl, R.F.; Capasso, F.; Gmachl, C.; Kosterev, A.A.; McManus, B.; Lewicki, R.; Pusharsky, M.; Wysocki, G.; Tittel, F.K. Quantum cascade lasers in chemical physics. *Chem. Phys. Lett.* **2010**, *487*, 1–18. [CrossRef]
48. Westberg, J.; Sterczewski, L.A.; Wysocki, G. Mid-infrared multiheterodyne spectroscopy with phase-locked quantum cascade lasers. *Appl. Phys. Lett.* **2017**, *110*, 141108. [CrossRef]
49. Gong, L.; Lewicki, R.; Griffin, R.J.; Flynn, J.H.; Lefer, B.L.; Tittel, F.K. Atmospheric ammonia measurements in Houston, TX using an external-cavity quantum cascade laser-based sensor. *Atmos. Chem. Phys.* **2011**, *11*, 9721–9733. [CrossRef]
50. Cui, X.; Dong, F.; Zhang, Z.; Sun, P.; Xia, H.; Fertein, E.; Chen, W. Simultaneous detection of ambient methane, nitrous oxide, and water vapor using an external-cavity quantum cascade laser. *Atmos. Environ.* **2018**, *189*, 125–132. [CrossRef]
51. Spearrin, R.M.; Goldenstein, C.S.; Schultz, I.A.; Jeffries, J.B.; Hanson, R.K. Simultaneous sensing of temperature, CO, and CO_2 in a scramjet combustor using quantum cascade laser absorption spectroscopy. *Appl. Phys. B-Lasers Opt.* **2014**, *117*, 689–698. [CrossRef]
52. Nadeem, F.; Mandon, J.; Khodabakhsh, A.; Cristescu, S.M.; Harren, F.J.M. Sensitive Spectroscopy of Acetone Using a Widely Tunable External-Cavity Quantum Cascade Laser. *Sensors* **2018**, *18*, 2050. [CrossRef]
53. Maity, A.; Pal, M.; Maithani, S.; Banik, G.D.; Pradhan, M. Wavelength modulation spectroscopy coupled with an external-cavity quantum cascade laser operating between 7.5 and 8 μm. *Laser Phys. Lett.* **2018**, *15*. [CrossRef]
54. Hugi, A.; Maulini, R.; Faist, J. External cavity quantum cascade laser. *Semicond. Sci. Technol.* **2010**, *2535*, 83001–83014. [CrossRef]
55. Zhou, W.; Wu, D.; McClintock, R.; Slivken, S.; Razeghi, M. High performance monolithic, broadly tunable mid-infrared quantum cascade lasers. *Optica* **2017**, *4*, 1228–1231. [CrossRef]
56. Rauter, P.; Capasso, F. Multi-wavelength quantum cascade laser arrays. *Laser Photonics Rev.* **2015**, *9*, 452–477. [CrossRef]
57. Razeghi, M.; Bandyopadhyay, N.; Bai, Y.; Lu, Q.; Slivken, S. Recent advances in mid infrared (3–5 μm) Quantum Cascade Lasers. *Opt. Mater. Express* **2013**, *3*, 1872–1884. [CrossRef]
58. Bandyopadhyay, N.; Bai, Y.; Tsao, S.; Nida, S.; Slivken, S.; Razeghi, M. Room temperature continuous wave operation of λ ~ 3–3.2 μm quantum cascade lasers. *Appl. Phys. Lett.* **2012**, *101*, 241110. [CrossRef]
59. Wolf, J.M.; Bismuto, A.; Beck, M.; Faist, J. Distributed-feedback quantum cascade laser emitting at 3.2 μm. *Opt. Express* **2014**, *22*, 2111–2118. [CrossRef]
60. Yang, R.Q. Infrared laser based on intersubband transitions in quantum wells. *Superlattices Microstruct.* **1995**, *17*, 77. [CrossRef]
61. Kim, M.; Canedy, C.L.; Bewley, W.W.; Kim, C.S.; Lindle, J.R.; Abell, J.; Vurgaftman, I.; Meyer, J.R. Interband cascade laser emitting at λ=3.75μm in continuous wave above room temperature. *Appl. Phys. Lett.* **2008**, *92*, 191110. [CrossRef]

62. Du, Z.; Luo, G.; An, Y.; Li, J. Dynamic spectral characteristics measurement of DFB interband cascade laser under injection current tuning. *Appl. Phys. Lett.* **2016**, *109*, 011903. [CrossRef]
63. Jose Gomes da Silva, I.; Tutuncu, E.; Nagele, M.; Fuchs, P.; Fischer, M.; Raimundo, I.M.; Mizaikoff, B. Sensing hydrocarbons with interband cascade lasers and substrate-integrated hollow waveguides. *Analyst* **2016**, *141*, 4432–4437. [CrossRef]
64. Li, J.; Du, Z.; Zhang, Z.; Song, L.; Guo, Q. Hollow waveguide-enhanced mid-infrared sensor for fast and sensitive ethylene detection. *Sens. Rev.* **2017**, *37*, 82–87. [CrossRef]
65. Li, C.; Zheng, C.; Dong, L.; Ye, W.; Tittel, F.K.; Wang, Y. Ppb-level mid-infrared ethane detection based on three measurement schemes using a 3.34-µm continuous-wave interband cascade laser. *Appl. Phys. B-Lasers Opt.* **2016**, *122*. [CrossRef]
66. Li, C.; Dong, L.; Zheng, C.; Tittel, F.K. Compact TDLAS based optical sensor for ppb-level ethane detection by use of a 3.34 µm room-temperature CW interband cascade laser. *Sens. Actuator B-Chem.* **2016**, *232*, 188–194. [CrossRef]
67. He, Q.; Zheng, C.; Lou, M.; Ye, W.; Wang, Y.; Tittel, F.K. Dual-feedback mid-infrared cavity-enhanced absorption spectroscopy for H_2CO detection using a radio-frequency electrically-modulated interband cascade laser. *Opt. Express* **2018**, *26*, 15436–15444. [CrossRef] [PubMed]
68. Du, Z.; Zhen, W.; Zhang, Z.; Li, J.; Gao, N. Detection of methyl mercaptan with a 3393-nm distributed feedback interband cascade laser. *Appl. Phys. B-Lasers Opt.* **2016**, *122*. [CrossRef]
69. Li, J.; Luo, G.; Du, Z.; Ma, Y. Hollow waveguide enhanced dimethyl sulfide sensor based on a 3.3 µm interband cascade laser. *Sens. Actuator B-Chem.* **2018**, *255*, 3550–3557. [CrossRef]
70. Risby, T.H.; Tittel, F.K. Current status of midinfrared quantum and interband cascade lasers for clinical breath analysis. *Opt. Eng.* **2010**, *49*, 11. [CrossRef]
71. Ghorbani, R.; Schmidt, F.M. ICL-based TDLAS sensor for real-time breath gas analysis of carbon monoxide isotopes. *Opt. Express* **2017**, *25*, 12743–12752. [CrossRef]
72. Girard, J.J.; Spearrin, R.M.; Goldenstein, C.S.; Hanson, R.K. Compact optical probe for flame temperature and carbon dioxide using interband cascade laser absorption near 4.2 µm. *Combust. Flame* **2017**, *178*, 158–167. [CrossRef]
73. Sonnenfroh, D.M. Interband cascade laser–based sensor for ambient CH_4. *Opt. Eng.* **2010**, *49*, 111118. [CrossRef]
74. Edlinger, M.V.; Scheuermann, J.; Weih, R.; Zimmermann, C.; Nähle, L.; Fischer, M.; Koeth, J.; Höfling, S.; Kamp, M. Monomode Interband Cascade Lasers at 5.2 µm for Nitric Oxide Sensing. *IEEE Photonics Technol. Lett.* **2014**, *26*, 480–482. [CrossRef]
75. Sprengel, S.; Andrejew, A.; Federer, F.; Veerabathran, G.K.; Boehm, G.; Amann, M.C. Continuous wave vertical cavity surface emitting lasers at 2.5 µm with InP-based type-II quantum wells. *Appl. Phys. Lett* **2015**, *106*, 151102. [CrossRef]
76. Andrejew, A.; Sprengel, S.; Amann, M.-C. GaSb-based vertical-cavity surface-emitting lasers with an emission wavelength at 3 µm. *Opt. Lett.* **2016**, *41*, 2799–2802. [CrossRef] [PubMed]
77. Bewley, W.W.; Canedy, C.L.; Kim, C.S.; Merritt, C.D.; Warren, M.V.; Vurgaftman, I.; Meyer, J.R.; Kim, M. Room-temperature mid-infrared interband cascade vertical-cavity surface-emitting lasers. *Appl. Phys. Lett.* **2016**, *109*, 151108. [CrossRef]
78. Veerabathran, G.K.; Sprengel, S.; Andrejew, A.; Amann, M.C. Room-temperature vertical-cavity surface-emitting lasers at 4 µm with GaSb-based type-II quantum wells. *Appl. Phys. Lett.* **2017**, *110*, 071104. [CrossRef]
79. Cossel, K.C.; Waxman, E.M.; Finneran, I.A.; Blake, G.A.; Ye, J.; Newbury, N.R. Gas-phase broadband spectroscopy using active sources: Progress, status, and applications [Invited]. *J. Opt. Soc. Am. B* **2017**, *34*, 104–129. [CrossRef] [PubMed]
80. Schliesser, A.; Picqué, N.; Hänsch, T.W. Mid-infrared frequency combs. *Nat. Photonics* **2012**, *6*, 440–449. [CrossRef]
81. Galli, I.; Cappelli, F.; Cancio, P.; Giusfredi, G.; Mazzotti, D.; Bartalini, S.; De Natale, P. High-coherence mid-infrared frequency comb. *Opt. Express* **2013**, *21*, 28877–28885. [CrossRef]
82. Droste, S.; Ycas, G.; Washburn Brian, R.; Coddington, I.; Newbury Nathan, R. Optical Frequency Comb Generation based on Erbium Fiber Lasers. *Nanophotonics* **2016**, *5*, 196. [CrossRef]

83. Cizmeciyan, M.N.; Cankaya, H.; Kurt, A.; Sennaroglu, A. Kerr-lens mode-locked femtosecond Cr2+:ZnSe laser at 2420 nm. *Opt. Lett.* **2009**, *34*, 3056–3058. [CrossRef]
84. Burghoff, D.; Kao, T.-Y.; Han, N.; Chan, C.W.I.; Cai, X.; Yang, Y.; Hayton, D.J.; Gao, J.-R.; Reno, J.L.; Hu, Q. Terahertz laser frequency combs. *Nat. Photonics* **2014**, *8*, 462. [CrossRef]
85. Bagheri, M.; Frez, C.; Sterczewski, L.A.; Gruidin, I.; Fradet, M.; Vurgaftman, I.; Canedy, C.L.; Bewley, W.W.; Merritt, C.D.; Kim, C.S.; et al. Passively mode-locked interband cascade optical frequency combs. *Sci. Rep.* **2018**, *8*, 3322. [CrossRef] [PubMed]
86. Sotor, J.; Martynkien, T.; Schunemann, P.G.; Mergo, P.; Rutkowski, L.; Soboń, G. All-fiber mid-infrared source tunable from 6 to 9 μm based on difference frequency generation in OP-GaP crystal. *Opt. Express* **2018**, *26*, 11756–11763. [CrossRef] [PubMed]
87. Zhu, F.; Hundertmark, H.; Kolomenskii, A.A.; Strohaber, J.; Holzwarth, R.; Schuessler, H.A. High-power mid-infrared frequency comb source based on a femtosecond Er:fiber oscillator. *Opt. Lett.* **2013**, *38*, 2360–2362. [CrossRef] [PubMed]
88. Jin, Y.; Cristescu, S.M.; Harren, F.J.M.; Mandon, J. Two-crystal mid-infrared optical parametric oscillator for absorption and dispersion dual-comb spectroscopy. *Opt. Lett.* **2014**, *39*, 3270–3273. [CrossRef]
89. Liu, P.; Wang, S.; He, P.; Zhang, Z. Dual-channel operation in a synchronously pumped optical parametric oscillator for the generation of broadband mid-infrared coherent light sources. *Opt. Lett.* **2018**, *43*, 2217–2220. [CrossRef] [PubMed]
90. Smolski, V.O.; Yang, H.; Gorelov, S.D.; Schunemann, P.G.; Vodopyanov, K.L. Coherence properties of a 2.6–7.5 μm frequency comb produced as a subharmonic of a Tm-fiber laser. *Opt. Lett.* **2016**, *41*, 1388–1391. [CrossRef] [PubMed]
91. Lu, Q.Y.; Razeghi, M.; Slivken, S.; Bandyopadhyay, N.; Bai, Y.; Zhou, W.J.; Chen, M.; Heydari, D.; Haddadi, A.; McClintock, R.; et al. High power frequency comb based on mid-infrared quantum cascade laser at λ ~ 9 μm. *Appl. Phys. Lett.* **2015**, *106*, 051105. [CrossRef]
92. Faist, J.; Villares, G.; Scalari, G.; Rösch, M.; Bonzon, C.; Hugi, A.; Beck, M. Quantum Cascade Laser Frequency Combs. *Nanophotonics* **2016**, *5*, 272. [CrossRef]
93. Dong, M.; Cundiff, S.T.; Winful, H.G. Physics of frequency-modulated comb generation in quantum-well diode lasers. *Phys. Rev. A* **2018**, *97*, 053822. [CrossRef]
94. Hugi, A.; Villares, G.; Blaser, S.; Liu, H.C.; Faist, J. Mid-infrared frequency comb based on a quantum cascade laser. *Nature* **2012**, *492*, 229. [CrossRef]
95. Jouy, P.; Wolf, J.M.; Bidaux, Y.; Allmendinger, P.; Mangold, M.; Beck, M.; Faist, J. Dual comb operation of λ ~ 8.2 μm quantum cascade laser frequency comb with 1 W optical power. *Appl. Phys. Lett.* **2017**, *111*, 141102. [CrossRef]
96. Villares, G.; Hugi, A.; Blaser, S.; Faist, J. Dual-comb spectroscopy based on quantum-cascade-laser frequency combs. *Nat. Commun.* **2014**, *5*, 5192. [CrossRef] [PubMed]
97. Kazakov, D.; Piccardo, M.; Wang, Y.; Chevalier, P.; Mansuripur, T.S.; Xie, F.; Zah, C.-e.; Lascola, K.; Belyanin, A.; Capasso, F. Self-starting harmonic frequency comb generation in a quantum cascade laser. *Nat. Photonics* **2017**, *11*, 789–792. [CrossRef]
98. Liu, L.; Zhang, X.; Xu, T.; Dai, Z.; Dai, S.; Liu, T. Simple and seamless broadband optical frequency comb generation using an InAs/InP quantum dot laser. *Opt. Lett.* **2017**, *42*, 1173–1176. [CrossRef] [PubMed]
99. Panapakkam, V.; Anthur, A.; Vujicic, V.; Gaimard, Q.; Merghem, K.; Aubin, G.; Lelarge, F.; Viktorov, E.; Barry, L.P.; Ramdane, A. Asymmetric corner frequency in the 1/f FM-noise PSD of optical frequency combs generated by quantum-dash mode-locked lasers. *Appl. Phys. Lett.* **2016**, *109*, 181102. [CrossRef]
100. Forrer, A.; Rösch, M.; Singleton, M.; Beck, M.; Faist, J.; Scalari, G. Coexisting frequency combs spaced by an octave in a monolithic quantum cascade laser. *Opt. Express* **2018**, *26*, 23167–23177. [CrossRef] [PubMed]
101. Martyniuk, P.; Rogalski, A. Performance limits of the mid-wave InAsSb/AlAsSb nBn HOT infrared detector. *Opt. Quantum Electron.* **2014**, *46*, 581–591. [CrossRef]
102. Piotrowski, A.; Piotrowski, J. Uncooled Infrared Detectors in Poland, History and Recent Progress. In Proceedings of the 26th European Conference on Solid-State Transducers (Eurosensors), Krakow, Poland, 9–12 September 2012; pp. 1506–1512.
103. Deng, Z.; Jeong, K.S.; Guyot-Sionnest, P. Colloidal Quantum Dots Intraband Photodetectors. *ACS Nano* **2014**, *8*, 11707–11714. [CrossRef]

104. El-Tokhy, M.S.; Mahmoud, I.I.; Konber, H.A. Comparative study between different quantum infrared photodetectors. *Opt. Quantum Electron.* **2009**, *41*, 933–956. [CrossRef]
105. Gueriaux, V.; l'Isle, N.B.d.; Berurier, A.; Huet, O.; Manissadjian, A.; Facoetti, H.; Marcadet, X.; Carras, M.; Trinité, V.; Nedelcu, A. Quantum well infrared photodetectors: Present and future. *Opt. Eng.* **2011**, *50*, 20. [CrossRef]
106. Asano, T.; Hu, C.; Zhang, Y.; Liu, M.; Campbell, J.C.; Madhukar, A. Design Consideration and Demonstration of Resonant-Cavity-Enhanced Quantum Dot Infrared Photodetectors in Mid-Infrared Wavelength Regime (3–5 μm). *IEEE J. Quantum Electron.* **2010**, *46*, 1484–1491. [CrossRef]
107. Dianat, P. A Scalable Low-Cost Manufacturing to Hybridize Infrared Detectors with Si Read-Out Circuits. In Proceedings of the 2018 IEEE Research and Applications of Photonics In Defense Conference (RAPID), Miramar Beach, FL, USA, 22–24 August 2018; pp. 1–3.
108. Zamiri, M.; Anwar, F.; Klein, B.A.; Rasoulof, A.; Dawson, N.M.; Schuler-Sandy, T.; Deneke, C.F.; Ferreira, S.O.; Cavallo, F.; Krishna, S. Antimonide-based membranes synthesis integration and strain engineering. *Proc. Natl. Acad. Sci. USA* **2017**, *114*, E1–E8. [CrossRef] [PubMed]
109. Singh, A.; Pal, R. Infrared Avalanche Photodiode Detectors. *Def. Sci. J.* **2017**, *67*, 159. [CrossRef]
110. Singh, A.; Srivastav, V.; Pal, R. HgCdTe avalanche photodiodes: A review. *Opt. Laser Technol.* **2011**, *43*, 1358–1370. [CrossRef]
111. Shen, C.; Zhang, Y.; Ni, J. Compact cylindrical multipass cell for laser absorption spectroscopy. *Chin. Opt. Lett.* **2013**, *11*, 091201. [CrossRef]
112. Mohamed, T.; Zhu, F.; Chen, S.; Strohaber, J.; Kolomenskii, A.A.; Bengali, A.A.; Schuessler, H.A. Multipass cell based on confocal mirrors for sensitive broadband laser spectroscopy in the near infrared. *Appl. Opt.* **2013**, *52*, 7145–7151. [CrossRef]
113. Kühnreich, B.; Höh, M.; Wagner, S.; Ebert, V. Direct single-mode fibre-coupled miniature White cell for laser absorption spectroscopy. *Rev. Sci. Instrum.* **2016**, *87*, 023111. [CrossRef] [PubMed]
114. Ren, W.; Jiang, W.; Tittel, F.K. Single-QCL-based absorption sensor for simultaneous trace-gas detection of CH4 and N2O. *Appl. Phys. B-Lasers Opt.* **2014**, *117*, 245–251. [CrossRef]
115. Liu, K.; Wang, L.; Tan, T.; Wang, G.; Zhang, W.; Chen, W.; Gao, X. Highly sensitive detection of methane by near-infrared laser absorption spectroscopy using a compact dense-pattern multipass cell. *Sens. Actuator B-Chem.* **2015**, *220*, 1000–1005. [CrossRef]
116. Guo, Y.; Sun, L. Compact optical multipass matrix system design based on slicer mirrors. *Appl. Opt.* **2018**, *57*, 1174–1181. [CrossRef]
117. Ofner, J.; Krüger, H.-U.; Zetzsch, C. Circular multireflection cell for optical spectroscopy. *Appl. Opt.* **2010**, *49*, 5001–5004. [CrossRef] [PubMed]
118. Manninen, A.; Tuzson, B.; Looser, H.; Bonetti, Y.; Emmenegger, L. Versatile multipass cell for laser spectroscopic trace gas analysis. *Appl. Phys. B-Lasers Opt.* **2012**, *109*, 461–466. [CrossRef]
119. Tuzson, B.; Mangold, M.; Looser, H.; Manninen, A.; Emmenegger, L. Compact multipass optical cell for laser spectroscopy. *Opt. Lett.* **2013**, *38*, 257–259. [CrossRef] [PubMed]
120. Jouy, P.; Mangold, M.; Tuzson, B.; Emmenegger, L.; Chang, Y.C.; Hvozdara, L.; Herzig, H.P.; Wägli, P.; Homsy, A.; Rooij, N.F.D. Mid-infrared spectroscopy for gases and liquids based on quantum cascade technologies. *Analyst* **2014**, *139*, 2039–2046. [CrossRef]
121. Knox, D.A.; King, A.K.; McNaghten, E.D.; Brooks, S.J.; Martin, P.A.; Pimblott, S.M. Novel utilisation of a circular multi-reflection cell applied to materials ageing experiments. *Appl. Phys. B-Lasers Opt.* **2015**, *119*, 55–64. [CrossRef]
122. Mangold, M.; Tuzson, B.; Hundt, M.; Jágerská, J.; Looser, H.; Emmenegger, L. Circular paraboloid reflection cell for laser spectroscopic trace gas analysis. *J. Opt. Soc. Am. A* **2016**, *33*, 913–919. [CrossRef]
123. Graf, M.; Looser, H.; Emmenegger, L.; Tuzson, B. Beam folding analysis and optimization of mask-enhanced toroidal multipass cells. *Opt. Lett.* **2017**, *42*, 3137–3140. [CrossRef]
124. Graf, M.; Emmenegger, L.; Tuzson, B. Compact, circular, and optically stable multipass cell for mobile laser absorption spectroscopy. *Opt. Lett.* **2018**, *43*, 2434–2437. [CrossRef]
125. Xiong, B.; Du, Z.; Liu, L.; Zhang, Z.; Li, J.; Cai, Q. Hollow-waveguide-based carbon dioxide sensor for capnography. *Chin. Opt. Lett.* **2015**, *13*, 111201–111204. [CrossRef]

126. Tütüncü, E.; Nägele, M.; Fuchs, P.; Fischer, M.; Mizaikoff, B. iHWG-ICL: Methane Sensing with Substrate-Integrated Hollow Waveguides Directly Coupled to Interband Cascade Lasers. *ACS Sens.* **2016**, *1*, 847–851. [CrossRef]
127. Gayraud, N.; Kornaszewski, Ł.W.; Stone, J.M.; Knight, J.C.; Reid, D.T.; Hand, D.P.; MacPherson, W.N. Mid-infrared gas sensing using a photonic bandgap fiber. *Appl. Opt.* **2008**, *47*, 1269–1277. [CrossRef] [PubMed]
128. Jin, W.; Xuan, H.F.; Ho, H.L. Sensing with hollow-core photonic bandgap fibers. *Meas. Sci. Technol.* **2010**, *21*, 094014. [CrossRef]
129. Mizaikoff, B. Mid-IR fiber-optic sensors. *Anal. Chem.* **2003**, *75*, 258A–267A. [CrossRef] [PubMed]
130. Charlton, C.M.; Thompson, B.T.; Mizaikoff, B. Hollow Waveguide Infrared Spectroscopy and Sensing. In *Frontiers in Chemical Sensors*; Springer: Berlin/Heidelberg, Germany, 2005; pp. 133–167.
131. Jin-Yi, L.; Zhen-Hui, D.; Wang, R.X.; Xu, Y.; Song, L.M.; Guo, Q.H. Applications of Hollow Waveguide in Spectroscopic Gas Sensing. *Spectrosc. Spectr. Anal.* **2017**, *37*, 2259–2266. [CrossRef]
132. Liu, N.; Sun, J.; Deng, H.; Ding, J.; Zhang, L.; Li, J. Recent progress on gas sensor based on quantum cascade lasers and hollow fiber waveguides. In Proceedings of the Fourth International Conference on Optical and Photonics Engineering, Chengdu, China, 26–30 September 2016; p. 5.
133. Wörle, K.; Seichter, F.; Wilk, A.; Armacost, C.; Day, T.; Godejohann, M.; Wachter, U.; Vogt, J.; Radermacher, P.; Mizaikoff, B. Breath Analysis with Broadly Tunable Quantum Cascade Lasers. *Anal. Chem.* **2013**, *85*, 2697–2702. [CrossRef] [PubMed]
134. Tütüncü, E.; Kokoric, V.; Szedlak, R.; Macfarland, D.; Zederbauer, T.; Detz, H.; Andrews, A.M.; Schrenk, W.; Strasser, G.; Mizaikoff, B. Advanced gas sensors based on substrate-integrated hollow waveguides and dual-color ring quantum cascade lasers. *Analyst* **2016**, *141*, 6202–6207. [CrossRef] [PubMed]
135. Tütüncü, E.; Kokoric, V.; Wilk, A.; Seichter, F.; Schmid, M.; Hunt, W.E.; Manuel, A.M.; Mirkarimi, P.; Alameda, J.B.; Carter, J.C.; et al. Fiber-Coupled Substrate-Integrated Hollow Waveguides: An Innovative Approach to Mid-infrared Remote Gas Sensors. *ACS Sens.* **2017**, *2*, 1287–1293. [CrossRef]
136. Kokoric, V.; Theisen, J.; Wilk, A.; Penisson, C.; Bernard, G.; Mizaikoff, B.; Gabriel, J.-C.P. Determining the Partial Pressure of Volatile Components via Substrate-Integrated Hollow Waveguide Infrared Spectroscopy with Integrated Microfluidics. *Anal. Chem.* **2018**, *90*, 4445–4451. [CrossRef]
137. Van Well, B.; Murray, S.; Hodgkinson, J.; Pride, R.; Strzoda, R.; Gibson, G.; Padgett, M. An open-path, hand-held laser system for the detection of methane gas. *J. Opt. A Pure Appl. Opt.* **2005**, *7*, S420–S424. [CrossRef]
138. Sun, J.; Ding, J.; Liu, N.; Yang, G.; Li, J. Detection of multiple chemicals based on external cavity quantum cascade laser spectroscopy. *Spectrochim. Acta Part A Mol. Spectrosc.* **2018**, *191*, 532–538. [CrossRef]
139. Chen, X.; Cheng, L.; Guo, D.; Kostov, Y.; Choa, F.S. Quantum cascade laser based standoff photoacoustic chemical detection. *Opt. Express* **2011**, *19*, 20251–20257. [CrossRef] [PubMed]
140. Puiu, A.; Fiorani, L.; Rosa, O.; Borelli, R.; Pistilli, M.; Palucci, A. Lidar/DIAL detection of acetone at 3.3 μm by a tunable OPO laser system. *Laser Phys.* **2014**, *24*, 085606. [CrossRef]
141. Daghestani, N.S.; Brownsword, R.; Weidmann, D. Analysis and demonstration of atmospheric methane monitoring by mid-infrared open-path chirped laser dispersion spectroscopy. *Opt. Express* **2014**, *22* (Suppl. 7), A1731–A1743. [CrossRef]
142. Jaworski, P.; Stachowiak, D.; Nikodem, M. Standoff detection of gases using infrared laser spectroscopy. In Proceedings of the SPIE Photonics Europe, Brussels, Belgium, 3–7 April 2016; p. 6.
143. Wang, Z.; Sanders, S.T. Toward single-ended absorption spectroscopy probes based on backscattering from rough surfaces: H_2O vapor measurements near 1350 nm. *Appl. Phys. B-Lasers Opt.* **2015**, *121*, 187–192. [CrossRef]
144. Andresen, B.F.; Lwin, M.; Fulop, G.F.; Corrigan, P.; Gross, B.; Norton, P.R.; Moshary, F.; Ahmed, S. Mid-infrared backscattering measurements of building materials using a quantum cascade laser. In Proceedings of the Conference on Infrared Technology and Applications XXXVI, Orlando, FL, USA, 5–9 April 2010; p. 766043.
145. Lambert-Girard, S.; Allard, M.; Piché, M.; Babin, F. Broadband and tunable optical parametric generator for remote detection of gas molecules in the short and mid-infrared. *Appl. Opt.* **2015**, *54*, 2594–2605. [CrossRef] [PubMed]

146. Brumfield, B.E.; Phillips, M.C. Standoff detection of turbulent chemical mixture plumes using a swept external cavity quantum cascade laser. *Opt. Eng.* **2017**, *57*. [CrossRef]
147. Peng, W.Y.; Goldenstein, C.S.; Mitchell Spearrin, R.; Jeffries, J.B.; Hanson, R.K. Single-ended mid-infrared laser-absorption sensor for simultaneous in situ measurements of H2O, CO2, CO, and temperature in combustion flows. *Appl. Opt.* **2016**, *55*, 9347–9359. [CrossRef] [PubMed]
148. Wallin, S.; Pettersson, A.; Östmark, H.; Hobro, A. Laser-based standoff detection of explosives: A critical review. *Anal. Bioanal. Chem.* **2009**, *395*, 259–274. [CrossRef]
149. Gaudio, P. Laser Based Standoff Techniques: A Review on Old and New Perspective for Chemical Detection and Identification. In *Cyber and Chemical, Biological, Radiological, Nuclear, Explosives Challenges: Threats and Counter Efforts*; Martellini, M., Malizia, A., Eds.; Springer International Publishing: Cham, Germany, 2017; pp. 155–177. [CrossRef]
150. Golston, L.M.; Tao, L.; Brosy, C.; Schäfer, K.; Wolf, B.; McSpiritt, J.; Buchholz, B.; Caulton, D.R.; Pan, D.; Zondlo, M.A.; et al. Lightweight mid-infrared methane sensor for unmanned aerial systems. *Appl. Phys. B-Lasers Opt.* **2017**, *123*. [CrossRef]
151. Roşca, S.; Suomalainen, J.; Bartholomeus, H.; Herold, M. Comparing terrestrial laser scanning and unmanned aerial vehicle structure from motion to assess top of canopy structure in tropical forests. *Interface Focus* **2018**, *8*, 20170038. [CrossRef]
152. Joanna, K. Basic Principles and Analytical Application of Derivative Spectrophotometry. In *Macro to Nano Spectroscopy*; Uddin, J., Ed.; IntechOpen: Rijeka, Croatia, 2012; pp. 253–268.
153. Patel, K.N.; Patel, J.K.; Rajput, G.C.; Rajgor, N.B. Derivative spectrometry method for chemical analysis: A review. *Der Pharm. Lett.* **2010**, *2*, 139–150.
154. Bosch Ojeda, C.; Sanchez Rojas, F. Recent applications in derivative ultraviolet/visible absorption spectrophotometry: 2009–2011: A review. *Microchem. J.* **2013**, *106*, 1–16. [CrossRef]
155. Czarnecki, M.A. Resolution Enhancement in Second-Derivative Spectra. *Appl. Spectrosc.* **2015**, *69*, 67–74. [CrossRef] [PubMed]
156. Xi-yang, L.; Nan, G.; Zhen-hui, D.; Jin-yi, L.; Chao, C.; Zong-hua, Z. Infrared Spectroscopy Quantitative Detection Method Based on Second Order Derivative Spectrum and Characteristic Absorption Window. *Spectrosc. Spectr. Anal.* **2017**, *37*, 1765–1770. [CrossRef]
157. Elzanfaly, E.S.; Hassan, S.A.; Salem, M.Y.; El-Zeany, B.A. Continuous Wavelet Transform, a powerful alternative to Derivative Spectrophotometry in analysis of binary and ternary mixtures: A comparative study. *Spectroc. Acta Part A-Mol. Biomol. Spectrosc.* **2015**, *151*, 945–955. [CrossRef] [PubMed]
158. Lee, W.; Choi, S.; Kim, T.K.; Kang, J.; Park, K.S.; Kim, Y.-H. Determination of Insulator-to-Semiconductor Transition in Sol-Gel Oxide Semiconductors Using Derivative Spectroscopy. *Materials* **2016**, *9*, 6. [CrossRef] [PubMed]
159. El-Sayed, A.-A.Y.; El-Salem, N.A. Recent Developments of Derivative Spectrophotometry and Their Analytical Applications. *Anal. Sci.* **2005**, *21*, 595–614. [CrossRef]
160. Hamid, D.; Frederic, L.; Brian, W.P.; Fabien, C. Application of spectral derivative data in visible and near-infrared spectroscopy. *Phys. Med. Biol.* **2010**, *55*, 3381. [CrossRef]
161. Deng, H.; Sun, J.; Li, P.; Liu, Y.; Yu, B.; Li, J. Sensitive detection of acetylene by second derivative spectra with tunable diode laser absorption spectroscopy. *Opt. Appl.* **2016**, *46*, 353–363. [CrossRef]
162. Du, Z.; Yan, Y.; Li, J.; Zhang, S.; Yang, X.; Xiao, Y. In situ, multiparameter optical sensor for monitoring the selective catalytic reduction process of diesel engines. *Sens. Actuator B-Chem.* **2018**, *267*, 255–264. [CrossRef]
163. Meng, Y.; Liu, T.; Liu, K.; Jiang, J.; Wang, R.; Wang, T.; Hu, H. A Modified Empirical Mode Decomposition Algorithm in TDLAS for Gas Detection. *IEEE Photonics J.* **2014**, *6*, 1–7. [CrossRef]
164. Wang, S.; Du, Z.; Yuan, L.; Ma, Y.; Wang, X.; Han, R.; Meng, S. Measurement of Atmospheric Dimethyl Sulfide with a Distributed Feedback Interband Cascade Laser. *Sensors* **2018**, *18*, 3216. [CrossRef] [PubMed]
165. Du, Z.; Wan, J.; Li, J.; Luo, G.; Gao, H.; Ma, Y. Detection of Atmospheric Methyl Mercaptan Using Wavelength Modulation Spectroscopy with Multicomponent Spectral Fitting. *Sensors* **2017**, *17*, 379. [CrossRef] [PubMed]
166. Foltynowicz, A.; Masłowski, P.; Ban, T.; Adler, F.; Cossel, K.C.; Briles, T.C.; Ye, J. Optical frequency comb spectroscopy. *Faraday Discuss.* **2011**, *150*. [CrossRef]
167. Thorpe, M.J.; Moll, K.D.; Jones, R.J.; Safdi, B.; Ye, J. Broadband Cavity Ringdown Spectroscopy for Sensitive and Rapid Molecular Detection. *Science* **2006**, *311*, 1595–1599. [CrossRef] [PubMed]

168. Coddington, I.; Swann, W.C.; Newbury, N.R. Coherent Multiheterodyne Spectroscopy Using Stabilized Optical Frequency Combs. *Phys. Rev. Lett.* **2008**, *100*, 013902. [CrossRef] [PubMed]
169. Lee, J.; Kim, Y.-J.; Lee, K.; Lee, S.; Kim, S.-W. Time-of-flight measurement with femtosecond light pulses. *Nat. Photonics* **2010**, *4*, 716. [CrossRef]
170. Mandon, J.; Guelachvili, G.; Picqué, N. Fourier transform spectroscopy with a laser frequency comb. *Nat. Photonics* **2009**, *3*, 99. [CrossRef]
171. Foltynowicz, A.; Masłowski, P.; Fleisher, A.J.; Bjork, B.J.; Ye, J. Cavity-enhanced optical frequency comb spectroscopy in the mid-infrared application to trace detection of hydrogen peroxide. *Appl. Phys. B-Lasers Opt.* **2012**, *110*, 163–175. [CrossRef]
172. Khodabakhsh, A.; Ramaiah-Badarla, V.; Rutkowski, L.; Johansson, A.C.; Lee, K.F.; Jiang, J.; Mohr, C.; Fermann, M.E.; Foltynowicz, A. Fourier transform and Vernier spectroscopy using an optical frequency comb at 3–5.4 µm. *Opt. Lett.* **2016**, *41*, 2541–2544. [CrossRef] [PubMed]
173. Bernhardt, B.; Ozawa, A.; Jacquet, P.; Jacquey, M.; Kobayashi, Y.; Udem, T.; Holzwarth, R.; Guelachvili, G.; Hänsch, T.W.; Picqué, N. Cavity-enhanced dual-comb spectroscopy. *Nat. Photonics* **2009**, *4*, 55. [CrossRef]
174. Bernhardt, B.; Sorokin, E.; Jacquet, P.; Thon, R.; Becker, T.; Sorokina, I.T.; Picqué, N.; Hänsch, T.W. Mid-infrared dual-comb spectroscopy with 2.4 µm Cr2+:ZnSe femtosecond lasers. *Appl. Phys. B-Lasers Opt.* **2010**, *100*, 3–8. [CrossRef]
175. Rieker, G.B.; Giorgetta, F.R.; Swann, W.C.; Kofler, J.; Zolot, A.M.; Sinclair, L.C.; Baumann, E.; Cromer, C.; Petron, G.; Sweeney, C.; et al. Frequency-comb-based remote sensing of greenhouse gases over kilometer air paths. *Optica* **2014**, *1*, 290–298. [CrossRef]
176. Ideguchi, T.; Poisson, A.; Guelachvili, G.; Picqué, N.; Hänsch, T.W. Adaptive real-time dual-comb spectroscopy. *Nat. Commun.* **2014**, *5*, 3375. [CrossRef]
177. Muraviev, A.V.; Smolski, V.O.; Loparo, Z.E.; Vodopyanov, K.L. Massively parallel sensing of trace molecules and their isotopologues with broadband subharmonic mid-infrared frequency combs. *Nat. Photonics* **2018**, *12*, 209–214. [CrossRef]
178. Ycas, G.; Giorgetta, F.R.; Baumann, E.; Coddington, I.; Herman, D.; Diddams, S.A.; Newbury, N.R. High-coherence mid-infrared dual-comb spectroscopy spanning 2.6 to 5.2 µm. *Nat. Photonics* **2018**, *12*, 202–208. [CrossRef]
179. Link, S.M.; Klenner, A.; Mangold, M.; Zaugg, C.A.; Golling, M.; Tilma, B.W.; Keller, U. Dual-comb modelocked laser. *Opt. Express* **2015**, *23*, 5521–5531. [CrossRef]
180. Link, S.M.; Klenner, A.; Keller, U. Dual-comb modelocked lasers: Semiconductor saturable absorber mirror decouples noise stabilization. *Opt. Express* **2016**, *24*, 1889–1902. [CrossRef] [PubMed]
181. Link, S.M.; Maas, D.J.H.C.; Waldburger, D.; Keller, U. Dual-comb spectroscopy of water vapor with a free-running semiconductor disk laser. *Science* **2017**, *356*, 1164. [CrossRef] [PubMed]
182. Thorpe, M.J.; Ye, J. Cavity-enhanced direct frequency comb spectroscopy. *Appl. Phys. B-Lasers Opt.* **2008**, *91*, 397–414. [CrossRef]
183. Thorpe, M.J.; Balslev-Clausen, D.; Kirchner, M.S.; Ye, J. Cavity-enhanced optical frequency comb spectroscopy: Application to human breath analysis. *Opt. Express* **2008**, *16*, 2387–2397. [CrossRef] [PubMed]
184. Thorpe, M.J.; Adler, F.; Cossel, K.C.; de Miranda, M.H.G.; Ye, J. Tomography of a supersonically cooled molecular jet using cavity-enhanced direct frequency comb spectroscopy. *Chem. Phys. Lett.* **2009**, *468*, 1–8. [CrossRef]
185. Spaun, B.; Changala, P.B.; Patterson, D.; Bjork, B.J.; Heckl, O.H.; Doyle, J.M.; Ye, J. Continuous probing of cold complex molecules with infrared frequency comb spectroscopy. *Nature* **2016**, *533*, 517. [CrossRef]
186. Changala, P.B.; Spaun, B.; Patterson, D.; Doyle, J.M.; Ye, J. Sensitivity and resolution in frequency comb spectroscopy of buffer gas cooled polyatomic molecules. *Appl. Phys. B-Lasers Opt.* **2016**, *122*, 292. [CrossRef]
187. Masłowski, P.; Cossel, K.C.; Foltynowicz, A.; Ye, J. Cavity-Enhanced Direct Frequency Comb Spectroscopy. In *Cavity-Enhanced Spectroscopy and Sensing*; Gagliardi, G., Loock, H.-P., Eds.; Springer: Berlin/Heidelberg, Germnay, 2014; pp. 271–321. [CrossRef]
188. Xiao, S.; Weiner, A.M. 2-D wavelength demultiplexer with potential for $\geq$ 1000 channels in the C-band. *Opt. Express* **2004**, *12*, 2895–2902. [CrossRef]
189. Diddams, S.A.; Hollberg, L.; Mbele, V. Molecular fingerprinting with the resolved modes of a femtosecond laser frequency comb. *Nature* **2007**, *445*, 627. [CrossRef] [PubMed]

190. Gohle, C.; Stein, B.; Schliesser, A.; Udem, T.; Hänsch, T.W. Frequency Comb Vernier Spectroscopy for Broadband, High-Resolution, High-Sensitivity Absorption and Dispersion Spectra. *Phys. Rev. Lett.* **2007**, *99*, 263902. [CrossRef]
191. Nugent-Glandorf, L.; Neely, T.; Adler, F.; Fleisher, A.J.; Cossel, K.C.; Bjork, B.; Dinneen, T.; Ye, J.; Diddams, S.A. Mid-infrared virtually imaged phased array spectrometer for rapid and broadband trace gas detection. *Opt. Lett.* **2012**, *37*, 3285–3287. [CrossRef]
192. Fleisher, A.J.; Bjork, B.J.; Bui, T.Q.; Cossel, K.C.; Okumura, M.; Ye, J. Mid-Infrared Time-Resolved Frequency Comb Spectroscopy of Transient Free Radicals. *J. Phys. Chem. Lett.* **2014**, *5*, 2241–2246. [CrossRef]
193. Nelson, D.D.; Shorter, J.H.; McManus, J.B.; Zahniser, M.S. Sub-part-per-billion detection of nitric oxide in air using a thermoelectrically cooled mid-infrared quantum cascade laser spectrometer. *Appl. Phys. B-Lasers Opt.* **2002**, *75*, 343–350. [CrossRef]
194. Weber, W.H.; Remillard, J.T.; Chase, R.E.; Richert, J.F.; Capasso, F.; Gmachl, C.; Hutchinson, A.L.; Sivco, D.L.; Baillargeon, J.N.; Cho, A.Y. Using a Wavelength-Modulated Quantum Cascade Laser to Measure NO Concentrations in the Parts-per-Billion Range for Vehicle Emissions Certification. *Appl. Spectrosc.* **2002**, *56*, 706–714. [CrossRef]
195. Martín-Mateos, P.; Jerez, B.; de Dios, C.; Acedo, P. Mid-infrared heterodyne phase-sensitive dispersion spectroscopy using difference frequency generation. *Appl. Phys. B-Lasers Opt.* **2018**, *124*. [CrossRef]
196. Song, F.; Zheng, C.; Yan, W.; Ye, W.; Wang, Y.; Tittel, F.K. Interband cascade laser based mid-infrared methane sensor system using a novel electrical-domain self-adaptive direct laser absorption spectroscopy (SA-DLAS). *Opt. Express* **2017**, *25*, 31876–31888. [CrossRef] [PubMed]
197. Zheng, C.; Ye, W.; Sanchez, N.P.; Li, C.; Dong, L.; Wang, Y.; Griffin, R.J.; Tittel, F.K. Development and field deployment of a mid-infrared methane sensor without pressure control using interband cascade laser absorption spectroscopy. *Sens. Actuator B-Chem.* **2017**, *244*, 365–372. [CrossRef]
198. Dong, L.; Li, C.; Sanchez, N.P.; Gluszek, A.K.; Griffin, R.J.; Tittel, F.K. Compact CH4 sensor system based on a continuous-wave, low power consumption, room temperature interband cascade laser. *Appl. Phys. Lett.* **2016**, *108*. [CrossRef]
199. Song, F.; Zheng, C.; Yu, D.; Zhou, Y.; Yan, W.; Ye, W.; Zhang, Y.; Wang, Y.; Tittel, F.K. Interband cascade laser-based ppbv-level mid-infrared methane detection using two digital lock-in amplifier schemes. *Appl. Phys. B-Lasers Opt.* **2018**, *124*. [CrossRef]
200. Manfred, K.M.; Ritchie, G.A.D.; Lang, N.; Röpcke, J.; van Helden, J.H. Optical feedback cavity-enhanced absorption spectroscopy with a 3.24 µm interband cascade laser. *Appl. Phys. Lett.* **2015**, *106*. [CrossRef]
201. Triki, M.; Nguyen Ba, T.; Vicet, A. Compact sensor for methane detection in the mid infrared region based on Quartz Enhanced Photoacoustic Spectroscopy. *Infrared Phys. Technol.* **2015**, *69*, 74–80. [CrossRef]
202. Zhu, F.; Bicer, A.; Askar, R.; Bounds, J.; Kolomenskii, A.A.; Kelessides, V.; Amani, M.; Schuessler, H.A. Mid-infrared dual frequency comb spectroscopy based on fiber lasers for the detection of methane in ambient air. *Laser Phys. Lett.* **2015**, *12*. [CrossRef]
203. Khodabakhsh, A.; Rutkowski, L.; Morville, J.; Johansson, A.C.; Soboń, G.; Foltynowicz, A. Continuous-Filtering Vernier Spectroscopy at 3.3 µm Using a Femtosecond Optical Parametric Oscillator. In Proceedings of the Conference on Lasers and Electro-Optics, San Jose, CA, USA, 14–19 May 2017; p. SW1L.5.
204. Vainio, M.; Merimaa, M.; Halonen, L. Frequency-comb-referenced molecular spectroscopy in the mid-infrared region. *Opt. Lett.* **2011**, *36*, 4122–4124. [CrossRef]
205. Silander, I.; Hausmaninger, T.; Ma, W.; Harren, F.J.; Axner, O. Doppler-broadened mid-infrared noise-immune cavity-enhanced optical heterodyne molecular spectrometry based on an optical parametric oscillator for trace gas detection. *Opt. Lett.* **2015**, *40*, 439–442. [CrossRef] [PubMed]
206. Li, C.; Dong, L.; Zheng, C.; Lin, J.; Wang, Y.; Tittel, K.F. Ppbv-Level Ethane Detection Using Quartz-Enhanced Photoacoustic Spectroscopy with a Continuous-Wave, Room Temperature Interband Cascade Laser. *Sensors* **2018**, *18*, 723. [CrossRef] [PubMed]
207. Razeghi, M.; Jahjah, M.; Lewicki, R.; Tittel, F.K.; Krzempek, K.; Stefanski, P.; So, S.; Thomazy, D. CW DFB RT diode laser based sensor for trace-gas detection of ethane using novel compact multipass gas absorption cell. In Proceedings of the Quantum Sensing and Nanophotonic Devices X, San Francisco, CA, USA, 3–7 February 2013.

208. Krzempek, K.; Lewicki, R.; Nähle, L.; Fischer, M.; Koeth, J.; Belahsene, S.; Rouillard, Y.; Worschech, L.; Tittel, F.K. Continuous wave, distributed feedback diode laser based sensor for trace-gas detection of ethane. *Appl. Phys. B-Lasers Opt.* **2012**, *106*, 251–255. [CrossRef]
209. Zhao, Z.-S.; Liao, Y.; Grattan, K.T.; Wang, W.; Jiang, D.; Zhang, W.; Wei, L.; Feng, Q.; Li, P.; Wang, Y.; et al. Research on propane leak detection system and device based on mid infrared laser. In Proceedings of the AOPC 2017: Fiber Optic Sensing and Optical Communications, Beijing, China, 4–6 June 2017.
210. Adler, F.; Masłowski, P.; Foltynowicz, A.; Cossel, K.C.; Briles, T.C.; Hartl, I.; Ye, J. Mid-infrared Fourier transform spectroscopy with a broadband frequency comb. *Opt. Express* **2010**, *18*, 21861–21872. [CrossRef] [PubMed]
211. Belyanin, A.A.; Smowton, P.M.; Lewicki, R.; Witinski, M.; Li, B.; Wysocki, G. Spectroscopic benzene detection using a broadband monolithic DFB-QCL array. In Proceedings of the Novel In-Plane Semiconductor Lasers XV, San Francisco, CA, USA, 15–18 February 2016.
212. Lassen, M.; Harder, D.B.; Brusch, A.; Nielsen, O.S.; Heikens, D.; Persijn, S.; Petersen, J.C. Photo-acoustic sensor for detection of oil contamination in compressed air systems. *Opt. Express* **2017**, *25*, 1806–1814. [CrossRef] [PubMed]
213. Lundqvist, S.; Kluczynski, P.; Weih, R.; von Edlinger, M.; Nähle, L.; Fischer, M.; Bauer, A.; Höfling, S.; Koeth, J. Sensing of formaldehyde using a distributed feedback interband cascade laser emitting around 3493nm. *Appl. Opt.* **2012**, *51*, 6009–6013. [CrossRef] [PubMed]
214. Tanaka, K.; Miyamura, K.; Akishima, K.; Tonokura, K.; Konno, M. Sensitive measurements of trace gas of formaldehyde using a mid-infrared laser spectrometer with a compact multi-pass cell. *Infrared Phys. Technol.* **2016**, *79*, 1–5. [CrossRef]
215. Dong, L.; Yu, Y.; Li, C.; So, S.; Tittel, F.K. Ppb-level formaldehyde detection using a CW room-temperature interband cascade laser and a miniature dense pattern multipass gas cell. *Opt. Express* **2015**, *23*, 19821–19830. [CrossRef]
216. Ren, W.; Luo, L.; Tittel, F.K. Sensitive detection of formaldehyde using an interband cascade laser near 3.6 µm. *Sens. Actuator B-Chem.* **2015**, *221*, 1062–1068. [CrossRef]
217. Tuzson, B.; Jagerska, J.; Looser, H.; Graf, M.; Felder, F.; Fill, M.; Tappy, L.; Emmenegger, L. Highly Selective Volatile Organic Compounds Breath Analysis Using a Broadly-Tunable Vertical-External-Cavity Surface-Emitting Laser. *Anal. Chem.* **2017**, *89*, 6377–6383. [CrossRef]
218. Borri, S.; Patimisco, P.; Galli, I.; Mazzotti, D.; Giusfredi, G.; Akikusa, N.; Yamanishi, M.; Scamarcio, G.; De Natale, P.; Spagnolo, V. Intracavity quartz-enhanced photoacoustic sensor. *Appl. Phys. Lett.* **2014**, *104*, 091114. [CrossRef]
219. Liu, X.; Zhang, G.; Huang, Y.; Wang, Y.; Qi, F. Two-dimensional temperature and carbon dioxide concentration profiles in atmospheric laminar diffusion flames measured by mid-infrared direct absorption spectroscopy at 4.2 µm. *Appl. Phys. B-Lasers Opt.* **2018**, *124*. [CrossRef]
220. Nwaboh, J.A.; Qu, Z.; Werhahn, O.; Ebert, V. Interband cascade laser-based optical transfer standard for atmospheric carbon monoxide measurements. *Appl. Opt.* **2017**, *56*, E84–E93. [CrossRef] [PubMed]
221. Dang, J.; Yu, H.; Zheng, C.; Wang, L.; Sui, Y.; Wang, Y. Development a low-cost carbon monoxide sensor using homemade CW-DFB QCL and board-level electronics. *Opt. Laser Technol.* **2018**, *101*, 57–67. [CrossRef]
222. Lee, D.D.; Bendana, F.A.; Schumaker, S.A.; Spearrin, R.M. Wavelength modulation spectroscopy near 5 µm for carbon monoxide sensing in a high-pressure kerosene-fueled liquid rocket combustor. *Appl. Phys. B-Lasers Opt.* **2018**, *124*. [CrossRef]
223. Diemel, O.; Pareja, J.; Dreizler, A.; Wagner, S. An interband cascade laser-based in situ absorption sensor for nitric oxide in combustion exhaust gases. *Appl. Phys. B-Lasers Opt.* **2017**, *123*. [CrossRef]
224. Shi, C.; Wang, D.; Wang, Z.; Ma, L.; Wang, Q.; Xu, K.; Chen, S.-C.; Ren, W. A Mid-Infrared Fiber-Coupled QEPAS Nitric Oxide Sensor for Real-Time Engine Exhaust Monitoring. *IEEE Sens. J.* **2017**, *17*, 7418–7424. [CrossRef]
225. Upadhyay, A.; Wilson, D.; Lengden, M.; Chakraborty, A.L.; Stewart, G.; Johnstone, W. Calibration-Free WMS Using a cw-DFB-QCL, a VCSEL, and an Edge-Emitting DFB Laser With In-Situ Real-Time Laser Parameter Characterization. *IEEE Photonics J.* **2017**, *9*, 1–17. [CrossRef]
226. Almodovar, C.A.; Spearrin, R.M.; Hanson, R.K. Two-color laser absorption near 5 µm for temperature and nitric oxide sensing in high-temperature gases. *J. Quant. Spectrosc. Radiat. Transf.* **2017**, *203*, 572–581. [CrossRef]

227. Wojtas, J.; Gluszek, A.; Hudzikowski, A.; Tittel, F.K. Mid-Infrared Trace Gas Sensor Technology Based on Intracavity Quartz-Enhanced Photoacoustic Spectroscopy. *Sensors* **2017**, *17*, 513. [CrossRef]
228. Sur, R.; Peng, W.Y.; Strand, C.; Mitchell Spearrin, R.; Jeffries, J.B.; Hanson, R.K.; Bekal, A.; Halder, P.; Poonacha, S.P.; Vartak, S.; et al. Mid-infrared laser absorption spectroscopy of NO2 at elevated temperatures. *J. Quant. Spectrosc. Radiat. Transf.* **2017**, *187*, 364–374. [CrossRef]
229. Sun, J.; Deng, H.; Liu, N.; Wang, H.; Yu, B.; Li, J. Mid-infrared gas absorption sensor based on a broadband external cavity quantum cascade laser. *Rev. Sci. Instrum.* **2016**, *87*, 123101. [CrossRef] [PubMed]
230. Gambetta, A.; Cassinerio, M.; Coluccelli, N.; Fasci, E.; Castrillo, A.; Gianfrani, L.; Gatti, D.; Marangoni, M.; Laporta, P.; Galzerano, G. Direct phase-locking of a 8.6-mum quantum cascade laser to a mid-IR optical frequency comb: Application to precision spectroscopy of N_2O. *Opt. Lett.* **2015**, *40*, 304–307. [CrossRef] [PubMed]
231. Nikodem, M.; Wysocki, G. Chirped laser dispersion spectroscopy for remote open-path trace-gas sensing. *Sensors* **2012**, *12*, 16466–16481. [CrossRef] [PubMed]
232. Razeghi, M.; Lewicki, R.; Sudharsanan, R.; Kosterev, A.A.; Thomazy, D.M.; Brown, G.J.; Risby, T.H.; Solga, S.; Schwartz, T.B.; Tittel, F.K. Real time ammonia detection in exhaled human breath using a distributed feedback quantum cascade laser based sensor. In Proceedings of the Quantum Sensing and Nanophotonic Devices VIII, San Francisco, CA, USA, 23–27 January 2011.
233. Li, J.; Yang, S.; Wang, R.; Du, Z.; Wei, Y. Ammonia detection using hollow waveguide enhanced laser absorption spectroscopy based on a 9.56 μm quantum cascade laser. In Proceedings of the Applied Optics and Photonics China (AOPC2017), Beijing, China, 4–6 January 2017; p. 8.
234. Plunkett, S.; Parrish, M.E.; Shafer, K.H.; Shorter, J.H.; Nelson, D.D.; Zahniser, M.S. Hydrazine detection limits in the cigarette smoke matrix using infrared tunable diode laser absorption spectroscopy. *Spectroc. Acta Part A-Mol. Biomol. Spectrosc.* **2002**, *58*, 2505–2517. [CrossRef]
235. Taslakov, M.; Simeonov, V.; Froidevaux, M.; van den Bergh, H. Open-path ozone detection by quantum-cascade laser. *Appl. Phys. B-Lasers Opt.* **2005**, *82*, 501–506. [CrossRef]
236. Ren, W.; Jiang, W.; Sanchez, N.P.; Patimisco, P.; Spagnolo, V.; Zah, C.-E.; Xie, F.; Hughes, L.C.; Griffin, R.J.; Tittel, F.K. Hydrogen peroxide detection with quartz-enhanced photoacoustic spectroscopy using a distributed-feedback quantum cascade laser. *Appl. Phys. Lett.* **2014**, *104*, 041117. [CrossRef]
237. Siciliani de Cumis, M.; Viciani, S.; Borri, S.; Patimisco, P.; Sampaolo, A.; Scamarcio, G.; De Natale, P.; D'Amato, F.; Spagnolo, V. Widely-tunable mid-infrared fiber-coupled quartz-enhanced photoacoustic sensor for environmental monitoring. *Opt. Express* **2014**, *22*, 28222–28231. [CrossRef]
238. Spagnolo, V.; Patimisco, P.; Borri, S.; Scamarcio, G.; Bernacki, B.E.; Kriesel, J. Part-per-trillion level SF6 detection using a quartz enhanced photoacoustic spectroscopy-based sensor with single-mode fiber-coupled quantum cascade laser excitation. *Opt. Lett.* **2012**, *37*, 4461–4463. [CrossRef]
239. Spagnolo, V.; Patimisco, P.; Borri, S.; Scamarcio, G.; Bernacki, B.E.; Kriesel, J. Mid-infrared fiber-coupled QCL-QEPAS sensor. *Appl. Phys. B-Lasers Opt.* **2013**, *112*, 25–33. [CrossRef]
240. Waclawek, J.P.; Moser, H.; Lendl, B. Compact quantum cascade laser based quartz-enhanced photoacoustic spectroscopy sensor system for detection of carbon disulfide. *Opt. Express* **2016**, *24*, 6559–6571. [CrossRef] [PubMed]
241. Cao, X.; Li, J.; Gao, H.; Du, Z.; Ma, Y. Simultaneous Determination of Carbon Disulfide, Carbon Monoxide, and Dinitrogen Oxide by Differential Absorption Spectroscopy Using a Distributed Feedback Quantum Cascade Laser. *Anal. Lett.* **2017**, *50*, 2342–2350. [CrossRef]
242. Wysocki, G.; McCurdy, M.; So, S.; Weidmann, D.; Roller, C.; Curl, R.F.; Tittel, F.K. Pulsed quantum-cascade laser-based sensor for trace-gas detection of carbonyl sulfide. *Appl. Opt.* **2004**, *43*, 6040–6046. [CrossRef] [PubMed]
243. Du, Z.; Li, J.; Gao, H.; Luo, G.; Cao, X.; Ma, Y. Ultrahigh-resolution spectroscopy for methyl mercaptan at the $\nu 2$ -band by a distributed feedback interband cascade laser. *J. Quant. Spectrosc. Radiat. Transf.* **2017**, *196*, 123–129. [CrossRef]
244. Dong, L.; Tittel, F.K.; Li, C.; Sanchez, N.P.; Wu, H.; Zheng, C.; Yu, Y.; Sampaolo, A.; Griffin, R.J. Compact TDLAS based sensor design using interband cascade lasers for mid-IR trace gas sensing. *Opt. Express* **2016**, *24*, A528–A535. [CrossRef] [PubMed]

245. Zheng, C.; Ye, W.; Sanchez, N.P.; Gluszek, A.K.; Hudzikowski, A.J.; Li, C.; Dong, L.; Griffin, R.J.; Tittel, F.K. Infrared Dual-Gas CH4/C2H6Sensor Using Two Continuous-Wave Interband Cascade Lasers. *IEEE Photonics Technol. Lett.* **2016**, *28*, 2351–2354. [CrossRef]
246. Ye, W.; Li, C.; Zheng, C.; Sanchez, N.P.; Gluszek, A.K.; Hudzikowski, A.J.; Dong, L.; Griffin, R.J.; Tittel, F.K. Mid-infrared dual-gas sensor for simultaneous detection of methane and ethane using a single continuous-wave interband cascade laser. *Opt. Express* **2016**, *24*, 16973–16985. [CrossRef]
247. Razeghi, M.; Ye, W.; Zheng, C.; Tittel, F.K.; Sanchez, N.P.; Gluszek, A.K.; Hudzikowski, A.J.; Lou, M.; Dong, L.; Griffin, R.J. A compact mid-infrared dual-gas CH_4/C_2H_6 sensor using a single interband cascade laser and custom electronics. In Proceedings of the Quantum Sensing and Nano Electronics and Photonics XIV, San Francisco, CA, USA, 29 January–2 February 2017; p. 1011134.
248. Krzempek, K.; Abramski, K.M.; Nikodem, M. All-fiber mid-infrared difference frequency generation source and its application to molecular dispersion spectroscopy. *Laser Phys. Lett.* **2017**, *14*. [CrossRef]
249. Sampaolo, A.; Csutak, S.; Patimisco, P.; Giglio, M.; Menduni, G.; Passaro, V.; Tittel, F.K.; Deffenbaugh, M.; Spagnolo, V. Methane, ethane and propane detection using a compact quartz enhanced photoacoustic sensor and a single interband cascade laser. *Sens. Actuator B-Chem.* **2019**, *282*, 952–960. [CrossRef]
250. Schiller, C.L.; Bozem, H.; Gurk, C.; Parchatka, U.; Königstedt, R.; Harris, G.W.; Lelieveld, J.; Fischer, H. Applications of quantum cascade lasers for sensitive trace gas measurements of CO, CH_4, N_2O and HCHO. *Appl. Phys. B-Lasers Opt.* **2008**, *92*, 419–430. [CrossRef]
251. Haakestad, M.W.; Lamour, T.P.; Leindecker, N.; Marandi, A.; Vodopyanov, K.L. Intracavity trace molecular detection with a broadband mid-IR frequency comb source. *J. Opt. Soc. Am. B* **2013**, *30*, 631–640. [CrossRef]
252. Dong, L.; Cao, Y.; Sanchez, N.P.; Griffin, R.J.; Tittel, F.K. Mid-infrared detection of atmospheric CH_4, N_2O and H_2O based on a single continuous wave quantum cascade laser. In Proceedings of the CLEO 2015, San Jose, CA, USA, 10–15 May 2015; p. AF2J.3.
253. Wei, C.; Pineda, D.I.; Paxton, L.; Egolfopoulos, F.N.; Spearrin, R.M. Mid-infrared laser absorption tomography for quantitative 2D thermochemistry measurements in premixed jet flames. *Appl. Phys. B-Lasers Opt.* **2018**, *124*. [CrossRef]
254. Shorter, J.H.; Nelson, D.D.; Barry McManus, J.; Zahniser, M.S.; Milton, D.K. Multicomponent Breath Analysis With Infrared Absorption Using Room-Temperature Quantum Cascade Lasers. *IEEE Sens. J.* **2009**, *10*, 76–84. [CrossRef] [PubMed]
255. Sun, K.; Tao, L.; Miller, D.J.; Khan, M.A.; Zondlo, M.A. Inline multi-harmonic calibration method for open-path atmospheric ammonia measurements. *Appl. Phys. B-Lasers Opt.* **2012**, *110*, 213–222. [CrossRef]
256. Ma, Y.; Lewicki, R.; Razeghi, M.; Tittel, F.K. QEPAS based ppb-level detection of CO and N_2O using a high power CW DFB-QCL. *Opt. Express* **2013**, *21*, 1008–1019. [CrossRef]
257. Reidl-Leuthner, C.; Ofner, J.; Tomischko, W.; Lohninger, H.; Lendl, B. Simultaneous open-path determination of road side mono-nitrogen oxides employing mid-IR laser spectroscopy. *Atmos. Environ.* **2015**, *112*, 189–195. [CrossRef]
258. Razeghi, M.; Yu, Y.; Sanchez, N.P.; Lou, M.; Zheng, C.; Wu, H.; Głuszek, A.K.; Hudzikowski, A.J.; Griffin, R.J.; Tittel, F.K. CW DFB-QCL- and EC-QCL-based sensor for simultaneous NO and NO_2 measurements via frequency modulation multiplexing using multi-pass absorption spectroscopy. In Proceedings of the Quantum Sensing and Nano Electronics and Photonics XIV, San Francisco, CA, USA, 29 January–2 February 2017.
259. Yu, Y.; Sanchez, N.P.; Yi, F.; Zheng, C.; Ye, W.; Wu, H.; Griffin, R.J.; Tittel, F.K. Dual quantum cascade laser-based sensor for simultaneous NO and NO_2 detection using a wavelength modulation-division multiplexing technique. *Appl. Phys. B-Lasers Opt.* **2017**, *123*, 164. [CrossRef]
260. Jagerska, J.; Jouy, P.; Tuzson, B.; Looser, H.; Mangold, M.; Soltic, P.; Hugi, A.; Bronnimann, R.; Faist, J.; Emmenegger, L. Simultaneous measurement of NO and NO_2 by dual-wavelength quantum cascade laser spectroscopy. *Opt. Express* **2015**, *23*, 1512–1522. [CrossRef] [PubMed]
261. Jágerská, J.; Jouy, P.; Hugi, A.; Tuzson, B.; Looser, H.; Mangold, M.; Beck, M.; Emmenegger, L.; Faist, J. Dual-wavelength quantum cascade laser for trace gas spectroscopy. *Appl. Phys. Lett.* **2014**, *105*. [CrossRef]
262. Chen, X.; Yang, C.-G.; Hu, M.; Shen, J.-K.; Niu, E.-C.; Xu, Z.-Y.; Fan, X.-L.; Wei, M.; Yao, L.; He, Y.-B.; et al. Highly-sensitive NO, NO 2, and NH 3 measurements with an open-multipass cell based on mid-infrared wavelength modulation spectroscopy. *Chin. Phys. B* **2018**, *27*. [CrossRef]

263. Chen, X.; Yang, C.; Hu, M.; Xu, Z.; Fan, X.; Wei, M.; Yao, L.; He, Y.; Kan, R. High sensitivity measurement of NO, NO_2 and NH_3 using MIR-QCL and time division multiplexing WMS technology. In Proceedings of the Hyperspectral Remote Sensing Applications and Environmental Monitoring and Safety Testing Technology, Beijing, China, 9–11 May 2016.
264. Lee, B.H.; Wood, E.C.; Zahniser, M.S.; McManus, J.B.; Nelson, D.D.; Herndon, S.C.; Santoni, G.W.; Wofsy, S.C.; Munger, J.W. Simultaneous measurements of atmospheric HONO and NO_2 via absorption spectroscopy using tunable mid-infrared continuous-wave quantum cascade lasers. *Appl. Phys. B-Lasers Opt.* **2010**, *102*, 417–423. [CrossRef]
265. Rawlins, W.T.; Hensley, J.M.; Sonnenfroh, D.M.; Oakes, D.B.; Allen, M.G. Quantum cascade laser sensor for SO_2 and SO_3 for application to combustor exhaust streams. *Appl. Opt.* **2005**, *44*, 6635–6643. [CrossRef] [PubMed]
266. Young, C.; Kim, S.S.; Luzinova, Y.; Weida, M.; Arnone, D.; Takeuchi, E.; Day, T.; Mizaikoff, B. External cavity widely tunable quantum cascade laser based hollow waveguide gas sensors for multianalyte detection. *Sens. Actuator B-Chem.* **2009**, *140*, 24–28. [CrossRef]

© 2019 by the authors. Licensee MDPI, Basel, Switzerland. This article is an open access article distributed under the terms and conditions of the Creative Commons Attribution (CC BY) license (http://creativecommons.org/licenses/by/4.0/).

Review

Laser Absorption Sensing Systems: Challenges, Modeling, and Design Optimization

Zhenhai Wang [1], Pengfei Fu [2] and Xing Chao [1,*]

[1] Center for Combustion Energy, Department of Energy and Power Engineering and Key Laboratory for Thermal Science and Power Engineering of Ministry of Education, Tsinghua University, Beijing 100084, China
[2] School of Aerospace Engineering, Tsinghua University, Beijing 100084, China
* Correspondence: chaox6@tsinghua.edu.cn; Tel.: +86-10-6278-4497

Received: 31 May 2019; Accepted: 3 July 2019; Published: 5 July 2019

Abstract: Laser absorption spectroscopy (LAS) is a promising diagnostic method capable of providing high-bandwidth, species-specific sensing, and highly quantitative measurements. This review aims at providing general guidelines from the perspective of LAS sensor system design for realizing quantitative species diagnostics in combustion-related environments. A brief overview of representative detection limits and bandwidths achieved in different measurement scenarios is first provided to understand measurement needs and identify design targets. Different measurement schemes including direct absorption spectroscopy (DAS), wavelength modulation spectroscopy (WMS), and their variations are discussed and compared in terms of advantages and limitations. Based on the analysis of the major sources of noise including electronic, optical, and environmental noises, strategies of noise reduction and design optimization are categorized and compared. This addresses various means of laser control parameter optimization and data processing algorithms such as baseline extraction, *in situ* laser characterization, and wavelet analysis. There is still a large gap between the current sensor capabilities and the demands of combustion and engine diagnostic research. This calls for a profound understanding of the underlying fundamentals of a LAS sensing system in terms of optics, spectroscopy, and signal processing.

Keywords: laser absorption spectroscopy (LAS); combustion sensing; direct absorption spectroscopy (DAS); wavelength modulation spectroscopy (WMS); design optimization; noise reduction algorithms

---

1. Introduction

1.1. Overview of Laser Absorption Sensor Developments and Applications

Laser absorption spectroscopy (LAS) has served as a useful tool for both fundamental and practical studies in energy and power systems since its earliest demonstration as a species diagnostic method in combustion-related environments over forty years ago [1,2]. In addition to the non-contact characteristics of most optically based methods (e.g., emission, Schlieren imaging, laser-induced fluorescence, particle image velocimetry), LAS benefits from the species-specific nature of molecular spectroscopy, therefore is capable of achieving highly quantitative and selective measurements of a number of important species parameters including gas composition, temperature, pressure, and velocity. The use of narrow-linewidth, tunable, low-power semiconductor lasers, on the other hand, offers the opportunity for real-time, time-resolved, and long-term stable field monitoring systems that can be realized under various measurement scenarios.

Up towards the early 2000s, gas sensors based on tunable diode laser absorption spectroscopy (TDLAS) were extensively applied in continuous emission monitoring and process controls and diffused in process industries as an accepted technique [3,4]. The maturity and commercialization of opto-electronic devices, thanks to the booming telecommunication industry, has allowed TDLAS sensors to evolve from laboratory-scale benchtop experimental tools to a robust industrial technology. With further reduction in cost and size, portable and field-deployable systems can be made. Using smart optical–mechanical designs and sophisticated signal processing schemes, some of these sensors are able to operate in the field without operator intervention for over 10 years [5].

Today, the design, control, and optimization of complex energy and propulsion systems poses unprecedented challenges for sensor technology. The extreme conditions associated with combustion that are typically unfriendly for optical systems, such as high temperature, pressure, and velocity, mechanical vibrations, high dust levels, and corrosiveness, make LAS still one of the best available contactless sensor techniques for harsh environments. However, extraction and interpretation of high-quality, quantitative information under such complicated, demanding conditions requires deeper understanding of the target environments and tailoring the sensor design to specific measurement goals. While state-of-the-art laser technology for novel single-mode mid-infrared and hyperspectral laser sources continues to offer much more freedom in choosing the spectral features and designing LAS sensors, much remains to be studied in improving their reliability, accuracy, and precision.

1.2. Scope and Organization of the Paper

There has been a number of excellent reviews and books on the development and application of LAS sensor technology. Even as early as in the late 1990s, several papers reviewed the fundamental theories and early successes of laser-based spectroscopic methods in gas dynamics and combustion flow measurements [6–8]. Nasim et al. illustrated the evolution of and confinement methods used in semiconductor diode lasers and discussed different techniques used to convert free-running diode lasers into true narrow linewidth tunable diode laser sources [9]. As lasers were much less mature at that time, these papers also reviewed the available laser light sources as an important part of the diagnostics development. More recently, several reviews reported the advances of the LAS technique and its demonstration in various disciplines. Hanson et al. presented an overview of combustion kinetics, propulsion, and combustion in practical systems and documented the impact of quantitative laser diagnostics methods [10]. Bolshov et al. highlighted the temperature, concentrations, and flow velocities in different combustion zones and emphasized the strategies and data processing algorithms for LAS measurements [11]. Goldenstein et al. provided a thorough review of the underlying fundamentals, design, and use of infrared LAS sensors for combustion gases and highlighted recent findings and some of the remaining measurement opportunities, challenges, and needs [12]. Liu et al. summarized the key principles and recent advances of line-of-sight (LOS) LAS techniques and then focused on spatially resolved gas sensing with LAS tomography [13].

More specifically, a few papers have focused on reviewing signal-to-noise-ratio (SNR) enhancement techniques to improve TDLAS performance. Li et al. summarized several commonly employed noise reduction schemes including signal averaging, modulation techniques, balanced detection, zero-background subtraction, and adaptive filtering [14]. Zhang et al. focused on the various mathematical algorithms applied in TDLAS signal processing to enhance the accuracy and resolution of the sensor [15].

As discussed above, sensing challenges vary largely with measurement targets and conditions. The maturity of semiconductor lasers and photodetectors allows many researchers in relevant fields to simply use LAS sensor as an established tool, whereas diagnosticians still strive to push the limit of this technique. While previous papers have provided comprehensive reviews on all major elements of LAS sensor systems including fundamental molecular spectroscopy, the development of opto-electronic devices, and various measurement schemes and applications, we refer the readers to other works for the fundamentals and will address the design perspectives of a LAS sensing system. Section 2 will start by summarizing and identifying the measurement challenges and design targets under different measurement conditions; Section 3 will follow by outlining the fundamentals of different diagnostic strategies and their underlying models and control factors; Section 4 will then provide a systematic analysis of approaches for noise reduction and signal optimization, so to extend the use of the LAS sensing technique to much more difficult measurement scenarios. In view of the urging demands in both the use and development of this sensor technology, a goal of this manuscript is to provide understanding and an overview of the key factors and considerations in the measurement targets, as well as a reference and guidelines to a strategic design procedure appropriate for obtaining optimized sensing results.

2. Sensor Design Targets and Measurement Challenges

The fundamental theory lying behind absorption spectroscopy is the Beer–Lambert law, presented in Equation (1), with multiple forms of expressions. The fractional transmission of light, given by the ratio between transmitted and incident light intensity, I_t [W/cm^2] and I_0 [W/cm^2], can be expressed as an exponential function of absorbance α_ν, with k_ν [1/cm] being the spectral absorption coefficient, L [cm] the absorption pathlength, n [molecule/cm^3] the number density of the absorbing species, σ_ν [cm^2/molecule] the absorption cross section, S [cm^{-2}/atm] the absorption linestrength of an individual transition line, ϕ_ν [cm] the frequency-dependent lineshape function, P [atm] the total pressure, and X_i the mole fraction of the absorbing species i. The subscript ν is used to signify the spectral dependence of the parameter on the light frequency ν:

$$\begin{aligned}(I_t/I_0)_\nu &= \exp(-\alpha_\nu) \\ &= \exp(-k_\nu L) \\ &= \exp(-n\sigma_\nu L) \\ &= \exp(-S\phi_\nu P X_i L)\end{aligned} \quad (1)$$

This rather simple but nontrivial relationship roots from the energy balance derivation in the equation of radiative transfer. In a typical absorption measurement, the total absorbance α_ν is the ultimate "signal" to optimize, and the performance of LAS species sensor is commonly evaluated through detection limits and detection bandwidth. The limits of detection can be quantified as the noise-equivalent absorbance (NEA) or the minimal detectable concentration (e.g., in ppm), and the measurement bandwidth (Hz) can be defined as half of the measurement temporal resolution taking into account the Nyquist criterion.

To directly compare the detectivity of LAS sensors under different measurement conditions, the pathlength-normalized detection limit (e.g., in ppm·m) is often used, on the basis of the fact that the absorbance is linearly proportional to the absorption pathlength. Further normalization with bandwidth renders a detection limit in the unit of ppm·m·Hz$^{-1/2}$ to account for the improvement of signal-to-noise ratio (SNR) with averaging. However, it needs to be noted that this inverse square-root dependence on averaging time can be applied if the signal carries only white noise. A more rigorous assessment of the validity of such normalization can be done with Allan variance analysis, which analyzes a sequence of data in the time domain. Although traditionally used as a measure of frequency stability of clocks and oscillators in metrology, the Allan variance analysis can also be adopted as a useful tool to identify and quantify different noise terms and indicate signal stability. As shown in the schematic of Figure 1, the root Allan variance, aka Allan deviation, of a time domain signal is computed as a function of different averaging times τ. By analyzing the characteristic regions and log–log scale slopes of the Allan deviation curve, different noise processes and modes may be identified.

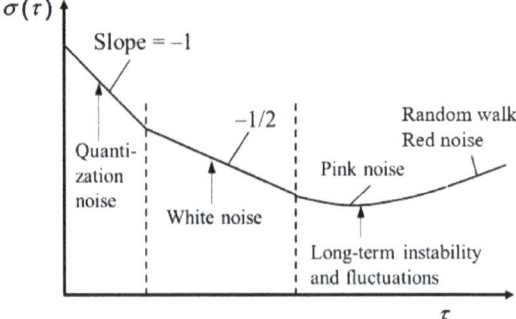

Figure 1. Schematic of different noise processes on an Allan deviation plot.

Even limiting our discussion to combustion-related LAS diagnostic studies, the demands and achieved performance of the laser sensors vary over a wide range. Figure 2 summarizes some representative results from a number of sensor development and implementation studies in terms of detection limits and bandwidths. It compares species concentration measurements in a variety of combustion systems including laboratory environments [1,16–22], shock tube studies [23–28], industry processes [29–33], and engines [34–46]. For the purpose of making a direct comparison between different studies, the reported absolute minimal detection limits for several commonly probed species in combustion, including CO, CO_2, H_2O, and C_2H_2, are plotted on the same diagram against detection bandwidths. It can be clearly observed that results from different applications naturally separate into different regimes: the well-controlled laboratory environments allow lower absorbance to be detected, where chemical kinetics studies in shock tubes typically demand for a much higher detection bandwidth than stabilized laboratory flames or gas cell; larger scale, more "realistic" environments such as industrial burners and engines present harsher conditions that lead to higher achievable minimal detection limits, but industrial process monitoring typically involves stringent emission control targets and trace species detection over continuous, long-term monitoring periods, whereas engine diagnostics are usually associated with high-velocity or turbulent flow and reaction conditions.

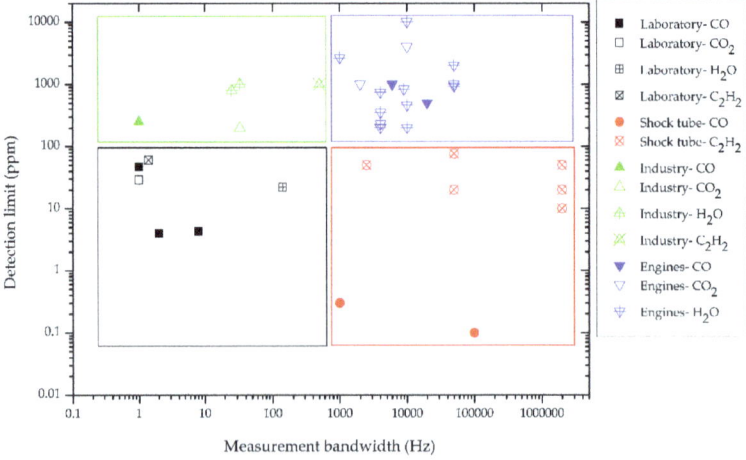

Figure 2. Representative detection limits and measurement bandwidths of laser absorption sensors in various combustion-related systems.

Despite its versatility in measuring temperature, pressure, and velocity, LAS is ultimately a species diagnostic method. It utilizes the "fingerprint" molecular spectra and infer the parameters to be measured on the basis of a correct interpretation of the recorded spectral feature. Figure 3 plots the absorption linestrengths over the infrared wavelength range of 0.5–6 µm of several of the representative C–O, C–H, N–O, N–H, and H–O compound gaseous species commonly studied in combustion. Today, with semiconductor laser technology, we are able to access almost any wavelength within this region, so that there is no longer much constraint in making an optimized selection of absorption transition, which can be solely based on appropriate strength level and minimal cross interference. For fairly harmonic molecules such as CO, the advantage of using a lower-order vibrational band at longer wavelength is more prominent in achieving lower detection limits as compared with a less harmonic molecule such as NH_3. One can also make the observation that high temperatures, often associated with combustion-related sensing, would by themselves present greater challenges for LAS sensors. The reasoning is twofold from the point of view of molecular spectroscopy: (1) the overall band strength is approximately inversely proportional to temperature, so that absorption is generally weaker at high temperatures, and (2) higher energy levels get more populated at higher temperatures, so that the spectra may get more crowded with hot bands and high J-number lines, especially when a multicomponent gas mixture is being studied. As will be reviewed in the following, many efforts have been devoted to reducing a correct absorption-free background and discriminate the molecular absorption by target species from a medley of various cross interference and noise processes.

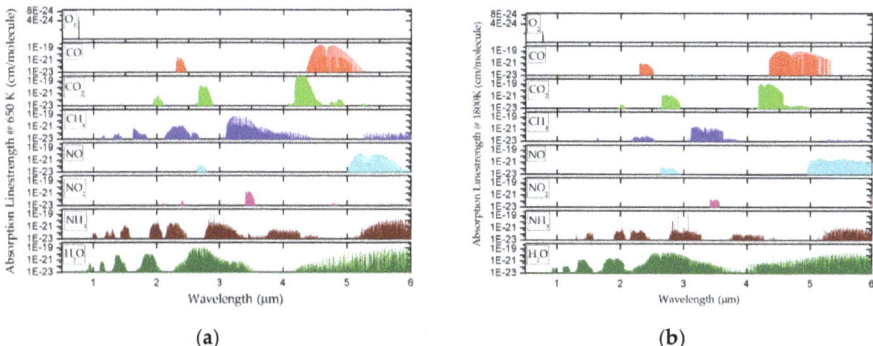

Figure 3. Absorption linestrengths for O_2, CO, CO_2, CH_4, NO, NO_2, NH_3, and H_2O at (**a**) 650 K, (**b**) 1800 K. Transitions with line strengths less than 1×10^{-23} cm/molecule are not shown. All data taken from HITRAN2016 [47].

3. Fundamentals of Different LAS Measurement Schemes

Absorption is based on signal difference between the light intensities before and after propagation through a medium. When the incident light source has a spectral linewidth much narrower than the absorption transition feature, the so-gained spectral information can be regarded as single-wavelength without the need for light dispersion elements which can potentially introduce a reduction in spectral resolution. The use of semiconductor lasers which can be easily and fast tuned with injection current and temperature also allows the access to a range of wavelengths and absorption feature profiles with a single laser device. A detailed description of typical laser sources employed in LAS including distributed feedback lasers (DFBs), vertical-cavity surface-emitting lasers (VCSELs), Fourier-domain mode-locked lasers (FDML), quantum cascade lasers (QCLs), interband cascade lasers (ICLs), and their corresponding features, such as linewidth, power, and tunability has been presented by Goldenstein et al. [12] and Liu et al. [13].

The two most commonly employed schemes in realizing a LAS sensor with wavelength-tunable, narrow-linewidth lasers include direct absorption spectroscopy (DAS) and wavelength modulation spectroscopy (WMS), and each can be further associated with fixed- or scanned-wavelength schemes.

For fixed-wavelength DAS, as the laser wavelength is not tuned or modulated, the detection bandwidth is solely limited by the detector bandwidth and the sampling rate of the data acquisition system. Therefore, this scheme is often used in applications where an extremely high temporal resolution at levels of 1 MHz or above is needed. However, it requires the absolute laser wavelength to be accurately known and precisely controlled. In contrast, the scanned-wavelength scheme is widely used for its higher robustness and better immunity against fluctuations in practical combustion system monitoring.

The basic principles of DAS and WMS have been detailed elsewhere [48]. Here, from a different perspective, we will give a brief summary of each technique and elucidate the fitting algorithms used to infer gas conditions.

3.1. Direct Absorption Spectrocopy

The direct absorption method infers gas properties directly from the fractional transmission of light I_t/I_0 using the Beer–Lambert relation. One of the key problems to solve in DAS is thus to determine the baseline, i.e., the light intensity I_0 before absorption by the medium. Certain sensor systems incorporate an additional reference signal with a non-resonant laser or a broadly tunable light source to correct for the baseline [12]. However, this is at the expense of added components and may introduce potential errors from additional optical etalons. The more commonly adopted approach is to mathematically fit the baseline using non-absorbing portions within a scan.

Two background correction strategies are often adopted, namely, the Levenberg–Marquardt (LM) fitting algorithm [49–51], and the advanced integrative (AI) fitting algorithm [52]. Both fitting algorithms involve an iterative process, and the major difference lies in the fitting speed and multi-line fitting capability. More details are summarized in Table 1 below, and here we illustrate a typical fitting procedure with an example of AI fitting algorithm.

AI fitting process:

Step 1: Background correction: a polynomial fit is applied to infer the background from the non-absorbing portions;

Step 2: Retrieval of line center position ν_0, which is estimated from the two symmetrical intervals from the maximum of background-corrected absorbance curve, thus no explicit initialization is needed;

Step 3: Retrieval of absorbance area: absorption linewidth is predetermined, and then the Voigt function is calculated;

Step 4: Voigt fit: several numerical approaches can be used to approximate the Voigt function [53,54] as no analytical form is available.

Most commonly, by using a low-order (order two to four) polynomial to fit the baseline and subsequently fitting the absorbance curve with a Voigt profile through an iterative process, the gas properties can be inferred. This works well for isolated or slightly overlapped features and if the baseline is reasonably smooth and the raw spectrum has a good SNR. However, when noises from harsh environments causing severe beam steering and baseline fluctuation carry frequency components comparable to the scanning/detection bandwidth, uncertainties in the DAS detection will significantly grow. Efforts to deal with such background drift will be detailed in Section 4.

Table 1. Comparison of fitting algorithms in direct absorption spectroscopy (DAS) and wavelength modulation spectroscopy (WMS) techniques. LM: Levenberg–Marquardt, AI: advanced integrative, SNR: signal-to-noise-ratio.

Technique	Parameter Initialization	Parameter Predetermined	Feature/Limitation
DAS LM algorithm	v_0, T, X	no	Easy to fit multiple absorption lines
DAS AI algorithm	no	Δv	Complicated to fit multiple absorption lines, a factor of 3-4 faster than the LM algorithm
2f-WMS	T, X	P HITRAN/HITEMP database [47]	Calibrated on the basis of database, hard to calculate absolute T
Calibration-free WMS-2f/1f with fixed-wavelength laser characterization	v_0, T, X	$i_1, i_2, \varphi_1, \varphi_2$ HITRAN/HITEMP database	Wavelength-dependent laser characteristics may lead to measurement errors
Calibration-free WMS-2f/1f with scanned-wavelength laser characterization	v_0, T, X	$v(t), {}^M I_0(t)$ HITRAN/HITEMP database	Non-linear laser intensity variation along measurement beam path difficult to quantify
Recovery of DAS lineshapes	v_0, T, X	$\varphi_1, X_{nf}, Y_{nf}$	Higher-order harmonics come with low SNR
DAS-calibrated WMS	no	P, T	Direct, on-the-fly calibration, increased precision and SNR for trace gas detection

3.2. 2f-WMS Method

By adding a high-frequency modulation to the laser injection current, the WMS scheme raises the signal detection band to a designated frequency range and later extracts the signal with band-pass filtering processes. This allows the method to have better noise rejection capability, and a 2~100 improvement in SNR has been demonstrated over DAS [55].

When the absorption features introduce distortions to the sinusoidally modulated laser intensity, harmonic components at integer multiples of the modulation frequency would appear. Due to the symmetry of an absorption feature about its line center, even orders of harmonic signals would peak near the line center position. For small modulation depths, the nonlinear laser intensity modulation can be neglected, and the peak of the second harmonic signal, or the 2f peak height, is a function of species concentration, pressure, and temperature and therefore can be taken as the target signal to measure [56,57]. The dependence on multiple non-precalibrated parameters is the major limitation of 2f-WMS method, however, for two-line thermometry, the 2f peak height ratio of the two selected lines can be directly reduced to linestrength ratio with the use of suitable modulation depth and line pair. Figure 4 shows the flow chart of the fitting procedure of the 2f-WMS method, including 2f peak ratio thermometry and species concentration extraction. To account for the unknown proportionality to environmentally dependent factors such as transmission losses and instantaneous laser intensity, the 2f peak magnitudes need to be pre-calibrated on the basis of HITRAN/HITEMP database line parameters with known pressure and temperature conditions and a nominal value of gas concentration.

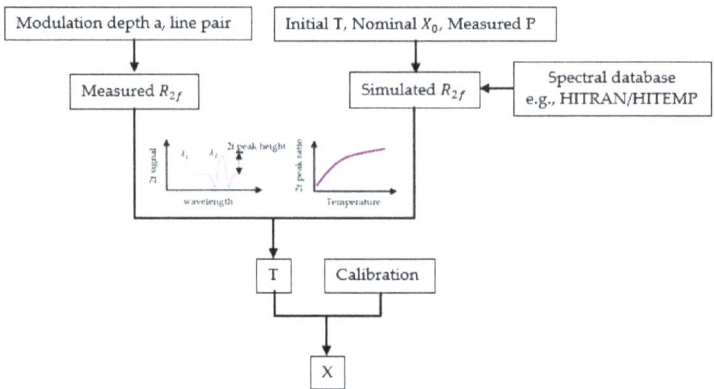

Figure 4. Flowchart of the 2*f*-WMS method.

3.3. Calibration-Free WMS-2f/1f Method

The calibration process needed for the WMS-2*f* method requires all environmental conditions to remain the same between the calibration stage and the actual measurement stage. Apparently, this luxury of having stable and known conditions is not always available, especially in practical applications where laser transmission fluctuations are much more difficult to predict or monitor. One solution is to normalize the 2*f* signal with the 1*f* component, referred to as the calibration-free WMS-2*f*/1*f* method.

The normalization process cancels out the unknown or changing laser intensity. However, to directly associate the measured WMS-2*f*/1*f* value with absolute gas conditions, laser intensity and frequency tuning parameters need to be accurately characterized. Typically, this laser characterization process identifies the first- and second-order laser intensity modulation indices (i_1, i_2) and intensity–frequency modulation phase shifts (φ_1, φ_2) at a specific center frequency ν_0, which are sufficient to describe the intensity and frequency modulation behavior of a DFB semiconductor laser.

Rieker et al. summarized potential sources of uncertainty in calibration-free WMS, with particular emphasis on the influence of pressure and optical depth in harsh environments [55]. Under such conditions, measurement uncertainties induced by pressure deviation between the simulation and the experimental conditions become more significant, and preference for stronger absorption features voids the optical-thin assumption so that larger measurement errors may result.

In view of such problems, scanned-wavelength WMS with larger scan ranges and modulation depths may be used. A modified calibration-free WMS scheme with an entirely different laser characterization strategy is proposed to account for the time-variant and wavelength-dependent change of the laser characteristic parameters [16,58,59]. The flow chart of this method is shown in Figure 5. This analysis scheme differs from previous WMS strategies in two apparent ways: (1) the use of measured intensity in real time avoids the need for a pre-determined analytic model to describe laser intensity, and can at the same time account for the wavelength-dependent transmission of optical components in the beam path, and (2) using the same data processing procedure for both simulation and measurement introduces equal contributions from any potential non-ideal performance of the lock-in and low-pass filter.

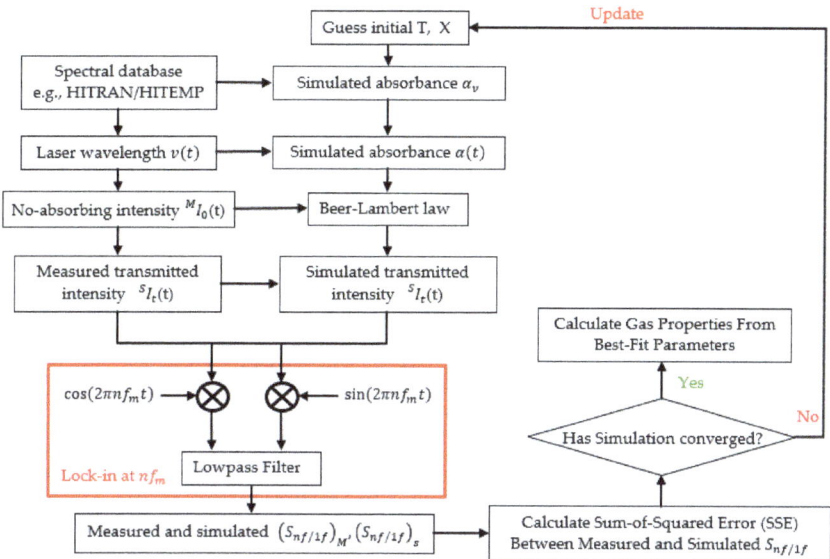

Figure 5. Flow chart of a calibration-free WMS scheme with scanned-wavelength laser characterization.

This method is applicable without constraints on optical depth or modulation index, so that pressure-broadened and blended absorption features can also be studied. However, nonlinear variations in laser characteristics due to calibration drift, temperature variations, and laser aging that add to the residual amplitude modulation (RAM) at different harmonic orders may introduce errors that mostly occur over longer terms.

3.4. Recovery of the Absorbance Profile Based On Higher-Order Harmonic Signals

While previous WMS approaches use the harmonic signals directly, a phasor decomposition method has been proposed to recover the absorbance profile from the first harmonic on the basis of RAM [60,61]. However, these earlier attempts only worked well with small modulation indices ($m < 0.2$) when the first-order harmonic shape is close to the first derivative of the absorbance profile. Peng et al. employed additional higher odd harmonics (3rd, 5th, ...) to enhance the recovery accuracy with large modulation indices [62]. Figure 6 shows the flow chart of the recovery and data reduction process. It is worth noting that while considering more harmonic components improves the model recovery accuracy, these high-order terms usually have low SNRs and are difficult to detect in practical measurements.

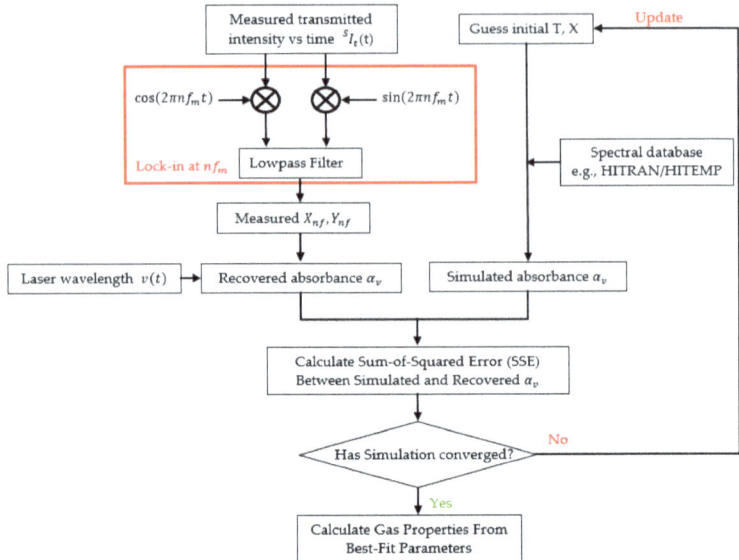

Figure 6. Flow chart of the recovery process of the absorbance profile based on higher-order harmonic signals.

3.5. DAS-Calibrated WMS

From the discussion above, one may conclude that DAS and WMS have their own advantages and limitations. While being more noise-immune in practical environments, calibration-free WMS methods rely on extensive laser characterization, and the nonlinear response to the sinusoidal modulation over the entire wavelength scan could lead to significant distortion in the 2f lineshape, especially with wide wavelength scans. Klein et al. have thus proposed a DAS-calibrated WMS method, which combines the simplicity and accuracy of the intrinsically calibration-free direct TDLAS (dTDLAS) with the enhanced precision of WMS-2f [63]. A rapid (125 Hz) time-division multiplexed scheme alternating between triangular scan and scanned modulation was used to realize quasi-simultaneous DAS and WMS measurements. The concept of this on-the-fly calibration is illustrated in Figure 7. An absorbance level of 0.1 was chosen as the decision criterion of whether DAS only or DAS-calibrated WMS will be used. This largely extends the dynamic range of a single sensor and is especially advantageous for in-field measurements without the need for reference gases or measurement interruption.

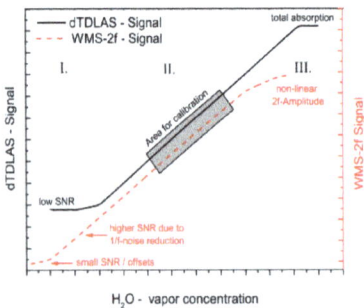

Figure 7. Schematic of the on-the-fly WMS calibration using time-multiplexed direct tunable diode laser absorption spectroscopy (dTDLAS) and WMS. Figure adapted from [63].

3.6. Summary of Measurement Schemes

A summary of various scanned-wavelength LAS techniques discussed above is presented in Table 1. For the purpose of comparison, fitting parameter initialization, parameter predetermination, as well as features and limitations for each algorithm are listed.

DAS is preferred for its calibration-free detection capability and simplicity, for applications where relatively isolated transitions are used. However, laser intensity fluctuations, excess noises, and baseline fitting errors put limitations to the use of this method especially for difficult environments. Thus, DAS is the method of choice for measurements with transitions of sufficient linestrength and narrow linewidth to allow laser access with wavelength scanning.

By contrast, the WMS method is more promising for measurement systems with small absorbance, high pressure broadening, or blended absorption features precluding a straightforward determination of the non-absorbing baseline. With $1f$ normalization, the calibration-free WMS-$2f/1f$ method can account for variations in laser intensity such as non-absorption losses due to light scattering or beam steering. This makes it advantageous for measurements in harsh environments involving high pressures, high opacity, high emission levels, and high temperatures.

The accuracy of DAS lineshape recovery from harmonic signals is limited by the low-SNR high-order terms, making this method unsuitable for practical measurements. A time-division multiplexed spectroscopic scheme is applied by alternating laser modulation and allows the enhancement of precision and dynamic range. This is an attractive strategy for in-field trace gas measurements without any need for reference gases or measurement interruption.

The LAS techniques discussed above typically provide a path-averaged parameter measurement over the zone of interest. Parameters including temperature, concentration, etc., are inferred by fitting measured and simulated spectral profiles. On the other hand, strategies have been developed to identify non-uniform characteristics in practical environments that commonly result from heat transfer, flow mixing, and combustion.

Two types of approaches including single line-of-sight (LOS) absorption and tomography have been proposed and experimentally demonstrated for the measurement of non-uniform zones. The first approach is based on either the *a priori* temperature (concentration) distribution profile or the probability density function along the LOS, namely, profile fitting and temperature binning, respectively [64]. It depends on simultaneous measurements of multiple absorption transitions with different temperature dependence. Tomography, on the other hand, relies on multiple LOS measurements and computational reconstruction algorithms. It reduces the requirements for spatially continuous optical access in traditional planar imaging and exhibits great potential for harsh and even optically dense environments [65]. However, this may complicate the optical system design and data processing of the experimental results.

Improvement in laser source and tomographic algorithms will contribute to greater accuracy in the temperature and concentration distribution reconstruction, using either of the two approaches above. Rieker et al. have extended the single LOS technique from advanced and expensive diode lasers with much broader tuning range (up to 10–15 cm^{-1}) to a dual-frequency comb spectrometer [66,67]. The temperature distribution reconstruction accuracies increase as more absorption lines are incorporated. In addition, the development of hyperspectral laser sources enables measurements of a large number of absorption transitions and thereby significantly reduces the number of projections [68,69]. A detailed summary of laser absorption tomography (LAT) algorithms has been given by Cai et al. [65].

4. Strategies for Noise Reduction and Design Optimization

4.1. Souces of Noise

For any sensing system, the noise and uncertainty in the obtained results need to be assessed to evaluate the performance of the sensor. High noise levels not only obscure the signal to be measured but also make the data reduction process difficult and time-consuming. In the following, we will first discuss the several major sources of noise in a LAS system, including laser noise, detector noise, analogue-to-digital conversion (ADC) noise, optical noise, and environmental noise [70,71]. Then, different approaches for noise reduction and signal optimization will be reviewed to guide future design considerations.

4.1.1. Laser and Detector Noise

The laser and detector excess noise are both frequency-dependent and known to have a pink noise spectrum. They are known as $1/f$ noise and are usually the dominant noise component at low frequencies from 1 to 10 kHz. The detector shot noise and thermal noise, on the other hand, are independent of frequency and known to have a white noise spectrum. Table 2 gives a brief summary and comparison of the different laser and detector noises.

Table 2. Comparison of laser and detector noises.

Type of noise	Cause	Feature	Frequency Range
Laser excess noise	Intensity fluctuation	Frequency dependent	1–10 kHz
Detector thermal noise	Thermal agitation of the charge carriers	Frequency independent	0.1–140 kHz
Detector shot noise	Discrete nature of electric charge	Frequency independent	0.1–140 kHz
Detector excess noise	Intensity fluctuation	Frequency dependent	1–10 kHz

4.1.2. Quantization Noise and ADC Resolution

The demands for accuracy and precision in trace species detection present great challenges for DAS. Therefore, high-speed and high-resolution ADC is required to recover the analogue detector signal with low quantization noise and to reduce the laser relative intensity noise (RIN, same as excess noise) through fast sampling. The frequency filtering feature of WMS has been widely adopted as it could improve the SNR by modulating and demodulating the absorption signal at an elevated frequency band. On the other hand, Lins et al. pointed out that as long as RIN is the dominating noise, the requirements for ADC resolution in $2f$-WMS and DAS setups are in fact similar [72]. As suggested by the simulation results, when the RIN is low and sufficient ADC resolution is provided, DAS can provide higher SNR than WMS. In this case, with the advances of modern electronics and digital signal processing, DAS can have more advantages over WMS [73].

4.1.3. Optical Interference Fringe Noise

In a multi-element optical system with coherent light source, many flat surfaces can lead to optical interference fringes, often referred to as optical etalons. These optical fringes often exhibit a free-spectral-range (FSR) comparable to the linewidth of the absorbing species, so that they are difficult to be distinguished from target absorption features.

Various means have been attempted to minimize the influence of optical etalons, including careful optical design [74–76], mechanical dithering of optics [77–79], balanced-ratio detection schemes [80,81], calibration-free WMS [82–84], and digital filtering techniques [85–89]. A summary of these approaches is given in Table 3 and briefly discussed in the following.

Table 3. Comparison of strategies to minimize interference fringe noise. AR: anti-reflective, EMD: empirical mode decomposition.

Strategy	Solutions	Feature
Optical design	Wedged or AR-coated windows; Optical isolators; Etalon immune distance	Difficult to minimize etalon in complex optical systems
Mechanical modulation	Dithering or rotating of mirrors; Brewster-plate spoiler	Increased system complexity; Limited detection bandwidth
Balanced detection schemes	Split laser beam into the sample beam and reference beam	Difficult to replicate all optical effects in reference path
WMS	Modulation index optimization; Use of higher harmonics	SNR sacrificed
Digital filtering technique	Wavelet transform; Kalman filtering; EMD	Increasing computation cost; Parameters need to be set carefully

Optical etalons essentially arise from the constructive and destructive interferences when light bounces back and forth between parallel surfaces, similar to the way optical cavity modes are formed. Optical designs to avoid etalons are therefore approaches to reduce reflections and to scramble these standing-wave modes. This includes using wedged or anti-reflective (AR) -coated windows, optical isolators, and placing the optical elements at "etalon-immune distances" to avoid etalon effects from optical feedback. Collimated laser beams can be intentionally diverged before passing through optical thins and re-collimated thereafter. In addition to these passive optical designs, an active control of the optical system can be applied to effectively reduce etalons. Mechanical modulations, such as dithering or rotation of the mirrors and use of a Brewster-plate spoiler, are active control approaches taken in high-performance optical systems. However, these do largely increase the system's cost and complexity and may limit the achievable detection bandwidth.

Certain laser devices exhibit unsuppressed cavity mode noise or fiber-coupling noise due to imperfect optical isolation, which appear as etalons on the output scans. Balanced-ratio detection schemes, also known as common-mode rejection schemes, are commonly employed to reduce such optical fringes. The laser beam is split into a measurement and a reference beam, and signals from the two detectors are fed into a balanced-ratio detection circuit that automatically cancels out the noises commonly present in the two signals. Etalons originated from the optical path external to the laser cannot generally be canceled in this way, since it would be difficult to achieve exactly the same beam path other than the portion being absorbed.

Etalon fringes often appear as low-frequency noises and can therefore be suppressed through the band-pass filtering process of WMS. On the other hand, fringe noises may add to the RAM and put additional constraint on selecting the optimal modulation parameters. This can lead to reduced SNR and larger uncertainties, especially since the optical alignment and etalon fringe profiles may change over time.

Digital filtering techniques are extensively investigated in absorption signal processing, including wavelet transform, Kalman filter, and empirical mode decomposition (EMD). These methods are purely from a standpoint of digital signal processing and frequency-domain analysis, thus require no additional optical or mechanical components. However, these techniques will inevitably increase the computation complexity, and the filter parameters need to be carefully designed on the basis of an appropriate understanding of the signal and noise characteristics. The extent of improvement that such techniques can achieve is ultimately limited by the quality of the light signal measured, so it is still of utmost importance to devote the best effort in designing a better optical system.

4.2. Strategies Based On Laser Control Parameter Optimization

The absorption signal can be enhanced by optimizing a number of laser control parameters including scanning amplitude, scan rate, modulation depth and frequency, and time delay constant of the lock-in amplifier.

A number of researchers have discussed parameter selection criteria for the WMS system. Werle et al. conducted a comprehensive analysis of the modulation degree so as to obtain maximum signal amplitude [8]; a Fourier series was utilized by Uehara et al. to analyze the relationship between modulated signals and optical penetration depth in frequency modulation [90]; Kluczyrrski et al. theoretically studied the effect of modulation frequency [91]; Neethu et al. studied the optimization of a number of modulation parameters in realizing a LAS oxygen sensing system [92].

Table 4 summaries the effect of different control parameters on the 2f-signal waveforms including maximum amplitude, SNR, peak width, and peak height ratio (PHR).

Table 4. Effects of different control parameters on 2f signals. PHR: peak height ratio.

	Scanning Amplitude	Scanning Frequency	Modulation Depth	Modulation Frequency	Time Constant
Maximum 2f	-[1]	↓[3]	↑↓	↑↓	↓
SNR	-	↑ or ↓	↑↓	↓	-
Peak width	↑[2]	↑	-	↓	-
PHR	↑	↓	-	↑ or –[4]	-

[1] no evident correlation, [2] increase, [3] decrease, [4] unchanged.

Due to the complex intercorrelation between various laser characteristic parameters, no explicit mathematical expressions generally exist to describe the dependence of the WMS signal on the laser control parameters. Therefore, the selection and optimization of these control parameters usually rely on empirical attempts and maximization of the target signal. The laser scan range, for example, is simply set to cover the target absorption feature and just enough non-absorption portions for baseline fitting, without leaving excessive margin which will decrease the detection bandwidth. For WMS in general, the modulation depth is chosen so that the WMS-2f signal at absorption line center is maximized. It has been demonstrated that for an isolated absorption feature, the highest peak value can be achieved by selecting the modulation depth to be 1.1 times the full-width-at-half-maximum (FWHM) Δv of the absorption feature [55].

In conventional scanned-wavelength WMS, a high-frequency sinusoidal wave is superimposed on a linear ramp of the laser injection current. As the laser tuning function can now readily be controlled digitally, variation of this tuning waveform can bring increased flexibility to signal optimization. Fried et al. employed a jump scanning function and dual fitting analysis to simultaneous optimize signals from two selected absorption profiles [93]. Chen et al. modified the linear ramp function so that the laser scanned more slowly near the line center ([94], Figure 8). The scan portion with stronger absorption thus occupied a larger fraction of the scan, resulting in the improvement of the overall SNR. Consequently, a factor of 1.8 improvement was observed with 2f detection, and an even higher enhancement by a factor of 3.3 was obtained with direct absorption measurements.

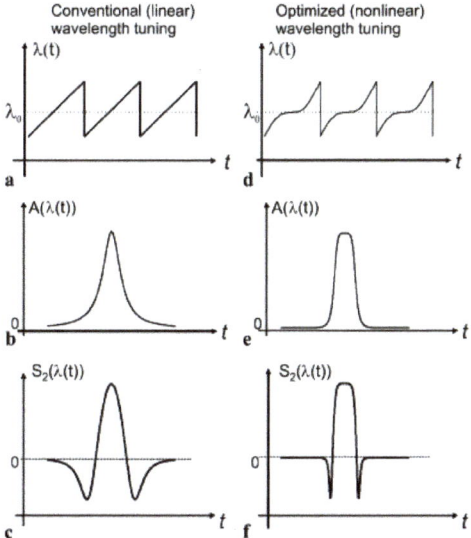

Figure 8. Schematic illustration of the wavelength scanning function: conventional (linear) (**a**) and optimized (nonlinear) (**d**). The resulting spectrum waveform is shown for direct spectroscopy in (**b**) for linear scanning and (**e**) for nonlinear scanning as well as for WMS (second harmonic detection) in (**c**,**f**). Figure adapted from [94].

When multi-species detection is attempted by using closely spaced absorption lines accessible with a single laser scan, the different widths and line profiles of the target species may impose significantly different optimization targets according to the optimal modulation depth criteria stated above. Du et al. demonstrated an *in situ*, multi-parameter LAS sensor during a selective catalytic reduction (SCR) process by accessing four H_2O spectral transitions and a group of NH_3 lines with a single diode laser [95]. As can be seen in Figure 9, WMS with varied modulation amplitude (WMS-VMA) and an optimized multispectral fitting algorithm was used to satisfy the optimization targets for both the H_2O and the NH_3 lines.

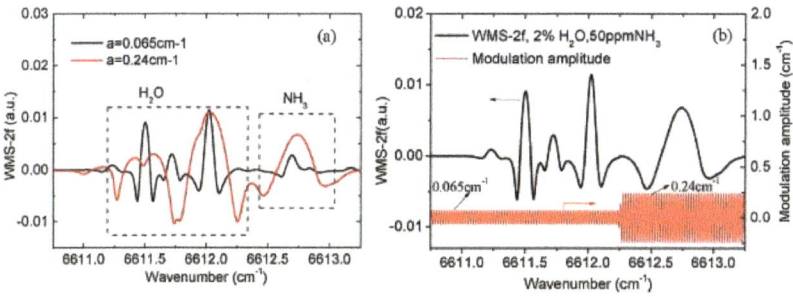

Figure 9. Simulated WMS-2*f* signal with (**a**) modulation amplitude 0.065 cm^{-1} and 0.24 cm^{-1}, respectively; (**b**) wavelength modulation spectroscopy with varied modulation amplitude (WMS-VMA). Figure adapted from [95].

WMS-2*f* detection is sensitive to the curvature of the signal. When blending different spectral features becomes more significant, absorption from the wings of adjacent lines would alter the curvature at the target line center, so that maximizing the line center 2*f* magnitude of an individual transition no longer renders the optimum of overall signal. In such cases, the optimal modulation depth a_m may largely deviate from the theoretical value of ~1.1 Δv. Peng et al. proposed a novel procedure for choosing an optimal WMS modulation depth in the presence of spectral interference by defining a new figure of merit to account for the spectral interference [96,97].

As shown in Equations (2) and (3), F and σ are defined to quantify the inverse magnitude of the WMS-2f signal and the sensitivity to spectral interference, respectively:

$$F(a_m) = \frac{S_{2f,\,max}}{S_{2f}(a_m)} \quad (2)$$

$$\sigma(a_m) = \left| \frac{S_{2f/1f,\,intf}(a_m, \overline{\phi}) - S_{2f/1f}(a_m)}{S_{2f/1f}(a_m)} \right| \quad (3)$$

where S_{2f} is the total interference-free WMS-2f signal at the target transition line center, $S_{2f,\,max}$ is the maximum possible interference-free WMS-2f signal, and $S_{2f/1f}$ is the WMS-2f/1f.

Ideally, a_m should be chosen so that both the inverse signal magnitude F and the perturbation from interference σ are simultaneously minimized. Such an a_m does not generally exist, so the strategy is to select $a_{m,\,opt} = \mathrm{argmin}(C)$, where the cost function $C = \sigma F$ is now the target of optimization. The so-obtained optimal modulation depth $a_{m,\,opt}$ is indicated by the black dashed line in Figure 10 at the minimum of C, where the value of F at this modulation depth is 1.5, indicating that 33% of the potential SNR is sacrificed.

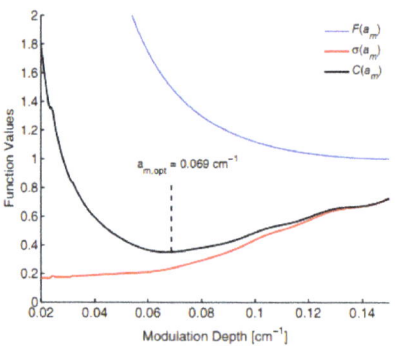

Figure 10. Inverse WMS-2f signal strength (F, blue line), NH$_3$ interference sensitivity (σ, red line), and cost function (C, red line) vs modulation depth for a 3 ppm NH$_3$ mixture in CH$_4$ – air $\phi = 0.6$ combustion exhaust at T = 600 K, P = 1 atm. The optimal modulation depth $a_{m,\,opt}$ is shown in the black/dashed line. Figure adapted from [96].

4.3. Strategies based On Signal Processing Schemes

4.3.1. Baseline Fitting for Blended Absorption Feature

As discussed above, using the AI or LM fitting algorithms to calculate the laser intensity baseline requires proper identification of the absorption-free flanks. However, for spectra with low SNR, distinguishing the non-absorption area using direct visual inspection (DVI) is hard. A new strategy was proposed by Li et al. using wavelet decomposition and iteration to remove the background drift [98]. The application of wavelet transform (WT) for TDLAS signal denoising is based on finding the optimal wavelet pairs to determine the baseline. Figure 11 shows effective nonlinear baseline correction and denoising using discrete wavelet transform (DWT). Compared with the commonly used DVI method, this DWT algorithm demonstrated potential for batch processing of TDLAS spectra. However, system cost and complexity increased.

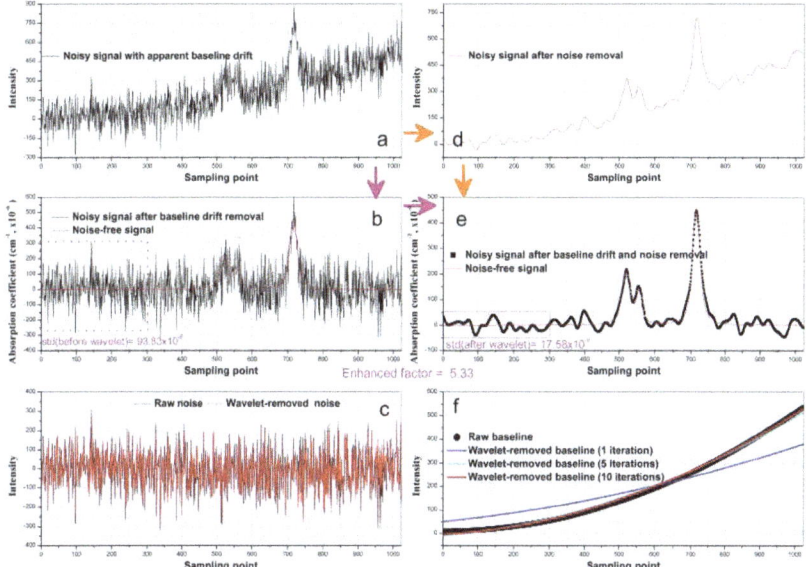

Figure 11. Nonlinear baseline correction and denoising using discrete wavelet transform (DWT). (**a**) Noisy signal with apparent baseline drift; (**b**) baseline drift-removed signal from (**a**) and noise-free signal; (**c**) raw noise and wavelet-removed noise; (**d**) denoised signal from (**a**); (**e**) baseline drift-removed signal from (**d**) and noise-free signal; (**f**) raw baseline and wavelet-removed baselines with different iterations. Figure adapted from [98].

Weisberger et al. have developed a blended-feature baseline fitting method (BFBL) using a more efficient iterative lookup table approach [99]. The baseline is estimated by coupling measured data with simulated fractional transmission at the peaks between absorption features known as baseline anchor points. Figure 12 illustrates the fitting procedure, and this technique was validated against measurements in a static heated cell and a wood-fired two-stage hydronic heater. The results indicated that it can be very useful with multiple overlapping absorption features.

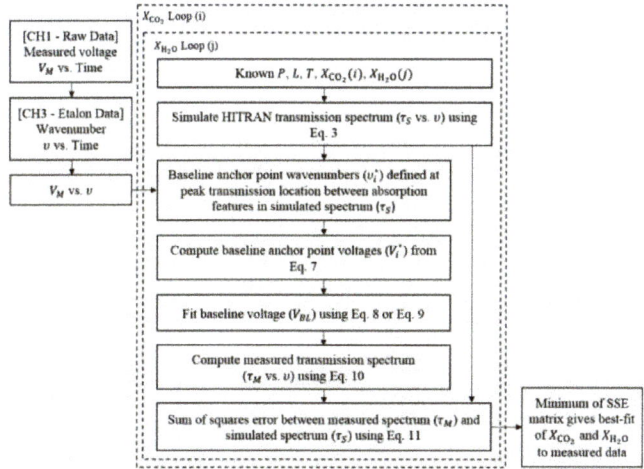

Figure 12. Flow chart for DAS baseline fitting data analysis. Figure adapted from [99].

4.3.2. *In Situ* Parameter Fitting

In Section 3, we have discussed the limitation of calibration-free WMS when it is important to perform accurate assessments of laser characteristics. It is often assumed that the magnitude of wavelength modulation is constant over the entire scanning range [58] and the differential current-to-wavelength tuning behavior is linear for small current variations around a bias point [100,101]. Such assumption may nevertheless introduce different degrees of uncertainty under different circumstances. Zhao et al. presented an improved methodology for assessing the wavelength response of a DFB laser by using a high-order empirical formula [102].

Although the laser characterization process is not difficult by itself, the task of accurately tracking its variations in real time is nontrivial. More recently, Upadhyay et al. have proposed a new calibration-free 2*f*-WMS technique to measure gas concentration and pressure without the need for laser pre-characterization [103–106]. Similar to the idea of DAS baseline fitting, the harmonic backgrounds, i.e., RAMs, are obtained by interpolating the non-absorbing wings of the X and Y components of the demodulated 1*f*, 2*f*, and 3*f* signals (Figure 13). The intensity modulation indices i_n can then be obtained from the RAMs, whereas the intensity–frequency modulation phase shifts φ_n can be obtained by taking the inverse tangent of the ratio of X and Ys.

The *in situ* and real-time characterization of relevant laser parameters ensures that the measurements are not affected by rapid non-absorbing laser intensity variations such as those due to light scattering, beam steering, vibrations, and window fouling or by slow variation effects such as temperature changes, calibration drift, and aging of the devices. However, even if it has been demonstrated that higher-order harmonic signals would have reduced spectral interference from neighboring lines, the requirement for non-absorbing baseline portions for the fitting still puts limitations to the applicable scope of this method.

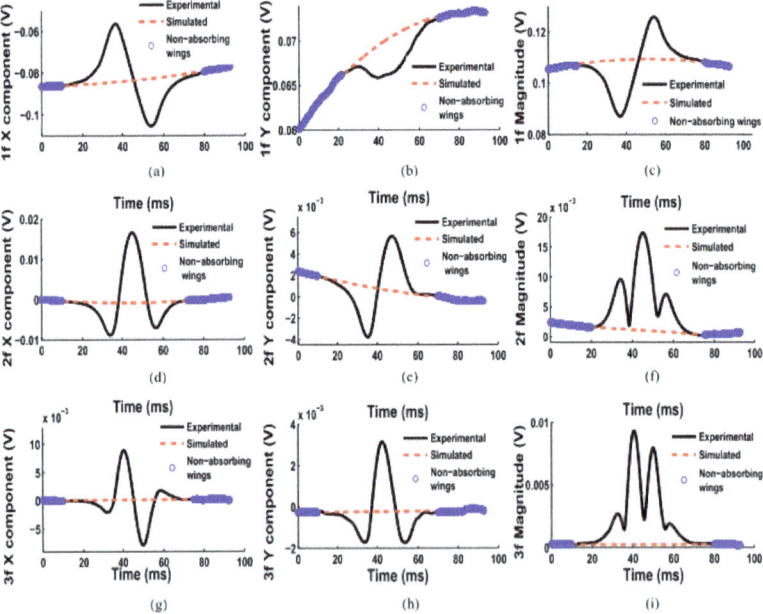

Figure 13. The 1650 nm DFB laser was modulated at m = 2.2, and the transmitted light through a 1% CH$_4$ sample at 1 bar pressure was demodulated by a lock-in amplifier (LIA) to obtain (**a**) 1*f* X-component along *I* H$_1$, (**b**) 1*f* Y-component orthogonal to *I* H$_1$, (**c**) magnitude of 1*f* Signal, (**d**) 2*f* X-component along *I* H$_2$, (**e**) 2*f* Y-component orthogonal to *I* H$_2$, (**f**) magnitude of 2*f* Signal, (**g**) 3*f* X-component along *I* H$_3$, (**h**) 3*f* Y-component orthogonal to *I* H$_3$, and (**i**) magnitude of 3*f* Signal. Figure adapted from [104].

4.3.3. Harmonic Wavelet Analysis of Modulated TDLAS Signals

Calibration-free WMS is essentially a signal processing technique to extract information about molecular absorption from a noisy environment with higher signal fidelity. In view of the unsolved issues arising from potential uncertainties in laser characteristic parameters, spectroscopic constants, as well as requirements for digital filter design in the digital lock-in amplification process [107–109], new signal analysis methodologies have been proposed to better analyze the time-dependent signal harmonics.

Duan et al. have employed signal processing techniques based on wavelet analysis of modulated signals obtained from TDLAS and demonstrated that wavelets have the potential to enable the detection of signal harmonics [110]. In addition to a pure frequency domain analysis, windowed Fourier analysis can provide information on the time dependence of the frequency components but will suffer from substantial inaccuracy when the window width decreases. As an alternative, wavelet analyses can provide more accurate information on this time-dependent evolution of different frequency components and have become popular in recent years.

Figure 14 shows the comparison of wavelet analyses of level 8, 9, and 10 with analytical predictions from a Voigt absorption profile. Each level corresponds to a specific value related to scaling in DWT. As the level increases, the frequency resolution of the wavelet analysis increases. With the shown correlation between the two, it is apparent that wavelet analysis performed well in extracting the 1f and 2f harmonics. This work demonstrated the potential of wavelet analyses in yielding a new methodology for the improvement of the conventional TDLAS system.

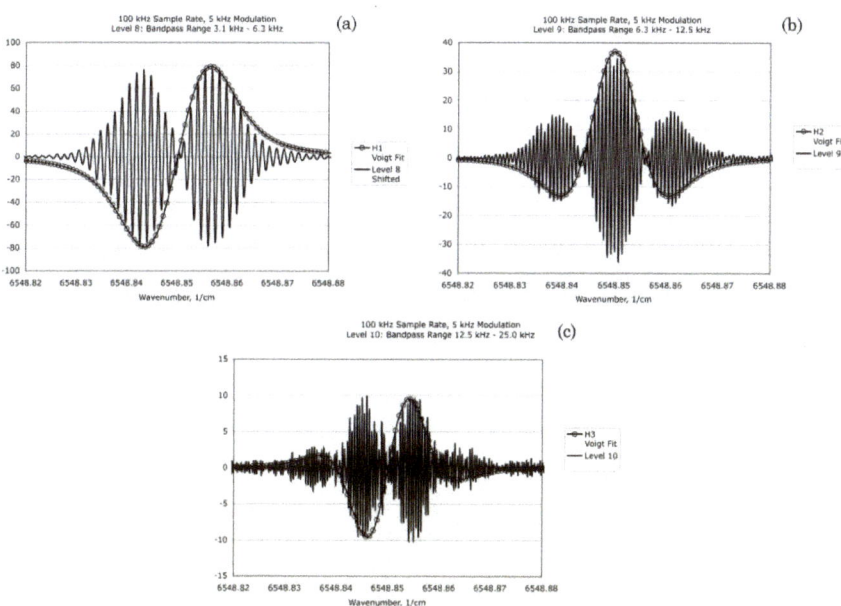

Figure 14. Comparison of results from the wavelet analyses with (**a**) H_1, (**b**) H_2, and (**c**) H_3 coefficients obtained with a Voigt absorption profile. Figure adapted from [110].

5. Conclusions and Future Outlook

Efforts in improving LAS sensor performance suitable for practical environments have never stopped attracting the attention of researchers and developers. By overviewing the achievements of the developed sensors, particularly with sensor deployment in laboratory studies, industrial processes, and engine diagnostics, it can be understood that different measurement applications pose largely different demands in terms of detection limits and bandwidths. In this review, we have attempted to establish a logical development from a system design perspective, for the purpose of better understanding measurement needs and providing guidelines and insights for designing a laser absorption sensing system. Understanding the different sources and processes of noise is the prerequisite for proposing appropriate strategies for SNR amelioration. Under typical measurement conditions, analogue and digital noise from opto-electronic devices, optical noise from the optical system, and excess noise from the environment are the most relevant disturbances that contribute to signal deterioration. In addition to designing a better optical system, strategies of noise reduction and design optimization play important roles in improving a sensor performance and achieving better adaptation to various environments. This can be done through either optimization of laser control and tuning parameters or various algorithms for data processing, such as baseline extraction, *in situ* laser characterization, and wavelet analysis. It is realized that the need for sensing technology development is highly fragmented and demand-driven. While each measurement strategy has its limitations and constraints in face of the challenges from complex environments, much needs to be done from an engineering design perspective and to profoundly comprehend the underlying scientific fundamentals in terms of optics, spectroscopy, and signal processing.

Author Contributions: Z.W. collected references, drafted and revised the paper; P.F. collected references and revised the paper; X.C. supervised the whole work, designed the paper and revised the draft.

Funding: This material is based on work supported by the National Science Foundation of China under Grant No. 51606111 and No. 61627804.

Conflicts of Interest: The author declares no conflict of interest.

Nomenclature

LAS	Laser absorption spectroscopy
DAS	Diode absorption spectroscopy
WMS	Wavelength modulation spectroscopy
TDLAS	Tunable diode laser absorption spectroscopy
LOS	Line-of-sight
SNR	Signal-to-noise ratio
I_t [W/cm^2]	Transmitted light intensity
I_0 [W/cm^2]	Incident light intensity
α_v	Absorbance
k_v [1/cm]	Spectral absorption coefficient
L [cm]	Absorption pathlength
n [molecule/cm^3]	Number density of the absorbing species
σ_v [cm^2/molecule]	Absorption cross-section
S [cm^{-2}/atm]	Absorption linestrength
ϕ_v [cm]	Frequency-dependent lineshape function
P [atm]	Pressure
X_i	Mole fraction of the absorbing species i
v	Light frequency
NEA	Noise equivalent absorbance
DFBs	Distributed feedback lasers
VCSELs	Vertical-cavity surface-emitting lasers
FDML	Fourier-domain mode locked
QCLs	Quantum cascade lasers
ICLs	Interband cascade lasers
LM	Levenberg–Marquardt
AI	Advanced integrative
HITRAN	High Resolution Transmission
HITEMP	High Temperature
R_{2f}	2f peak ratio
T	Temperature

X_0	Nominal concentration
a	Modulation depth
i_1,	1st-order laser intensity modulation indices
i_2	2nd-order laser intensity modulation indices
φ_1	1st-order laser intensity–frequency modulation phase shifts
φ_2	2nd-order laser intensity-frequency modulation phase shifts
ν_0	Specific center frequency
$\nu(t)$	Laser wavelength
$\alpha(t)$	Simulated absorbance
$^M I_0(t)$	No-absorbing intensity
$^S I_t(t)$	Measured transmitted intensity
$^S I_t(t)$	Simulated transmitted intensity
f_m	Modulation frequency
m	Modulation index
$(S_{nf/1f})_M$	Measured 1f-normalized nf signal
$(S_{nf/1f})_s$	Simulated 1f-normalized nf signal
SSE	Sum-of-squared error
RAM	Residual amplitude modulation
X_{nf}	Measured X component of nf signal
Y_{nf}	Measured Y component of nf signal
LAT	Laser absorption tomography
ADC	Analogue-to-digital conversion
RIN	Relative intensity noise
EMD	Empirical mode decomposition
PHR	Peak height ratio
$\Delta\nu$	Full-width-at-half-maximum
FWHM	Full-width-at-half-maximum
SCR	Selective catalytic reduction
WMS-VMA	Wavelength modulation spectroscopy with varied modulation amplitude
a_m	Modulation depth
F	Inverse WMS-2f signal strength
σ	Interference sensitivity
C	Cost function
S_{2f}	Interference-free WMS-2f signal at the target transition line center
$S_{2f,\ max}$	Maximum possible interference-free WMS-2f signal
$a_{m,\ opt}$	Optimal modulation depth
DVI	Direct visual inspection
WT	Wavelet transform
DWT	Discrete wavelet transform
BFBL	Blended-feature baseline fitting method
i_n	n^{th}-order laser intensity modulation indices
φ_n	n^{th}-order laser intensity–frequency modulation phase shifts
LIA	Lock-in amplifier

References

1. Hanson, R.K.; Falcone, P.K. Temperature measurement technique for high temperature gases using a tunable diode laser. *Appl. Opt.* **1978**, *17*, 2477–2480. [CrossRef] [PubMed]
2. Ried, J.; Shewchun, J.; Garside, B.K.; Balik, E.A. High sensitivity pollution detection employing tunable diode lasers. *Appl. Opt.* **1978**, *17*, 300–307. [CrossRef] [PubMed]
3. Druy, M.A. From laboratory technique to process gas sensor: The maturation of tunable diode laser absorption spectroscopy. *Spectroscopy* **2006**, *21*, 14–18.
4. Lackner, M. Tunable diode laser absorption spectroscopy in the process industries: A review. *Rev. Chem. Eng.* **2007**, *23*, 65–147. [CrossRef]

5. Available online: www.oxigraf.com (accessed on 3 July 2019).
6. Wolfrum, J. *Lasers in Combustion: From Basic Theory to Practical Devices*; Elsevier: Amsterdam, The Netherlands, 1998; Volume 27, pp. 1–41.
7. Allen, M.G. Diode laser absorption sensors for gas-dynamic and combustion flows. *Meas. Sci. Technol.* **1998**, *9*, 545–562. [CrossRef] [PubMed]
8. Werle, P. A review of recent advances in semiconductor laser based gas monitors. *Spectrochim. Acta Part A Mol. Biomol. Spectrosc.* **1998**, *54*, 197–236. [CrossRef]
9. Nasim, H.; Jamil, Y. Recent advancements in spectroscopy using tunable diode lasers. *Laser Phys. Lett.* **2013**, *10*, 043001. [CrossRef]
10. Hanson, R.K. Applications of quantitative laser sensors to kinetics, propulsion and practical energy systems. *Proc. Combust. Inst.* **2011**, *33*, 1–40. [CrossRef]
11. Bolshov, M.A.; Kuritsyn, Y.A.; Romanovskii, Y.V. Tunable diode laser spectroscopy as a technique for combustion diagnostics. *Spectrochim. Acta Part B Atomic Spectrosc.* **2015**, *106*, 45–66. [CrossRef]
12. Goldenstein, C.S.; Spearrin, R.M.; Jeffries, J.B.; Hanson, R.K. Infrared laser-absorption sensing for combustion gases. *Prog. Energy Combust. Sci.* **2017**, *60*, 132–176. [CrossRef]
13. Liu, C.; Xu, L. Laser absorption spectroscopy for combustion diagnosis in reactive flows: A review. *Appl. Spectrosc. Rev.* **2018**, *54*, 1–44. [CrossRef]
14. Li, J.; Yu, B.; Zhao, W.; Chen, W. A review of signal enhancement and noise reduction techniques for tunable diode laser absorption spectroscopy. *Appl. Spectrosc. Rev.* **2014**, *49*, 666–691. [CrossRef]
15. Zhang, T.; Kang, J.; Meng, D.; Wang, H.; Mu, Z.; Zhou, M.; Chen, C. Mathematical Methods and Algorithms for Improving Near-Infrared Tunable Diode-Laser Absorption Spectroscopy. *Sensors* **2018**, *18*, 4295. [CrossRef] [PubMed]
16. Chao, X.; Jeffries, J.B.; Hanson R, K. In situ absorption sensor for NO in combustion gases with a 5.2 µm quantum-cascade laser. *Proc. Combust. Inst.* **2011**, *33*, 725–733. [CrossRef]
17. Wagner, S.; Fisher, B.T.; Fleming, J.W.; Ebert, V. TDLAS-based in situ measurement of absolute acetylene concentrations in laminar 2D diffusion flames. *Proc. Combust. Inst.* **2009**, *32*, 839–846. [CrossRef]
18. Wagner, S.; Klein, M.; Kathrotia, T.; Riedel, U.; Dreizler, A.; Ebert, V. In situ TDLAS measurement of absolute acetylene concentration profiles in a non-premixed laminar counter-flow flame. *Appl. Phys. B* **2012**, *107*, 585–589. [CrossRef]
19. Nau, P.; Koppmann, J.; Lackner, A.; Kohse-Höinghaus, K.; Brockhinke, A. Quantum cascade laser-based MIR spectrometer for the determination of CO and CO_2 concentrations and temperature in flames. *Appl. Phys. B* **2015**, *118*, 361–368. [CrossRef]
20. Wagner, S.; Klein, M.; Kathrotia, T.; Riedel, U.; Kissel, T.; Dreizler, A.; Ebert, V. Absolute, spatially resolved, in situ CO profiles in atmospheric laminar counter-flow diffusion flames using 2.3 µm TDLAS. *Appl. Phys. B* **2012**, *109*, 533–540. [CrossRef]
21. Wunderle, K.; Wagner, S.; Pasti, I.; Pieruschka, R.; Rascher, U.; Schurr, U.; Ebert, V. Distributed feedback diode laser spectrometer at 2.7 µm for sensitive, spatially resolved H_2O vapor detection. *Appl. Opt.* **2009**, *48*, B172–B182. [CrossRef]
22. Seidel, A.; Wagner, S.; Dreizler, A.; Ebert, V. Robust, spatially scanning, open-path TDLAS hygrometer using retro-reflective foils for fast tomographic 2-D water vapor concentration field measurements. *Atmos. Meas. Tech.* **2015**, *8*, 2061–2068. [CrossRef]
23. Vanderover, J.; Wang, W.; Oehlschlaeger, M.A. A carbon monoxide and thermometry sensor based on mid-IR quantum-cascade laser wavelength-modulation absorption spectroscopy. *Appl. Phys. B* **2011**, *103*, 959–966. [CrossRef]
24. Stranic, I.; Hanson, R.K. Laser absorption diagnostic for measuring acetylene concentrations in shock tubes. *J. Quant. Spectrosc. Radiat. Transf.* **2014**, *142*, 58–65. [CrossRef]
25. Utsav, K.C.; Nasir, E.F.; Farooq, A. A mid-infrared absorption diagnostic for acetylene detection. *Appl. Phys. B* **2015**, *120*, 223–232.
26. Spearrin, R.M.; Li, S.; Davidson, D.F.; Jeffries, J.B.; Hanson, R.K. High-temperature iso-butene absorption diagnostic for shock tube kinetics using a pulsed quantum cascade laser near 11.3 µm. *Proc. Combust. Inst.* **2015**, *35*, 3645–3651. [CrossRef]

27. Sun, K.; Wang, S.; Sur, R.; Chao, X.; Jeffries, J.B.; Hanson, R.K. Sensitive and rapid laser diagnostic for shock tube kinetics studies using cavity-enhanced absorption spectroscopy. *Opt. Express* **2014**, *22*, 9291–9300. [CrossRef]
28. Sun, K.; Wang, S.; Sur, R.; Chao, X.; Jeffries, J.B.; Hanson, R.K. Time-resolved in situ detection of CO in a shock tube using cavity-enhanced absorption spectroscopy with a quantum-cascade laser near 4.6 µm. *Opt. Express* **2014**, *22*, 24559–24565. [CrossRef]
29. Ebert, V.; Fitzer, J.; Gerstenberg, I.; Pleban, K.U.; Pitz, H.; Wolfrum, J.; Jochem, M.; Martin, J. Simultaneous laser-based in situ detection of oxygen and water in a waste incinerator for active combustion control purposes. In Proceedings of the Symposium (International) on Combustion, Boulder, Colorado, USA, 2–7 August 1998.
30. Teichert, H.; Fernholz, T.; Ebert, V. Simultaneous in situ measurement of CO, H_2O, and gas temperatures in a full-sized coal-fired power plant by near-infrared diode lasers. *Appl. Opt.* **2003**, *42*, 2043–2051. [CrossRef] [PubMed]
31. Ebert, V.; Teichert, H.; Strauch, P.; Kolb, T.; Seifert, H.; Wolfrum, J. Sensitive in situ detection of CO and O_2 in a rotary kiln-based hazardous waste incinerator using 760 nm and new 2.3 µm diode lasers. *Proc. Combust. Inst.* **2005**, *30*, 1611–1618. [CrossRef]
32. Sun, K.; Sur, R.; Jeffries, J.B.; Hanson, R.K.; Clark, T.; Anthony, J.; Northington, J. Application of wavelength-scanned wavelength-modulation spectroscopy H_2O absorption measurements in an engineering-scale high-pressure coal gasifier. *Appl. Phys. B* **2014**, *117*, 411–421. [CrossRef]
33. Nau, P.; Kutne, P.; Eckel, G.; Meier, W.; Hotz, C.; Fleck, S. Infrared absorption spectrometer for the determination of temperature and species profiles in an entrained flow gasifier. *Appl. Opt.* **2017**, *56*, 2982–2990. [CrossRef]
34. Witzel, O.; Klein, A.; Wagner, S.; Meffert, C.; Schulz, C.; Ebert, V. High-speed tunable diode laser absorption spectroscopy for sampling-free in-cylinder water vapor concentration measurements in an optical IC engine. *Appl. Phys. B* **2012**, *109*, 521–532. [CrossRef]
35. Witzel, O.; Klein, A.; Meffert, C.; Wagner, S.; Kaiser, S.; Schulz, C.; Ebert, V. VCSEL-based, high-speed, in situ TDLAS for in-cylinder water vapor measurements in IC engines. *Opt. Express* **2013**, *21*, 19951–19965. [CrossRef] [PubMed]
36. Witzel, O.; Klein, A.; Meffert, C.; Schulz, C.; Kaiser, S.A.; Ebert, V. Calibration-free, high-speed, in-cylinder laser absorption sensor for cycle-resolved, absolute H_2O measurements in a production IC engine. *Proc. Combust. Inst.* **2015**, *35*, 3653–3661. [CrossRef]
37. Jatana, G.S.; Magee, M.; Fain, D.; Naik, S.V.; Shaver, G.M.; & Lucht, R.P. Simultaneous high-speed gas property measurements at the exhaust gas recirculation cooler exit and at the turbocharger inlet of a multicylinder diesel engine using diode-laser-absorption spectroscopy. *Appl. Opt.* **2015**, *54*, 1220–1231. [CrossRef] [PubMed]
38. Spearrin, R.M.; Goldenstein, C.S.; Schultz, I.A.; Jeffries, J.B.; Hanson, R.K. Simultaneous sensing of temperature, CO, and CO_2 in a scramjet combustor using quantum cascade laser absorption spectroscopy. *Appl. Phys. B* **2014**, *117*, 689–698. [CrossRef]
39. Schultz, I.A.; Goldenstein, C.S.; Strand, C.L.; Jeffries, J.B.; Hanson, R.K. Hypersonic scramjet testing via TDLAS measurements of temperature and column density in a reflected shock tunnel. In Proceedings of the 52nd Aerospace Sciences Meeting, National Harbor, MD, USA, 13–17 January 2014.
40. Caswell, A.W.; Roy, S.; An, X.; Sanders, S.T.; Schauer, F.R.; Gord, J.R. Measurements of multiple gas parameters in a pulsed-detonation combustor using time-division-multiplexed Fourier-domain mode-locked lasers. *Appl. Opt.* **2013**, *52*, 2893–2904. [CrossRef] [PubMed]
41. Caswell, A.; Roy, S.; An, X.; Sanders, S.; Hoke, J.; Schauer, F.; Gord, J. High-bandwidth H_2O absorption sensor for measuring pressure, enthalpy, and mass flux in a pulsed-detonation combustor. In Proceedings of the 50th AIAA Aerospace Sciences Meeting including the New Horizons Forum and Aerospace Exposition, Nashville, TN, USA, 9–12 January 2012.
42. Goldenstein, C.S.; Spearrin, R.M.; Jeffries, J.B.; Hanson, R.K. Infrared laser absorption sensors for multiple performance parameters in a detonation combustor. *Proc. Combust. Inst.* **2015**, *35*, 3739–3747. [CrossRef]
43. Goldenstein, C.S.; Schultz, I.A.; Spearrin, R.M.; Jeffries, J.B.; Hanson, R.K. Diode Laser Measurements of Temperature and H_2O for Monitoring Pulse Detonation Combustor Performance. In Proceedings of the ICDERS Conference, Taipei, Taiwan, 28 July–2 August 2013.

44. Spearrin, R.M.; Goldenstein, C.S.; Jeffries, J.B.; Hanson, R.K. Fiber-coupled 2.7 µm laser absorption sensor for CO_2 in harsh combustion environments. *Meas. Sci. Technol.* **2013**, *24*, 055107. [CrossRef]
45. Spearrin, R.M.; Goldenstein, C.S.; Jeffries, J.B.; Hanson, R.K. Quantum cascade laser absorption sensor for carbon monoxide in high-pressure gases using wavelength modulation spectroscopy. *Appl. Opt.* **2014**, *53*, 1938–1946. [CrossRef]
46. Goldenstein, C.S.; Almodóvar, C.A.; Jeffries, J.B.; Hanson, R.K.; Brophy, C.M. High-bandwidth scanned-wavelength-modulation spectroscopy sensors for temperature and H_2O in a rotating detonation engine. *Meas. Sci. Technol.* **2014**, *25*, 105104. [CrossRef]
47. Gordon, I.E.; Rothman, L.S.; Hill, C.; Kochanov, R.V.; Tan, Y.; Bernath, P.F.; Birk, M.; Boudon, V.; Campargue, A.; Chance, K.V.; et al. The HITRAN2016 molecular spectroscopic database. *J. Quant. Spectrosc. Radiat. Transf.* **2017**, *203*, 3–69. [CrossRef]
48. Chao, X. *Development of Laser Absorption Sensors for Combustion Gases*; Stanford University: Stanford, CA, USA, 2012.
49. Sanders, S.T.; Wang, J.; Jeffries, J.B.; Hanson, R.K. Diode-laser absorption sensor for line-of-sight gas temperature distributions. *Appl. Opt.* **2001**, *40*, 4404–4415. [CrossRef] [PubMed]
50. Liu, X.; Jeffries, J.B.; Hanson R, K. Measurements of spectral parameters of water-vapour transitions near 1388 and 1345 nm for accurate simulation of high-pressure absorption spectra. *Meas. Sci. Technol.* **2007**, *18*, 1185. [CrossRef]
51. Li, H.; Farooq, A.; Jeffries, J.B.; Hanson, R.K. Diode laser measurements of temperature-dependent collisional-narrowing and broadening parameters of Ar-perturbed H_2O transitions at 1391.7 and 1397.8 nm. *J. Quant. Spectrosc. Radiat. Transf.* **2008**, *109*, 132–143. [CrossRef]
52. Skrotzki, J.; Habig, J.C.; Ebert, V. Integrative fitting of absorption line profiles with high accuracy, robustness, and speed. *Appl. Phys. B* **2014**, *116*, 393–406. [CrossRef]
53. Abrarov, S.M.; Quine, B.M.; Jagpal R, K. A simple interpolating algorithm for the rapid and accurate calculation of the Voigt function. *J. Quant. Spectrosc. Radiat. Transf.* **2009**, *110*, 376–383. [CrossRef]
54. Pagnini, G.; Mainardi, F. Evolution equations for the probabilistic generalization of the Voigt profile function. *J. Comput. Appl. Math.* **2010**, *233*, 1590–1595. [CrossRef]
55. Rieker, G.B.; Jeffries, J.B.; Hanson R, K. Calibration-free wavelength-modulation spectroscopy for measurements of gas temperature and concentration in harsh environments. *Appl. Opt.* **2009**, *48*, 5546–5560. [CrossRef]
56. Liu, J.T.C. *Near-Infrared Diode Laser Absorption Diagnostics for Temperature and Species in Engines*; Stanford University: Stanford, CA, USA, 2004.
57. Zhou, X. *Diode Laser Absorption Sensors for Combustion Control*; Stanford University: Stanford, CA, USA, 2005.
58. Sun, K.; Chao, X.; Sur, R.; Goldenstein, C.S.; Jeffries, J.B.; Hanson, R.K. Analysis of calibration-free wavelength-scanned wavelength modulation spectroscopy for practical gas sensing using tunable diode lasers. *Meas. Sci. Technol.* **2013**, *24*, 125203. [CrossRef]
59. Goldenstein, C.S.; Strand, C.L.; Schultz, I.A.; Sun, K.; Jeffries, J.B.; Hanson, R.K. Fitting of calibration-free scanned-wavelength-modulation spectroscopy spectra for determination of gas properties and absorption lineshapes. *Appl. Opt.* **2014**, *53*, 356–367. [CrossRef]
60. Stewart, G.; Johnstone, W.; Bain, J.R.P.; Ruxton, K.; Duffin, K. Recovery of absolute gas absorption line shapes using tunable diode laser spectroscopy with wavelength modulation—Part I: Theoretical analysis. *J. Lightwave Technol.* **2011**, *29*, 811–821.
61. Bain, J.R.P.; Johnstone, W.; Ruxton, K.; Stewart, G.; Lengden, M.; Duffin, K. Recovery of absolute gas absorption line shapes using tunable diode laser spectroscopy with wavelength modulation—Part 2: Experimental investigation. *J. Lightwave Technol.* **2011**, *29*, 987–996. [CrossRef]
62. Peng, Z.; Ding, Y.; Che, L.; Yang, Q. Odd harmonics with wavelength modulation spectroscopy for recovering gas absorbance shape. *Opt. Express* **2012**, *20*, 11976–11985.
63. Klein, A.; Witzel, O.; Ebert, V. Rapid, time-division multiplexed, direct absorption-and wavelength modulation-spectroscopy. *Sensors* **2014**, *14*, 21497–21513. [CrossRef] [PubMed]
64. Liu, X.; Jeffries, J.B.; Hanson R, K. Measurement of non-uniform temperature distributions using line-of-sight absorption spectroscopy. *AIAA J.* **2007**, *45*, 411–419. [CrossRef]
65. Cai, W.; Kaminski C, F. Tomographic absorption spectroscopy for the study of gas dynamics and reactive flows. *Prog. Energy Combust. Sci.* **2017**, *59*, 1–31. [CrossRef]

66. Rieker, G.B.; Giorgetta, F.R.; Swann, W.C.; Kofler, J.; Zolot, A.M.; Sinclair, L.C.; Baumann, E.; Cromer, C.; Petron, G.; Tans, P.P.; et al. Frequency-comb-based remote sensing of greenhouse gases over kilometer air paths. *Optica* **2014**, *1*, 290–298. [CrossRef]
67. Malarich, N.A.; Rieker G, B. Resolving gas temperature distributions with single-beam dual-comb absorption spectroscopy. In Proceedings of the Conference on Lasers and Electro-Optics (CLEO), San Jose, CA, USA, 14–19 May 2017.
68. Ma, L.; Cai, W.; Caswell, A.W.; Kraetschmer, T.; Sanders, S.T.; Roy, S.; Gord, J.R. Tomographic imaging of temperature and chemical species based on hyperspectral absorption spectroscopy. *Opt. Express* **2009**, *17*, 8602–8613. [CrossRef] [PubMed]
69. Ma, L.; Li, X.; Sanders, S.T.; Caswell, A.W.; Roy, S.; Plemmons, D.H.; Gord, J.R. 50-kHz-rate 2D imaging of temperature and H_2O concentration at the exhaust plane of a J85 engine using hyperspectral tomography. *Opt. Express* **2013**, *21*, 1152–1162. [CrossRef] [PubMed]
70. Mohammad, I.L.; Anderson, G.T.; Chen, Y. Noise estimation technique to reduce the effects of 1/f noise in Open Path Tunable Diode Laser Absorption Spectrometry (OP-TDLAS). In Proceedings of the Sensors for Extreme Harsh Environments. International Society for Optics and Photonics, Baltimore, MD, USA, 5–9 May 2014.
71. Werle, P.W.; Mazzinghi, P.; D'Amato, F.; De Rosa, M.; Maurer, K.; Slemr, F. Signal processing and calibration procedures for in situ diode-laser absorption spectroscopy. *Spectrochim. Acta Part A Mol. Biomol. Spectrosc.* **2004**, *60*, 1685–1705. [CrossRef]
72. Lins, B.; Zinn, P.; Engelbrecht, R.; Schmauss, B. Simulation-based comparison of noise effects in wavelength modulation spectroscopy and direct absorption TDLAS. *Appl. Phys. B* **2010**, *100*, 367–376. [CrossRef]
73. Zhu, J.; Cowie, A.; Kosterev, A.; Wyatt, D. Advantages of Using Direct Absorption Method for TDLAS Measurement. Available online: https://zh.scribd.com/document/234114018/Paper-Advantages-of-Using-Direct-Absorption-Method-for-TDLAS-Measurement-r1 (accessed on 3 July 2019).
74. Hodgkinson, J.; Tatam, R.P. Optical gas sensing: A review. *Meas. Sci. Technol.* **2012**, *24*, 012004. [CrossRef]
75. Hodgkinson, J.; Masiyano, D.; Tatam, R.P. Gas cells for tunable diode laser absorption spectroscopy employing optical diffusers. Part 1: Single and dual pass cells. *Appl. Phys. B* **2010**, *100*, 291–302. [CrossRef]
76. Ehlers, P.; Johansson, A.C.; Silander, I.; Foltynowicz, A.; Axner, O. Use of etalon-immune distances to reduce the influence of background signals in frequency-modulation spectroscopy and noise-immune cavity-enhanced optical heterodyne molecular spectroscopy. *JOSA B* **2014**, *31*, 2938–2945. [CrossRef]
77. Bomse, D.S.; Stanton, A.C.; Silver, J.A. Frequency modulation and wavelength modulation spectroscopies: Comparison of experimental methods using a lead-salt diode laser. *Appl. Opt.* **1992**, *31*, 718–731. [CrossRef] [PubMed]
78. Chou, S.I.; Baer, D.S.; Hanson, R.K. Diode laser absorption measurements of CH_3Cl and CH_4 near 1.65 µm. *Appl. Opt.* **1997**, *36*, 3288–3293. [CrossRef] [PubMed]
79. Webster, C.R. Brewster-plate spoiler: A novel method for reducing the amplitude of interference fringes that limit tunable-laser absorption sensitivities. *JOSA B* **1985**, *2*, 1464–1470. [CrossRef]
80. Masiyano, D.; Hodgkinson, J.; Schilt, S.; Tatam, R.P. Self-mixing interference effects in tunable diode laser absorption spectroscopy. *Appl. Phys. B* **2009**, *96*, 863–874. [CrossRef]
81. Persson, L.; Andersson, F.; Andersson, M.; Svanberg, S. Approach to optical interference fringes reduction in diode laser absorption spectroscopy. *Appl. Phys. B* **2007**, *87*, 523–530. [CrossRef]
82. Kluczynski, P.; Lindberg, Å.M.; Axner, O. Characterization of background signals in wavelength-modulation spectrometry in terms of a Fourier based theoretical formalism. *Appl. Opt.* **2001**, *40*, 770–782. [CrossRef]
83. Xiong, B.; Du, Z.; Li, J. Modulation index optimization for optical fringe suppression in wavelength modulation spectroscopy. *Rev. Sci. Instrum.* **2015**, *86*, 113104. [CrossRef] [PubMed]
84. Sun, K.; Chao, X.; Sur, R.; Jeffries, J.B.; Hanson, R.K. Wavelength modulation diode laser absorption spectroscopy for high-pressure gas sensing. *Appl. Phys. B* **2013**, *110*, 497–508. [CrossRef]
85. Xia, H.; Dong, F.; Zhang, Z.; Tu, G.; Pang, T.; Wu, B.; Wang, Y. Signal analytical processing based on wavelet transform for tunable diode laser absorption spectroscopy. In Proceedings of the Advanced Sensor Systems and Applications IV. International Society for Optics and Photonics, Beijing, China, 2 November 2010.
86. Li, C.; Guo, X.; Ji, W.; Wei, J.; Qiu, X.; Ma, W. on fringe removal of tunable diode laser multi-pass spectroscopy by wavelet transforms. *Opt. Quant. Electron.* **2018**, *50*, 275. [CrossRef]

87. Niu, M.; Han, P.; Song, L.; Hao, D.; Zhang, J.; Ma, L. Comparison and application of wavelet transform and Kalman filtering for denoising in δ 13 CO_2 measurement by tunable diode laser absorption spectroscopy at 2.008 µm. *Opt. Express* **2017**, *25*, A896–A905. [CrossRef] [PubMed]
88. Tian, G.; Li, J. Tunable diode laser spectrometry signal de-noising using discrete wavelet transform for molecular spectroscopy study. *Opt. Appl.* **2013**, *43*. [CrossRef]
89. Yang, R.; Bi, Y.; Zhou, Q.; Dong, X.; Lv, T. A background reduction method based on empirical mode decomposition for tunable diode laser absorption spectroscopy system. *Optik* **2018**, *158*, 416–423. [CrossRef]
90. Uehara, K. Dependence of harmonic signals on sample-gas parameters in wavelength-modulation spectroscopy for precise absorption measurements. *Appl. Phys. B Lasers Opt.* **1998**, *67*, 517–523. [CrossRef]
91. Kluczynski, P.; Axner, O. Theoretical description based on Fourier analysis of wavelength-modulation spectrometry in terms of analytical and background signals. *Appl. Opt.* **1999**, *38*, 5803–5815. [CrossRef]
92. Neethu, S.; Verma, R.; Kamble, S.S.; Radhakrishnan, J.K.; Krishnapur, P.P.; Padaki, V.C. Validation of wavelength modulation spectroscopy techniques for oxygen concentration measurement. *Sens. Actuators B Chem.* **2014**, *192*, 70–76. [CrossRef]
93. Fried, A.; Henry, B.; Drummond, J.R. Tunable diode laser ratio measurements of atmospheric constituents by employing dual fitting analysis and jump scanning. *Appl. Opt.* **1993**, *32*, 821–827. [CrossRef]
94. Chen, J.; Hangauer, A.; Strzoda, R.; Amann, M.C. Tunable diode laser spectroscopy with optimum wavelength scanning. *Appl. Phys. B* **2010**, *100*, 331–339. [CrossRef]
95. Du, Z.; Yan, Y.; Li, J.; Zhang, S.; Yang, X.; Xiao, Y. In situ, multiparameter optical sensor for monitoring the selective catalytic reduction process of diesel engines. *Sens. Actuators B Chem.* **2018**, *267*, 255–264. [CrossRef]
96. Peng, W.Y.; Sur, R.; Strand, C.L.; Spearrin, R.M.; Jeffries, J.B.; Hanson, R.K. High-sensitivity in situ QCLAS-based ammonia concentration sensor for high-temperature applications. *Appl. Phys. B* **2016**, *122*, 188. [CrossRef]
97. Sur, R.; Peng, W.Y.; Strand, C.; Spearrin, R.M.; Jeffries, J.B.; Hanson, R.K.; Bekal, A.; Halder, P.; Poonacha, S.P.; Vartak, S.; et al. Mid-infrared laser absorption spectroscopy of NO2 at elevated temperatures. *J. Quant. Spectrosc. Radiat. Transf.* **2017**, *187*, 364–374. [CrossRef]
98. Li, J.; Yu, B.; Fischer, H. Wavelet transform based on the optimal wavelet pairs for tunable diode laser absorption spectroscopy signal processing. *Appl. Spectrosc.* **2015**, *69*, 496–506. [CrossRef] [PubMed]
99. Weisberger, J.M.; Richter, J.P.; Parker, R.A.; DesJardin, P.E. Direct absorption spectroscopy baseline fitting for blended absorption features. *Appl. Opt.* **2018**, *57*, 9086–9095. [CrossRef]
100. Chen, J.; Hangauer, A.; Strzoda, R.; Amann, M.C. VCSEL-based calibration-free carbon monoxide sensor at 2.3 µm with in-line reference cell. *Appl. Phys. B* **2011**, *102*, 381–389. [CrossRef]
101. Qu, Z.; Ghorbani, R.; Valiev, D.; Schmidt, F.M. Calibration-free scanned wavelength modulation spectroscopy–application to H_2O and temperature sensing in flames. *Opt. Express* **2015**, *23*, 16492–16499. [CrossRef]
102. Zhao, G.; Tan, W.; Hou, J.; Qiu, X.; Ma, W.; Li, Z.; Axner, O. Calibration-free wavelength-modulation spectroscopy based on a swiftly determined wavelength-modulation frequency response function of a DFB laser. *Opt. Express* **2016**, *24*, 1723–1733. [CrossRef]
103. Upadhyay, A.; Lengden, M.; Wilson, D.; Humphries, G.S.; Crayford, A.P.; Pugh, D.G.; Johnstone, W. A new RAM normalized 1f-WMS technique for the measurement of gas parameters in harsh environments and a comparison with 2f/1f. *IEEE Photonics J.* **2018**, *10*, 1–11. [CrossRef]
104. Upadhyay, A.; Wilson, D.; Lengden, M.; Chakraborty, A.L.; Stewart, G.; Johnstone, W. Demonstration of Calibration-Free WMS Measurement of Gas Parameters with in-Situ Real-Time Characterization of Laser Parameters Using CW-DFB-QCL, VCSEL and DFB Lasers. 2016. Available online: https://core.ac.uk/download/pdf/77036446.pdf (accessed on 3 July 2019).
105. Upadhyay, A.; Wilson, D.; Lengden, M.; Chakraborty, A.L.; Stewart, G.; Johnstone, W. Calibration-free WMS using a cw-DFB-QCL, a VCSEL, and an edge-emitting DFB laser with in-situ real-time laser parameter characterization. *IEEE Photonics J.* **2017**, *9*, 1–17. [CrossRef]
106. Upadhyay, A.; Chakraborty, A.L. Calibration-free 2f WMS with in situ real-time laser characterization and 2f RAM nulling. *Opt. Lett.* **2015**, *40*, 4086–4089. [CrossRef] [PubMed]
107. Zhai, B.; He, Q.; Huang, J.; Zheng, C.; Wang, Y. Design and realization of harmonic signal orthogonal lock-in amplifier used in infrared gas detection. *Acta Photonica Sin.* **2014**, *43*, 1125001. [CrossRef]

108. Jiang, C.; Liu, Y.; Yu, B.; Yin, S.; Chen, P. TDLAS-WMS second harmonic detection based on spectral analysis. *Rev. Sci. Instrum.* **2018**, *89*, 083106. [CrossRef] [PubMed]
109. Du, Y.; Peng, Z.; Ding, Y. Wavelength modulation spectroscopy for recovering absolute absorbance. *Opt. Express* **2018**, *26*, 9263–9272. [CrossRef] [PubMed]
110. Duan, H.; Gautam, A.; Shaw, B.D.; Cheng, H. Harmonic wavelet analysis of modulated tunable diode laser absorption spectroscopy signals. *Appl. Opt.* **2009**, *48*, 401–407. [CrossRef] [PubMed]

© 2019 by the authors. Licensee MDPI, Basel, Switzerland. This article is an open access article distributed under the terms and conditions of the Creative Commons Attribution (CC BY) license (http://creativecommons.org/licenses/by/4.0/).

Review

Recent Developments in Modulation Spectroscopy for Methane Detection Based on Tunable Diode Laser

Fei Wang, Shuhai Jia *, Yonglin Wang and Zhenhua Tang

Department of Mechanical Engineering, Xi'an Jiaotong University, Xi'an 710049, China
* Correspondence: shjia@mail.xjtu.edu.cn; Tel.: +86-131-5206-8353

Received: 30 April 2019; Accepted: 11 July 2019; Published: 15 July 2019

Abstract: In this review, methane absorption characteristics mainly in the near-infrared region and typical types of currently available semiconductor lasers are described. Wavelength modulation spectroscopy (WMS), frequency modulation spectroscopy (FMS), and two-tone frequency modulation spectroscopy (TTFMS), as major techniques in modulation spectroscopy, are presented in combination with the application of methane detection.

Keywords: methane; tunable diode laser; wavelength modulation spectroscopy; frequency modulation spectroscopy; two-tone frequency modulation spectroscopy

1. Introduction

Due to global warming and climate change, the monitoring and detection of atmospheric gas concentration has come to be of great value. Although the average background level of methane (CH_4) (~1.89 ppm) in the earth's atmosphere is roughly 200 times lower than that of CO_2 (~400 ppm), the contribution of CH_4 to the greenhouse effect per mole is 25 times larger than CO_2 [1,2]. Therefore, a fast, accurate, and precise monitoring of trace greenhouse gas of CH_4 is essential. There are some methods for detecting methane, including chemical processes [3–6] and optical spectroscopy [7–9]. Optical spectroscopy for detecting gases is based on the Beer-Lambert law [10–12], in which the light attenuation is related to the effective length of the sample in an absorbing medium, and to the concentration of absorbing species, respectively. By this theory, the emission wavelength of the narrow-linewidth diode laser is scanned over the target gas absorption line, and tunable diode laser absorption spectroscopy (TDLAS) has become an effective technique for the rapid and online analysis of gas component concentration, due to the advantage of high spectrum resolution [13–22]. Direct detection and wavelength modulation spectroscopy are the most common sensing methods of TDLAS [23,24]. Comparatively speaking, wavelength/frequency modulation spectroscopy is less vulnerable to the effects of background noise and more suitable for detecting trace gases. Moreover, modulation spectroscopy is widely used for the detection of various gases with the advantage of high signal-to-noise ratio (SNR) [25,26]. Wavelength modulation spectroscopy (WMS), frequency modulation spectroscopy (FMS), and two-tone frequency modulation spectroscopy (TTFMS) are the main techniques in modulation absorption spectroscopy. Since each technique has its strengths and weaknesses, they have also been applied for detecting methane depending on the situation.

In this review, after introducing methane absorption characteristics and recent progress of tunable diode lasers (TDLs), recent advances in methane detection using modulation spectroscopy are presented.

2. Methane Absorption Lines

The CH_4 molecule has a spherical top, and belongs to the tetrahedral point family. It exhibits four fundamental vibration modes: v_1 = 2913 cm^{-1}, v_2 = 1533.3 cm^{-1}, v_3 = 3018.9 cm^{-1} and

$v_4 = 1305.9$ cm^{-1} [27]. Of these, the two bending vibrations are v_2 (asymmetric) and v_4 (symmetric), while v_1 (symmetric) and v_3 (asymmetric) are the two stretching vibrations [28]. The successive resonance spacing is about 1500 cm^{-1} [29]. In the near-infrared region (1100–1800 nm), the $2v_3$ band near 1670 nm and the $v_2 + 2v_3$ band near 1300 nm are primary overtone rotational-vibrational combination bands.

Based on the Beer-Lambert law (Equation (1)) [10,11], many laser absorption spectroscopy techniques are applied to measure gas concentration, including WMS, FMS, and TTFMS.

$$\tau(v) = \frac{I_t(v)}{I_o(v)} = e^{-\sigma(v)xL} = e^{-\alpha(v)} \tag{1}$$

where, $\tau(v)$ represents the transmittance of light, $I_o(v)$ and $I_t(v)$ represent the incident and transmitted intensities of a certain wavelength, respectively, $\sigma(v)$ represents the cross-section of gas absorption at a certain wavelength, x represents the gas concentration, and L represents the length of the path that the light travels through absorbing media, $\alpha(v)$ represents the transmission coefficient. In absorption spectroscopy, the detection sensitivity is related to the length of the absorption path and the absorption line intensity of molecule, respectively [2].

According to differences in measurement area and monitoring technique, different frequency bands involving methane transitions from near infrared to mid infrared have been applied [30]. With the development of near-infrared light sources and fiber technology, the corresponding test system in the near infrared region is more mature [31–33]. Washenfelder et al. [34] investigated that the absorption strengths of the $2v_3$ band were suitable for providing high sensitivity for ground-based high-resolution spectrometry in the near-infrared spectrum. The absorption intensity of CH$_4$ in the $2v_3$ band is more than four orders of magnitude stronger than that of H$_2$O and CO$_2$, which can be safely neglected [35]. The related spectroscopic parameters of $2v_3$ band, including line positions, line intensities, line widths, line shifts, and line couplings are certain, as described in reference [36–41].

3. Tunable Diode Laser

Semiconductor diode lasers are mainly made of gallium arsenic (GaAs), aluminum (GaAlAs), indium phosphite (InGaAlP, InGaAs or InGaP) and lead salt [42,43]. Moreover, diode lasers are frequently used for modulation spectroscopy because of fast tunability of laser wavelength and fast response times [44,45]. Compared with the spatial characteristics, the spectral characteristics of lasers, such as linewidth and tunability, are more valuable in modulation spectroscopy. To ensure high-quality measurement, narrow line widths, single frequency emission and the inherent stability of the laser are of importance.

The longitudinal mode spacing is relatively narrow, and the oscillation bands of the laser exist in many longitudinal modes. Therefore, the distributed Bragg reflection (DBR) and distributed feedback (DFB) lasers are common diode lasers in the near-infrared spectral region, in which the feedback necessary for the lasing action is distributed throughout the cavity length, and the longitudinal mode selection is improved [46–48]. Such lasers covering the absorption bands of CH$_4$ at 1650 nm have been applied for detecting methane gas [49]. Recently, quantum well (QW) DFB lasers have generated the emission wavelengths of 2.6 μm [50] and 3.4 μm [51,52]. Nevertheless, the overtone and combination bands of many target molecules are in the near infrared region, which is strong enough to get ppm, even ppb detection levels [13].

Depending on the laser materials, commercial diode lasers can be classified into two generic groups: gallium arsenide-based lasers with wavelengths below 3 μm, and lead salt-based lasers with wavelengths above 3 μm, which are usually fabricated from semiconductor materials in groups of III-V and IV-VI, respectively [53]. Among them, the laser beams of lead-salt diode lasers can range from 3 to 30 μm, covering the fundamental transitions of most atmospheric trace gases, which is in the mid-infrared region. Therefore, lead-salt diode lasers are appropriate for spectroscopic gas detection in theory. However, some drawbacks, such as instability in single mode operation, low output power, and high costs, result in their limited practical application in comparison to GaAs lasers [54,55].

Mid-infrared wavelength band covers the fundamental bands of most gas molecules. Therefore, many researchers are carrying out studies with the aim of improving the performance of mid-infrared lasers or develop new type lasers for higher sensitivity. For example, a diode laser with the external cavity (EC) (e.g., Littrow or Littman-Metcalf configuration) has been proposed to to be tuned over a broad spectral range, which is called an external cavity diode laser (ECDL) [56]. However, mode hops limit more improvements of spectral range in a single scan [57]. By contrast, Interband cascade lasers (ICLs) [58] in the 2.5–4 μm wavelength range and quantum cascade lasers (QCLs) [59] in the 4–12 μm wavelength range can provide continuous-wave (CW) output power levels with low input powers. Due to the use of interband transitions, the laser action of ICLs can be obtained at lower electrical input powers than QCLs. However, the phonon interactions in ICLs typically occur on a much slow time scale, which is slower than the longitudinal optical phonon interactions in QCLs. QCLs are widely used as the convenient spectroscopic source to trace gas analysis. In addition, a significant advance in mid-infrared spectroscopic detection has been made, due to their room temperature operation, inherently narrow linewidth and high output power [60–63]. It is worth pointing out that the linewidth reduction of QCLs is achieved by frequency locking to resonant cavities [64]. Recently, an EC configuration is adopted to further improve the performance of QCLs [65,66], and rapid wavelength modulation can be achieved by directly modulating the injection current of QCL chip. A wide tuning range between 7960 and 8840 nm can be performed by continuous-wave operation of EC QCL [67].

4. Application of Modulation Spectroscopy for Methane Detection

Due to the advantage of high spectral resolution, TDLAS has been widely used to measure methane absorptions and concentrations in the applications of molecular spectroscopy [68], natural gas leak detection [69], and trace greenhouse gas monitoring [70] and so on. With TDLAS, the laser frequency is scanned across the absorption line of methane gas, by tuning the output wavelength of the diode laser. Many methane sensor systems are developed utilizing modulation spectroscopy [71,72] to improve the limits of methane detection. The modulation spectroscopy technique used for TDLAS has been reviewed in detail elsewhere [29,44,73,74], and this section aims to bring relevant methane detection based on modulation spectroscopy up to date with recent developments.

4.1. Wavelength Modulation Spectroscopy Applied for Methane Detection

4.1.1. Principle of Wavelength Modulation Spectroscopy

In a typical TDLAS with WMS, the laser frequency is modulated by applying a low-frequency ramp and a high-frequency sine wave [23,70]. After interacting with the absorption line of the target gas, the modulated light generates the signals at different harmonics of modulation frequency [75]. At a fixed harmonic, the amount of signal attenuation is proportional to the gas absorption, and the modulated wavelength is demodulated by a phase-sensitive detector [76]. The demodulated wavelength at a particular harmonic nf ($n = 2, 3, 4$, etc.) is usually directed for the detection, and the detection bandwidth is shifted to higher frequencies, where $1/f$ noise is smaller. The output spectrum of the laser modulated by a radio frequency is shown in Figure 1 [77].

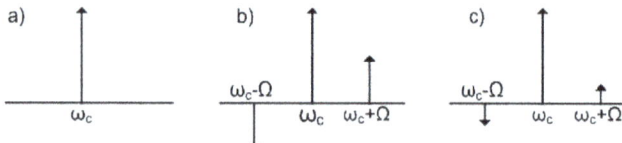

Figure 1. The spectral output of the laser at a radio frequency: (**a**) unmodulated; (**b**) modulated with no absorption; (**c**) modulated with absorption. w_c and $w_c \pm \Omega$ represent the carrier frequency and the side band frequencies, respectively; Ω represents the frequency of the laser. Reproduced with permission from [77], the Institute of Electrical and Electronics Engineers, 2008.

4.1.2. Near Infrared Methane Detection Systems Based on WMS

Based on TDLs with WMS, researchers have successfully studied many high-performance methane detection systems in the near infrared region [71,72,78,79]. Typically, a DFB laser is often used as a light source. A portable methane detection device was developed by using a DFB diode laser centered at 1.654 μm [71]. The second harmonic wavelength modulation spectroscopy (2f-WMS) was adopted. Within the detection range of 0–10^6 ppm, the relative detection error was less than 7%. Liu et al. [72] also designed a trace methane gas sensor. The frequency modulation technology was applied in WMS to move the bandwidth of detection from low frequency to high frequency, to reduce 1/f noise. Zheng et al. [78] described a portable CH_4 detection sensor. The temperature of the DFB laser was controlled by a software-based proportion-integration-differentiation (PID) algorithm. In addition, the measurement range was from 0 to 100%. Then, the absorption length was added from 0.2 m to 0.4 m, and an open gas sensing probe was set in an improved system [79]. In the lab group's previous reports, the two mid-infrared detection sensors had been developed [80,81]. Though the near-infrared sensor has a longer effective path length of the gas cell, the mid-infrared sensor has a higher MDL of 5 ppm. This is due to the fact that the absorption line intensity at 3.31 μm is stronger over two orders than that at 1.65 μm. The wavelet-denoising (WD)-assisted wavelength modulation technique is successfully suppressing the noise in mid-infrared detection sensors. A sequential multipoint sensor applying 2f-WMS technique was firstly developed by Shemshad [82]. The sensor did not use any multiplexing techniques to distribute the laser intensity among a multitude of gas cells.

In some systems, the DFB laser is replaced with a vertical cavity laser (VCL) as a light source [83]. Paige et al. [69] developed a portable natural gas leak detector based on a VCL. The detector could measure methane concentrations from ambient methane levels (1.8 ppm) to pure gas. In the detection process, the response time of the detector was 1–2 s, and the detection precision was below 1%.

Due to the low cost of near-infrared laser and the development of optical communications networks, some instruments in near infrared have been expanded into the commercialized application. The LI-7700 Methane Analyzer (LI-COR Biosciences Lincoln, NE, USA) was developed for detecting methane by eddy covariance method, which had the advantages of light weight, open path, and low power requirement [84]. A tunable diode laser centered at 1.65 μm and a Herriott cell with a 30 m effective path length were employed. The wavelength was modulated across the absorption band at the sub-MHz frequency. Pressure- and temperature-induced changes in line shape and population distribution, changing in laser power and mirror reflectivity as well, were compensated by using computational fitting algorithms. This ensured that measurements remained accurate over a wide range of pressure and temperature conditions. The Laser Gas™ iQ2 analyzer (NEO, AS) was the first all-in-one TDLAS analyzer to measure up to four gases (O_2, CO, CH_4, H_2O) and temperature depending on configuration, which eliminated the need for multiple units for combustion analysis. Then the Laser Gas™ II Open Path (OP) Monitor (NEO, AS), a compact and high-performance gas analyzer, was created for long-distance monitoring in ambient air.

4.1.3. Mid Infrared Methane Detection Systems Based on WMS

Methane detection systems based on the mid-infrared laser have been researched for quite some time. However, traditional mid-infrared sensors have some drawbacks, such as the low power of the laser and the instability of a laser in single mode operation. QCL, as a convenient mid-infrared laser source, is widely employed for methane monitoring at room temperature. A small in situ sensor system was developed to monitor the concentration of CH_4 and N_2O in real time [85]. The light source was a QCL operating at 7.83 μm. The wavelength was scanned over the absorption lines of CH_4 and N_2O at different operating temperatures, achieving simultaneous dual-species detection. The sampling volume of the multipass cell was only 225 mL, resulting in the compact size of the system. Moreover, an external cavity quantum cascade laser (EC-QCL), covering the absorption lines of four atmospheric greenhouse gases, was applied in a sensor system [86]. Additionally, gallium antimonite (GaSb)-based ICLs have been used for mid-infrared sensing [87,88]. In addition, the commercial availability of ICLs was achieved in 2009. Ye et al. [87] demonstrated a mid-infrared dual-gas detection system (CH_4 and C_2H_6) based on a single ICL [87]. However, the power consumption of ~250 W was relatively high. The weight of the oil-free vacuum pump and pressure controller and readout in the system was heavy. To address the limitations, new portable sensor systems were developed [88,89], in which competitive performances were revealed compared with other reported portable or handheld sensor devices. Mid-infrared sources based on DFG effects are also employed for methane detection. Commercial and off-the-shelf near-infrared lasers are purchased to act as the mixing sources. Armstrong et al. [90] described a mid-infrared methane detection system using the difference frequency generation (DFG) process in a periodically poled lithium niobate (PPLN) crystal. The DFG system was used to implement TDLAS with WMS. The pump wavelength and the signal wavelength were provided by a fiber Bragg grating diode laser and a DFB diode laser, respectively. The fundamental absorption line of methane located around 3.4 μm was addressed. Since then, several researchers have reported more absorption detection applications using DFG processes. A system utilizing a dual-wavelength amplifier for DFG process was firstly presented [91]. As shown in Figure 2, the dual-wavelength amplifier is used to amplify both of pump wavelength and signal wavelength, and simple 2f methane detection is carried out. However, the detectable concentration of methane at the same absorption line is less than achieved by Armstrong I. et al. [90]. The minimum detectable methane concentration was at the level of 26 ppbv for an open-path interaction length of 8 m [92]. Zhao et al. [93] described a single-frequency CW difference–DFG source, which was tunable from 3.1 to 3.6 μm. The output power of the source can reach tens of milliwatts. Therefore, the wideband-tunable mid-infrared source has the potential in the application of trace gas detection.

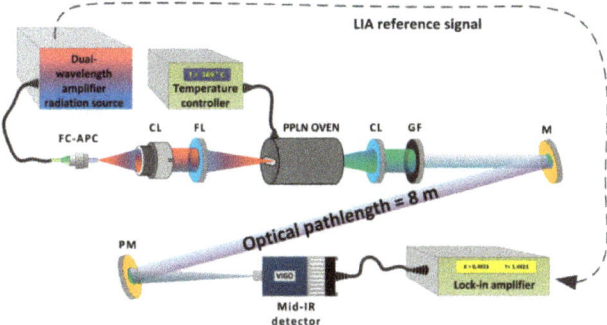

Figure 2. The methane detection system based on a dual-wavelength amplifier. Reproduced with permission from [91], Optical Society of America, 2013.

4.1.4. Multi-Mode Diode Laser Applied for Methane Detection Based on WMS

In the above research, almost every light source falls into a single-mode diode laser. The multi-mode diode laser is also used as a light source in some systems. A tunable multi-mode diode laser with a central wavelength of 1318 nm was applied for remote detection of methane by Gao et al. [94]. In the experiment, the multi-mode laser modes merely depended on the input current and temperature, and the tuning range was continual. The reliability of the data analysis process was certificated. Later, Cai et al. [95] reported what was probably the first application of measuring CH_4 and CO by using a multi-mode ECDL emitting around 2.33 µm, as shown in Figure 3. Correlation spectroscopy was used for signals identification and quantitative analysis. Although it is easier to obtain the signal within the tuning range with a multi-mode diode laser than with a single-mode diode laser due to its larger covering area, the mode-jump and mode competition, an intrinsic property of the multi-mode laser, becomes an obstacle to further applications in trace gas detection.

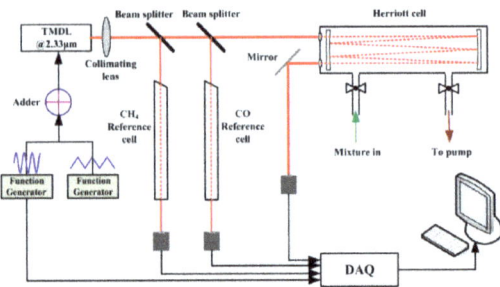

Figure 3. The experimental setup used for the simultaneous measurement of CH_4 and CO. Reproduced with permission from [95], Optical Society of America, 2016.

4.1.5. Optimization of Methane Detection Based on WMS

Apart from high detection sensitivity, the capacity of suppressing $1/f$ noise makes WMS a widely used technique [89,96]. However, optical interference fringe, the primary cause of background fluctuation, has not been effectively suppressed in WMS. Many methods, including the adoption of post-detection filtering and the improvement of the detection system, have been attempted to optimize the optical interference fringe in methane sensors [97]. For example, recently, a method of combining dual tone modulation (DTM) with vibration reflector (VR) was introduced, which decreased the standard deviation (STD) value of the background signal to 0.0924 ppm [98].

Other than the optical interference fringe, some kinds of MPCs, as well as the modulation spectroscopy technique, have also been under study to improve the detection sensitivity, by increasing the effective optical path while at the same time keeping the small size of methane sensor. White cells [99], Herriott cells [100] and Chernin [101] are the three most common MPCs applied for detecting gas, and variations on these have been developed. Liu et al. [2] studied a novel compact dense-pattern multipass cell (DP-MPC). The cell was used to detect atmospheric methane in TDLAS with WMS. Then, a confocal MPC was developed to build a compact and portable methane sensor [102]. The MPC was mainly comprised of confocal mirrors, of which the radii of curvature were 500 mm. Compared to the reported sensors, the sensitivity of ambient methane was improved. Additionally, many other MPCs are also used for the detection of atmospheric trace species, such as a multiple-reflection optical cell with three mirrors [103], the multipass cell formed by two twisted cylindrical mirrors [104] and circular multireflection cell [105]. Some parameters of novel MPCs are listed in Table 1.

Table 1. Some parameters of novel multipass cells.

Title of Multipass Cells	Components	Path Length	Detection Limit	Taken from Refs.
Dense-pattern multipass cell	Two silver-coated concave spherical mirrors	26.4 m	<79 ppb	[2]
Confocal multipass cell	A confocal configuration of six mirrors	290 m	1.2 ppb	[102]
Multiple-reflection optical cell	Three mirrors	140.3 m	128 ppb	[103]
Multiple-pass optical cell	Two twisted cylindrical high-reflectivity mirrors	29.5 m	-	[104]
Circular multireflection cell	A polished spherical reflecting surface	105 cm	300 ppm	[105]

4.2. Frequency Modulation Spectroscopy Applied for Methane Detection

4.2.1. Principle of Frequency Modulation Spectroscopy

FMS is an offshoot of WMS, and the difference between WMS and FMS lies in the magnitude of the modulation frequency. When modulation frequency is larger than the width of the absorption line of interest, the technique is named by frequency modulation (FM). Compared with that of WMS in the $1/f$ noise dominated region (10 kHz), the modulation frequency of FMS is the 100 kHz range, which is in a shot noise limited domain [29]. When detector quantum noise becomes the limiting factor for the sensitivity of the detection system, the detection limit can be described by Equation (2).

$$SNR = \frac{P_{signal}}{(P_{1/f}^2 + P_{sn}^2 + P_{tn}^2)^{1/2}} \quad (2)$$

wherein, $P_{1/f}$ represents the detector excess noise, P_{sn} represents the shot noise of the detector, P_{tn} represents the detector thermal noise, and P_{signal} represents the signal proportional to the incident laser power on the detector.

4.2.2. Methane Detection Systems Based on FMS

Generally, FMS can be applied for the situation where WMS is applied in reference [106]. Compared with that of WMS, processing electronics or detector of FMS should have a broader bandwidth in order to generate higher frequencies [107]. This is a crucial limitation of FMS. Werle et al. [108] determined that the absorbance of detection using one-tone FMS in a 1-Hz bandwidth could be 10^{-8} when the laser-induced shot noise of detector exceeded thermal noise. However, the power of many lead-salt diode lasers is not sufficient to produce shot-noise-limited spectroscopy. Frequency modulation of GaAlAs lasers was first demonstrated by Rickett et al. [109] in 1980. The method of FM-TDLAS was used by Gulluk et al. [110] to measure CO_2, CH_4, N_2O, and CO in air samples of a few cm^3. The lead-salt diode laser was tuned at a typical frequency of 1 kHz, and the rf signal of 100–195 MHz was superimposed on the laser current. Then, Pavone et al. [24] used a GaAlAs diode laser (DL) at 886 nm to obtain the sensitivities of three detection techniques in methane sensor in order to more clearly compare WMS and FMS with TTFMS. The laser was tuned on a third overtone methane transition for measuring the minimum detectable absorption. The modulation frequency was 1 kHz, 100 MHz, and 390 ± 5 MHz for WMS, FMS and TTFMS, and the minimum detectable absorptions were 4.5×10^{-7}, 9.7×10^{-8} and 6.4×10^{-8}, respectively.

Based on FMS, TDLs offer remote detection of methane gas the opportunity to operate over distances of 10 m or more in high sensitivity [111]. In these systems, the laser beam aimed through the probed region is collected after one-way transmission or further reflection from a topographic target [112,113]. The ambient methane detection using the FM technique at a frequency of 5.35 MHz was reported by Uehara et al. [112]. Then, Iseki et al. [113] described a portable methane detection sensor with a 1.65-μm InGaAsP DFB laser. The tuning range of wavelength was 14 cm^{-1}.

In addition to conventional lasers, new types of lasers such as QCL are also used for FMS in methane detection. Gagliardi et al. [114] developed a novel laser spectrometer, which relied on a QCL for detecting methane and nitrous oxide. Moreover, the research group firstly and thoroughly applied single-tone FM technique to QCL, detecting the same components under low pressure, as shown in Figure 4 [65]. Two QCLs of 8.06 μm and 7.3 μm are applied for FMS on N_2O and TTFMS on CH_4, respectively. For methane gas, the minimum detectable concentration can reach up to 400 ppt Hz$^{-1/2}$. Due to the tunability and sensitivity characteristics of the system, a mixture of gases can be monitored. As they summarized, QCL is an appropriate choice for frequency modulation spectroscopy in the mid-infrared region.

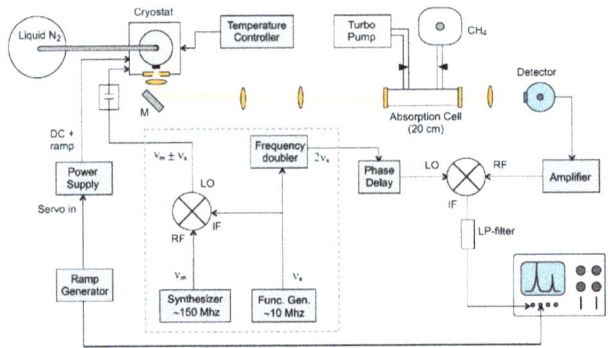

Figure 4. Schematic of the experimental set-up based on the FM technique. Reproduced with permission from [65], Springer, 2006.

4.2.3. Optimization of Methane Detection Based on FMS

Some researchers [115–117] have tried to improve the performance of methane sensors based on QCLs in several ways. For example, a fast optical modulation was achieved by introducing an fs NIR pulse train in a typical QCL, which was different from the traditional modulation method by temperature tuning or current injection [115]. The FM was obtained at frequencies up to 1.67 GHz, which was a benefit to get higher sensitivity of detection. Eichholz et al. [116] found out that the FM technique with QCLs at terahertz (THz) frequencies was suitable for high-resolution molecular spectroscopy. Then DFB QCLs, as the radiation source, were used to set up THz spectrometers [116,117]. The investigated molecular parameters such as transition frequency and pressure broadening of CH_3OH were presented.

4.3. Two-Tone Frequency Modulation Spectroscopy Applied for Methane Detection

4.3.1. Principle of Two-Tone Frequency Modulation Spectroscopy

The frequency modulation spectroscopy is called single-tone FMS (i.e., Standard FM) or two-tone FMS, depending on the number of modulation tones [107]. The modulation of lasers in TTFMS is completed by a pair of closely spaced frequencies simultaneously, which are $\omega_1 = \omega_c + \Omega/2$ and $\omega_1 = \omega_c - \Omega/2$, respectively. Wherein ω_c is the center frequency and Ω is the difference frequency. TTFMS has higher modulation frequencies in comparison with FMS. In the TTFM technique, the advantages of standard FM are used in conjunction with the benefits of a considerable reduction in detection bandwidth, with additional improvement in SNR [107].

4.3.2. Methane Detection Systems Based on TTFMS

In 1982, the TTFM technique was first proposed by Janik et al. [118]. Then modulation frequencies increased from the kHz range to the tens of GHz range due to using a CW dye laser source and an electro-optic modulator [119]. In the same year, the frequency of hundreds of MHz was obtained by using a lead-salt diode laser [120]. Modugno et al. [121] employed a DFB diode laser to develop a TTFMS spectrometer for monitoring methane. A balanced homodyne detection technique was adopted so that the spectrometer had high sensitivity. The modulation frequency and the rf frequency were 2 GHz and 5 MHz, respectively. The sensitivity of the spectrometer was up to $7(2) \times 10^{-8}$ at a 1-Hz bandwidth. Recently, the TTFMS technique has been used in conjunction with a DFG radiation source. Maddaloni et al. [122] developed a portable DFG spectrometer, as shown in Figure 5. Compared with direct absorption, SNR has been enhanced by a factor of 100 by using TTFMS. Gagliardi et al. [65] described another portable spectrometer based on QCLs and TTFMS for monitoring CH_4. The wavelength tunability vs. temperature was 2 GHz/K for QCL emitting at 7.3 µm. The output SNR was enhanced about six times than that of direct absorption.

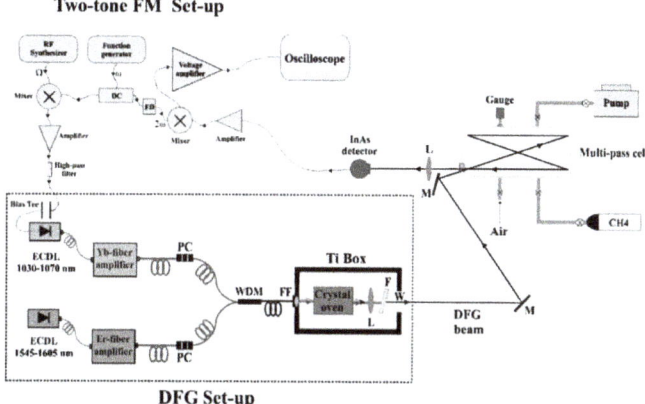

Figure 5. Schematic drawing of the experimental set-up of a portable spectrometer. Reproduced with permission from [122], Springer, 2006.

4.3.3. Optimization of Methane Detection Based on TTFMS

In absorption systems based on TTFM technique, optical interference fringe is also a limiting factor for detection sensitivity. A convenient fringe suppression method was employed to improve the sensitivity in a CH_4 detection system, as shown in Figure 6 [70]. Modulation depth optimization and TTFM technology is applied in this system. In addition, the MDL at 1.654 µm is enhanced to 130 ppb.m for a 50-min period. Observed detection sensitivity is given in the conclusion in Table 2.

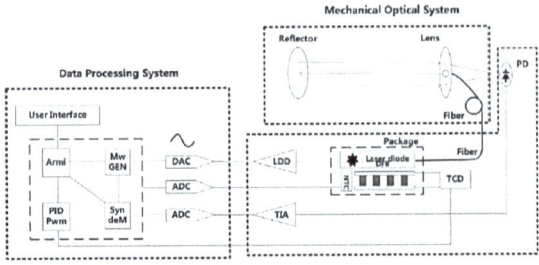

Figure 6. Frame diagram of the CH$_4$ detection system. Reproduced with permission from [70], Elsevier Science, 2017.

Table 2. Methane measurement characteristics for modulation spectroscopy techniques.

Spectroscopic Technique	Spectral Wavelength (nm)	Detection Limit/Measured Concentration	Source Type	Sample/Path Length	References
Wavelength Modulation	1318	25 ppm.m	DL	80 cm gas cell	[94]
	1650	1 ppm	VCL	107 cm gas cell	[69]
	1651	1.2 ppb	DFB	290 m gas cell	[102]
	1653	79 ppb	DFB	26.4 m gas cell	[2]
	1654	130 ppb.m [a]	DFB	Open path	[70]
	1654	11 ppm	DFB	0.2 m gas cell	[71]
	1654	1.4 ppm	DFB	76 m gas cell	[72]
	1654	11 ppm	DFB	0.2 m gas cell	[78]
	1654	12 ppm.m [b]	DFB	40 cm gas cell	[79]
	1654	5 ppb	DL	30 m open cell	[84]
	2330	81 ppb	ECDL	100 m gas cell	[95]
	3100	5 ppm	DFB	7.5 cm gas cell	[81]
	3291	2.1 ppmv [c]	ICL	54.6 m gas cell	[88]
	3291	5 ppbv	ICL	54.6 m gas cell	[89]
	3334	17.4 ppbv	ICL	54.6 m gas cell	[87]
	3403	1.31 ppm.m	DFB	30 mm gas cell	[90]
	3403	26 ppbv	DFB	Open path	[91]
	7800	2.2 ppbv [d]	QCL	57.6 m gas cell	[86]
Frequency Modulation	1654	450 ppb.m	DFB	0.2 m gas cell	[113]
	3357	20.3 ppbv	DL	25 cm gas cell	[110]
	7658	20 ppm	QCL	20 cm gas cell	[114]
Two-Tone Frequency Modulation	886	-	DL	1.5 m gas cell	[24]
	3314	3 ppb Hz$^{-1/2}$ [e]	ECDL	13 m gas cell	[122]
	3428	30 ppt Hz$^{-1/2}$	ECDL	13 m gas cell	[122]
	7300	-	QCL	20 cm gas cell	[65]

[a] Parts per billion meters (length normalized concentration unit); [b] Parts per million meters; [c] Parts per million by volume; [d] Parts per billion by volume; [e] Parts per billion by the negative square root of Hertz.

5. Conclusions

The characteristics of modulation spectroscopy techniques, based on TDLs for detecting methane in the last decade, have been typically reviewed and codified, as shown in Table 2. Recent developments in semiconductor lasers and modulation spectroscopy techniques for methane detection have been described throughout this article. Moreover, some other trends are becoming visible.

Firstly, although QCLs, as convenient mid-infrared laser sources, have not been widely used for commercial detection systems, they may find broad application in mid-infrared modulation spectroscopies in the near future. Then, other wideband-tunable room temperature mid-infrared sources and corresponding mid-infrared technology, such as DFG sources, optical fiber, and detection instruments, will be further developed for trace gas detection. Except for the development of TDLs and mid-infrared sources, the optimization of optical interference fringe and the formation of multipass cells are observable factors in improving the detection performance. Better sensitivity for methane detection system will be expected when new designs of lasers and multipass cells and methods of suppressing noise are put in operation.

Author Contributions: Conceptualization, F.W., and S.J.; investigation, F.W., S.J., Y.W., and Z.T.; Funding acquisition, S.J.; writing—original draft preparation, F.W. and S.J.

Funding: This work was supported by the National Natural Science Foundation of China (NSFC) (Grant No. 51575437), the NSFC—Shanxi Provincial Government coal-based low carbon joint fund (Grant No. U1510114), the Key Science and Technology Program of Shaanxi Province of China (Grant No. 2014K07-02) and XJTU University Funding (Grant No. PY3A048).

Conflicts of Interest: The authors declare no conflict of interest.

References

1. Rodhe, H. A comparison of the contribution of various gases to the greenhouse effect. *Science* **1990**, *248*, 1217–1219. [CrossRef] [PubMed]
2. Liu, K.; Wang, L.; Tan, T.; Wang, G.; Zhang, W.; Chen, W.; Gao, X. Highly sensitive detection of methane by near-infrared laser absorption spectroscopy using a compact dense-pattern multipass cell. *Sens. Actuators B* **2015**, *220*, 1000–1005. [CrossRef]
3. Karpov, E.E.; Karpov, E.F.; Suchkov, A.; Mironov, S.; Baranov, A.; Sleptsov, V.; Calliari, L. Energy efficient planar catalytic sensor for methane measurement. *Sens. Actuators A* **2013**, *194*, 176–180. [CrossRef]
4. Nagai, D.; Nishibori, M.; Itoh, T.; Kawabe, T.; Sato, K.; Shin, W. Ppm level methane detection using micro-thermoelectric gas sensors with Pd/Al$_2$O$_3$ combustion catalyst films. *Sens. Actuators B* **2015**, *206*, 488–494. [CrossRef]
5. Suzuki, T.; Kunihara, K.; Kobayashi, M.; Tabata, S.; Higaki, K.; Ohnishi, H. A micromachined gas sensor based on a catalytic thick film/SnO$_2$ thin film bilayer and thin film heater: Part 1: CH$_4$ sensing. *Sens. Actuators B* **2005**, *109*, 190–193. [CrossRef]
6. Laan, S.V.D.; Neubert, R.E.M.; Meijer, H.A.J. A single gas chromatograph for accurate atmospheric mixing ratio measurements of CO$_2$, CH$_4$, N$_2$O, SF$_6$ and CO. *Atmos. Meas. Tech.* **2009**, *2*, 549–559. [CrossRef]
7. Liu, D.; Fu, S.; Tang, M.; Shum, P.; Liu, D. Comb filter-based fiber-optic methane sensor system with mitigation of cross gas sensitivity. *J. Lightwave Technol.* **2012**, *30*, 3103–3109. [CrossRef]
8. Lin, H.; Liang, Z.; Li, E.; Yang, M.; Zhai, B. Analysis and design of an improved light interference methane sensor. In Proceedings of the 11th IEEE International Conference on Control & Automation (ICCA), Taiwan, China, 18–20 June 2014.
9. Ma, Y.; He, Y.; Tong, Y.; Yu, X.; Tittel, F.K. Quartz-tuning-fork enhanced photothermal spectroscopy for ultra-high sensitive trace gas detection. *Opt. Express* **2018**, *26*, 32103–32110. [CrossRef]
10. Leigh, R.J.; Corlett, G.K.; Friess, U.; Monks, P.S. Concurrent multiaxis differential optical absorption spectroscopy system for the measurement of tropospheric nitrogen dioxide. *Appl. Opt.* **2006**, *45*, 7504. [CrossRef]
11. Rustgi, O.P. Absorption Cross Sections of Argon and Methane between 600 and 170 Å. *JOSA* **1964**, *54*, 464–465. [CrossRef]
12. Swinehart, D.F. The beer-lambert law. *J. Chem. Educ.* **1962**, *39*, 333–335. [CrossRef]
13. Kireev, S.V.; Shnyrev, S.L. On-line monitoring of odorant in natural gas mixtures of different composition by the infrared absorption spectroscopy method. *Laser Phys. Lett.* **2018**, *15*. [CrossRef]
14. Zheng, W.; Zheng, C.; Yao, D.; Yang, S.; Dang, P.; Wang, Y. Development of a mid-infrared interband cascade laser methane sensor. *Acta Opt. Sin.* **2018**, *38*. [CrossRef]
15. Willer, U.; Saraji, M.; Khorsandi, A.; Geiser, P.; Schade, W. Near- and mid-infrared laser monitoring of industrial processes, environment and security applications. *Opt. Laser Eng.* **2006**, *44*, 699–710. [CrossRef]
16. Mappé, I.; Joly, L.; Durry, G.; Thomas, X.; Decarpenterie, T.; Cousin, J.; Dumelie, N.; Roth, E.; Chakir, A.; Grillon, P.G. A quantum cascade laser absorption spectrometer devoted to the in situ measurement of atmospheric N$_2$O and CH$_4$ emission fluxes. *Rev. Sci. Instrum.* **2013**, *84*, 222. [CrossRef]
17. Crosson, E.R. A cavity ring-down analyzer for measuring atmospheric levels of methane, carbon dioxide, and water vapor. *Appl. Phys. B* **2008**, *92*, 403–408. [CrossRef]

18. Berman, E.S.F.; Fladeland, M.; Liem, J.; Kolyer, R.; Gupta, M. Greenhouse gas analyzer for measurements of carbon dioxide, methane, and water vapor aboard an unmanned aerial vehicle. *Sens. Actuators B* **2012**, *169*, 128–135. [CrossRef]
19. Grossel, A.; Zéninari, V.; Parvitte, B.; Joly, L.; Courtois, D. Optimization of a compact photoacoustic quantum cascade laser spectrometer for atmospheric flux measurements: Application to the detection of methane and nitrous oxide. *Appl. Phys. B* **2007**, *88*, 483–492. [CrossRef]
20. Schiff, H.I.; Mackay, G.I.; Bechara, J. The use of tunable diode laser absorption spectroscopy for atmospheric measurements. *Res. Chem. Intermed.* **1994**, *20*, 525–556. [CrossRef]
21. Kamieniak, J.; Randviir, E.P.; Banks, C.E. Cheminform abstract: The latest developments in the analytical sensing of methane. *Trac Trends Anal. Chem.* **2016**, *47*, 146–157. [CrossRef]
22. He, Y.; Ma, Y.; Tong, Y.; Yu, X.; Tittel, F.K. Ultra-high sensitive light-induced thermoelastic spectroscopy sensor with a high Q-factor quartz tuning fork and a multipass cell. *Opt. Lett.* **2019**, *44*, 1904–1907. [CrossRef] [PubMed]
23. Behera, A.; Wang, A. Calibration-free wavelength modulation spectroscopy: Symmetry approach and residual amplitude modulation normalization. *Appl. Opt.* **2016**, *55*, 4446. [CrossRef] [PubMed]
24. Bomse, D.S.; Stanton, A.C.; Silver, J.A. Frequency modulation and wavelength modulation spectroscopies: Comparison of experimental methods using a lead-salt diode laser. *Appl. Opt.* **1992**, *31*, 718–731. [CrossRef] [PubMed]
25. Rojas, D.; Jung, P.; Axner, O. An investigation of the 2f-wavelength modulation technique for detection of atoms under optically thin as well as thick conditions. *Spectrochim. Acta Part B* **1997**, *52*, 1663–1686. [CrossRef]
26. Williams, R.M.; Kelly, J.F.; Sharpe, S.W.; Hartman, J.S.; Gmachl, C.F.; Capasso, F.; Sivco, D.L.; Baillargeon, J.N.; Cho, A.Y. Spectral and modulation performance of quantum cascade lasers with application to remote sensing. *Proc. SPIE* **1999**, *3758*, 11–22.
27. Chan, K.; Ito, H.; Inaba, H. Absorption measurement of $v_2 + 2v_3$ band of CH_4 at 1.33 µm using an InGaAsP light emitting diode. *Appl. Opt.* **1983**, *22*, 3802–3804. [CrossRef] [PubMed]
28. Schilt, S.; Besson, J.P.; Thévenaz, L. Near-infrared laser photoacoustic detection of methane: The impact of molecular relaxation. *Appl. Phys. B* **2006**, *82*, 319–328. [CrossRef]
29. Shemshad, J.; Aminossadati, S.M.; Kizil, M.S. A review of developments in near infrared methane detection based on tunable diode laser. *Sens. Actuators B* **2012**, *171–172*, 77–92. [CrossRef]
30. Tran, H.; Hartmann, J.M.; Toon, G.; Brown, L.R.; Frankenberg, C.; Warneke, T.; Spietz, P.; Hase, F. The $2v_3$ band of CH_4 revisited with line mixing: Consequences for spectroscopy and atmospheric retrievals at 1.67 µm. *J. Quant. Spectrosc. Radiat. Transf.* **2010**, *111*, 1344–1356. [CrossRef]
31. Gagliardi, G.; Gianfrani, L. Trace-gas analysis using diode lasers in the near-IR and long-path techniques. *Opt. Laser Eng.* **2002**, *37*, 509–520. [CrossRef]
32. Chan, K.; Ito, H.; Inaba, H. Optical remote monitoring of CH_4 gas using low-loss optical fiber link and InGaAsP light-emitting diode in 1.33-µm region. *Appl. Phys. Lett.* **1983**, *43*, 634–636. [CrossRef]
33. Chan, K.; Ito, H.; Inaba, H.; Furuya, T. 10 km-long fibre-optic remote sensing of CH_4 gas by near infrared absorption. *Appl. Phys. B* **1985**, *38*, 11–15. [CrossRef]
34. Washenfelder, R.A.; Wennberg, P.O.; Toon, G.C. Tropospheric methane retrieved from ground-based near-IR solar absorption spectra. *Geophys. Res. Lett.* **2003**, *30*, 2226. [CrossRef]
35. Gordon, I.E.; Rothman, L.S.; Hill, C.; Kochanov, R.V.; Tan, Y.; Bernath, P.F.; Birk, M.; Boudon, V.; Campargue, A.; Chance, K.V. The HITRAN2016 Molecular Spectroscopic Database. *J. Quant. Spectrosc. Radiat. Transf.* **2013**, *130*, 4–50. [CrossRef]
36. Gharavi, M.; Buckley, S.G. Diode laser absorption spectroscopy measurement of linestrengths and pressure broadening coefficients of the methane $2v_3$ band at elevated temperatures. *J. Mol. Spectrosc.* **2005**, *229*, 78–88. [CrossRef]
37. Frankenberg, C.; Warneke, T.; Butz, A.; Aben, I.; Hase, F.; Spietz, P.; Brown, L.R. Pressure broadening in the $2v_3$ band of methane and its implication on atmospheric retrievals. *Atmos. Chem. Phys.* **2008**, *8*, 5061–5075. [CrossRef]
38. Pine, A.S. N_2 and Ar broadening and line mixing in the P and R branches of the v_3 band of CH_4. *J. Quant. Spectrosc. Radiat. Transf.* **1997**, *57*, 157–176. [CrossRef]
39. Moorhead, J.G. The near infrared absorption spectrum of methane. *Phys. Rev.* **1932**, *39*, 83–88. [CrossRef]

40. Pine, A.S. Self, N$_2$, O$_2$, H$_2$, Ar, and He broadening in the v_3 band Q branch of CH$_4$. *J. Chem. Phys.* **1992**, *97*, 773–785. [CrossRef]
41. Tran, H.; Flaud, P.M.; Fouchet, T.; Gabard, T.; Hartmann, J.M. Model, software and database for line-mixing effects in the ν$_3$ and ν$_4$ bands of CH$_4$ and tests using laboratory and planetary measurements-II: H$_2$ (and He) broadening and the atmospheres of Jupiter and Saturn. *J. Quant. Spectrosc. Radiat. Transf.* **2006**, *101*, 284–305. [CrossRef]
42. Mantz, A.W. A review of spectroscopic applications of tunable semiconductor lasers. *Spectrochim. Acta, Part A* **1995**, *51*, 2211–2236. [CrossRef]
43. Agrawal, G.P.; Dutta, N.K. Laser structures and their performance. In *Long-Wavelength Semiconductor Lasers*; Van Nostrand Reinhold Company: New York, NY, USA, 1986; pp. 172–219.
44. Song, K.; Jung, E.C. Recent Developments in modulation spectroscopy for trace gas detection using tunable diode lasers. *Appl. Spectrosc. Rev.* **2003**, *38*, 395–432. [CrossRef]
45. Silveira, J.P.; Grasdepot, F. CH$_4$ optical sensor using a 1.31 μm DFB laser diode. *Sens. Actuators B* **1995**, *25*, 603–606. [CrossRef]
46. Kogelnik, H.; Shank, C.V. Erratum: Stimulated emission in a periodic structure. *Appl. Phys. Lett.* **1971**, *18*, 408. [CrossRef]
47. Scifres, D.R.; Burnham, R.D.; Streifer, W. Distributed-feedback single heterojunction GaAs diode laser. *Appl. Phys. Lett.* **1974**, *25*, 203–206. [CrossRef]
48. Goldenstein, C.S.; Spearrin, R.M.; Jeffries, J.B.; Hanson, R.K. Infrared laser-absorption sensing for combustion gases. *Prog. Energy Combust. Sci.* **2017**, *60*, 132–176. [CrossRef]
49. Shimose, Y.; Aizawa, M.; Nagai, H. Remote sensing of methane gas by differential absorption measurement using a wavelength tunable DFB LD. *IEEE Photonics Technol. Lett.* **1991**, *3*, 86–87. [CrossRef]
50. Guenter, J.K.; Amann, M.C.; Arafin, S.; Lei, C.; Vizbaras, K. Single mode and tunable GaSb-based VCSELs for wavelengths above 2 μm. *Proc. SPIE* **2011**, *7952*. [CrossRef]
51. Naehle, L.; Belahsene, S.; Edlinger, M.V.; Fischer, M.; Boissier, G.; Grech, P.; Narcy, G.; Vicet, A.; Rouillard, Y.; Koeth, J.; et al. Continuous-wave operation of type-i quantum well DFB laser diodes emitting in 3.4 μm wavelength range around room temperature. *Electron. Lett.* **2011**, *47*, 46. [CrossRef]
52. Hosoda, T.; Kipshidze, G.; Shterengas, L.; Belenky, G. Diode lasers emitting near 3.44 μm in continuous-wave regime at 300K. *Electron. Lett.* **2010**, *46*, 1455–1457. [CrossRef]
53. Tacke, M. New developments and applications of tunable IR lead salt lasers. *Infrared Phys. Technol.* **1995**, *36*, 447–463. [CrossRef]
54. Werle, P.; Slemr, F.; Gehrtz, M.; Bräuchle, C. Quantum-limited FM-spectroscopy with a lead-salt diode laser. *Appl. Phys. B* **1989**, *49*, 99–108. [CrossRef]
55. Cassidy, D.T.; Reid, J. Atmospheric pressure monitoring of trace gases using tunable diode lasers. *Appl. Opt.* **1982**, *21*, 1185–1190. [CrossRef] [PubMed]
56. Zorabedian, P. Tunable external-cavity semiconductor lasers. In *Tunable Lasers Handbook*; Duarte, F.J., Ed.; Academic Press: San Diego, CA, USA, 1995; pp. 349–442.
57. Liu, A.Q.; Zhang, X.M. A review of mems external-cavity tunable lasers. *J. Micromech. Microeng.* **2007**, *17*, R1–R13. [CrossRef]
58. Miller, J.H.; Bakhirkin, Y.A.; Ajtai, T.; Tittel, F.K.; Hill, C.J.; Yang, R.Q. Detection of formaldehyde using off-axis integrated cavity output spectroscopy with an interband cascade laser. *Appl. Phys. B* **2006**, *85*, 391–396. [CrossRef]
59. Li, J.; Parchatka, U.; Fischer, H. A formaldehyde trace gas sensor based on a thermoelectrically cooled CW-DFB quantum cascade laser. *Anal. Methods* **2014**, *6*, 5483–5488. [CrossRef]
60. Jacoby, M. Quantum cascade lasers. *Chem. Eng. News* **2013**, *88*, 42–43. [CrossRef]
61. Hancock, G.; Helden, J.H.V.; Peverall, R.; Ritchie, G.A.D.; Walker, R.J. Direct and wavelength modulation spectroscopy using a cw external cavity quantum cascade laser. *Appl. Phys. Lett.* **2009**, *94*. [CrossRef]
62. Kosterev, A.; Wysocki, G.; Bakhirkin, Y.; So, S.; Lewicki, R.; Fraser, M.; Tittel, F.; Curl, R.F. Application of quantum cascade lasers to trace gas analysis. *Appl. Phys. B* **2008**, *90*, 165–176. [CrossRef]
63. Tittel, F.K.; Wysocki, G.; Kosterev, A.A.; Bakhirkin, Y.A. Semiconductor Laser Based Trace Gas Sensor Technology: Recent Advances and Applications. In *Mid-Infrared Coherent Sources and Applications*; Ebrahim-Zadeh, M., Sorokina, I.T., Eds.; Springer: Houten, The Netherlands, 2007; pp. 467–493.

64. Taubman, M.S.; Myers, T.L.; Cannon, B.D.; Williams, R.M.; Federico, C.; Claire, G.; Sivco, D.L.; Cho, A.Y. Frequency stabilization of quantum-cascade lasers by use of optical cavities. *Opt. Lett.* **2002**, *27*, 2164–2166. [CrossRef]
65. Borri, S.; Bartalini, S.; Natale, P.D.; Inguscio, M.; Gmachl, C.; Capasso, F.; Sivco, D.L.; Cho, A.Y. Frequency modulation spectroscopy by means of quantum-cascade lasers. *Appl. Phys. B* **2006**, *85*, 223. [CrossRef]
66. Rao, G.N.; Andreas, K. External cavity tunable quantum cascade lasers and their applications to trace gas monitoring. *Appl. Opt.* **2011**, *50*, 100–115. [CrossRef] [PubMed]
67. Mohan, A.; Wittmann, A.; Hugi, A.; Blaser, S.; Faist, J. Room-temperature continuous-wave operation of an external-cavity quantum cascade laser. *Opt. Lett.* **2007**, *32*, 2792–2794. [CrossRef] [PubMed]
68. Kormann, R.; Fischer, H.; Gurk, C.; Helleis, F.; Th, K.; Kowalski, K.; Nigstedt, R.K.; Parchatka, U.; Wagner, V. Application of a multi-laser tunable diode laser absorption spectrometer for atmospheric trace gas measurements at sub-ppbv levels. *Spectrochim. Acta Part A* **2002**, *58*, 2489–2498. [CrossRef]
69. Paige, M.E.; Silver, J.A.; Petroski, D.; Bomse, D.S. Portable natural gas leak detector for survey inspections. *Proc. SPIE* **2006**, *6378*. [CrossRef]
70. Liang, W.; Bi, Y.; Zhou, Q.; Dong, X.; Lv, T.; Liang, W.; Bi, Y.; Zhou, Q.; Dong, X.; Lv, T. Developing CH_4 detection limit at λ = 1.654 μm by suppressing optical interference fringes in wavelength modulation spectroscopy. *Sens. Actuators B* **2017**, *255*, 2614–2620. [CrossRef]
71. Huang, J.Q.; Zheng, C.T.; Gao, Z.L.; Ye, W.L.; Wang, Y.D. Near-infrared methane detection device using wavelength-modulated distributed feedback diode laser around 1.654 μm. *Spectrosc. Lett.* **2014**, *47*, 197–205. [CrossRef]
72. Liu, Y.; Wu, J.N.; Chen, M.M.; Yang, X.H.; Chen, C. The trace methane sensor based on TDLAC-WMS. *Spectrosc. Spectral Anal.* **2016**, *36*, 279–282.
73. Hodgkinson, J.; Tatam, R.P. Optical gas sensing: A review. *Meas. Sci. Technol.* **2013**, *24*, 012004. [CrossRef]
74. Werle, P.; Slemr, F.; Maurer, K.; Kormann, R.; Mücke, R.; Jänker, B. Near- and mid-infrared laser-optical sensors for gas analysis. *Opt. Lasers Eng.* **2002**, *37*, 101–114. [CrossRef]
75. Schilt, S.; Thévenaz, L.; Robert, P. Wavelength modulation spectroscopy: Combined frequency and intensity laser modulation. *Appl. Opt.* **2003**, *42*, 6728. [CrossRef] [PubMed]
76. Reid, J.; Labrie, D. Second-harmonic detection with tunable diode lasers-Comparison of experiment and theory. *Appl. Phys. B* **1981**, *26*, 203–210. [CrossRef]
77. Magalhaes, F.; Carvalho, J.P.; Ferreira, L.A.; Araujo, F.M. Methane detection system based on Wavelength Modulation Spectroscopy and hollow-core fibres. In Proceedings of the SENSORS, Lecce, Italy, 26–29 October 2008; pp. 1277–1280.
78. Zheng, C.T.; Huang, J.Q.; Ye, W.L.; Lv, M.; Dang, J.M.; Cao, T.S.; Chen, C.; Wang, Y.D. Demonstration of a portable near-infrared CH_4 detection sensor based on tunable diode laser absorption spectroscopy. *Infrared Phys. Technol.* **2013**, *61*, 306–312. [CrossRef]
79. Li, B.; Zheng, C.; Liu, H.; He, Q.; Ye, W.; Zhang, Y.; Pan, J.; Wang, Y. Development and measurement of a near-infrared CH_4 detection system using 1.654 μm wavelength-modulated diode laser and open reflective gas sensing probe. *Sens. Actuators B* **2016**, *225*, 188–198. [CrossRef]
80. Ye, W.L.; Zheng, C.T.; Yu, X.; Zhao, C.X.; Song, Z.W.; Wang, Y.D. Design and performances of a mid-infrared CH_4 detection device with novel three-channel-based LS-FTF self-adaptive denoising structure. *Sens. Actuators B* **2011**, *155*, 37–45. [CrossRef]
81. Zheng, C.T.; Ye, W.L.; Huang, J.Q.; Cao, T.S.; Lv, M.; Dang, J.M.; Wang, Y.D. Performance improvement of a near-infrared CH_4 detection device using wavelet-denoising-assisted wavelength modulation technique. *Sens. Actuators B* **2014**, *190*, 249–258. [CrossRef]
82. Shemshad, J. Design of a fibre optic sequential multipoint sensor for methane detection using a single tunable diode laser near 1666 nm. *Sens. Actuators B* **2013**, *186*, 466–477. [CrossRef]
83. Lackner, M.; Totschnig, G.; Winter, F.W.; Ortsiefer, M.; Rosskopf, J. Demonstration of methane spectroscopy using a vertical-cavity surface-emitting laser at 1.68 μm with up to 5 MHz repetition rate. *Meas. Sci. Technol.* **2002**, *14*, 101. [CrossRef]
84. McDermitt, D.; Burba, G.; Xu, L.; Anderson, T.; Komissarov, A.; Riensche, B.; Schedlbauer, J.; Starr, G.; Zona, D.; Oechel, W.; et al. A new low-power, open-path instrument for measuring methane flux by eddy covariance. *Appl. Phys. B* **2011**, *102*, 391–405. [CrossRef]

85. Ren, W.; Jiang, W.; Tittel, F.K. QCL Based Absorption Sensor for Simultaneous Trace-Gas Detection of CH_4 and N_2O. *Appl. Phys. B* **2014**, *117*, 245–251. [CrossRef]
86. Yu, Y.; Sanchez, N.P.; Griffin, R.J.; Tittel, F.K. CW EC-QCL-based sensor for simultaneous detection of H_2O, HDO, N_2O and CH_4 using multi-pass absorption spectroscopy. *Opt. Express* **2016**, *24*, 10391–10401. [CrossRef] [PubMed]
87. Ye, W.; Li, C.; Zheng, C.; Sanchez, N.P.; Gluszek, A.K.; Hudzikowski, A.J.; Dong, L.; Griffin, R.J.; Tittel, F.K. Mid-infrared dual-gas sensor for simultaneous detection of methane and ethane using a single continuous-wave interband cascade laser. *Opt. Express* **2016**, *24*, 16973–16985. [CrossRef] [PubMed]
88. Zheng, C.; Ye, W.; Sanchez, N.P.; Li, C.; Dong, L.; Wang, Y.; Griffin, R.J.; Tittel, F.K. Development and field deployment of a mid-infrared methane sensor without pressure control using interband cascade laser absorption spectroscopy. *Sens. Actuators B* **2017**, *244*, 365–372. [CrossRef]
89. Dong, L.; Tittel, F.K.; Li, C.; Sanchez, N.P.; Wu, H.; Zheng, C.; Yu, Y.; Sampaolo, A.; Griffin, R.J. Compact TDLAS based sensor design using interband cascade lasers for mid-IR trace gas sensing. *Opt. Express* **2016**, *24*, A528–A535. [CrossRef] [PubMed]
90. Armstrong, I.; Johnstone, W.; Duffin, K.; Lengden, M.; Chakraborty, A.L.; Ruxton, K. Detection of in the mid-IR using difference frequency generation with tunable diode laser spectroscopy. *J. Lightwave Technol.* **2010**, *28*, 1435–1442. [CrossRef]
91. Krzempek, K.; Sobon, G.; Abramski, K.M. DFG-based mid-IR generation using a compact dual-wavelength all-fiber amplifier for laser spectroscopy applications. *Opt. Express* **2013**, *21*, 20023–20031. [CrossRef] [PubMed]
92. Krzempek, K.; Soboń, G.; Dudzik, G.; Sotor, J.; Abramski, K.M. Difference frequency generation of Mid-IR radiation in PPLN crystals using a dual-wavelength all-fiber amplifier. *Int. Soc. Opt. Photonics* **2014**, *8964*, 271–283.
93. Zhao, J.; Jia, F.; Feng, Y.; Nilsson, J. Continuous-wave 3.1–3.6 μm difference-frequency generation of dual wavelength-tunable fiber sources in PPMgLN-based rapid-tuning design. *IEEE J. Sel. Top. Quantum Electron.* **2017**, *99*, 1. [CrossRef]
94. Gao, Q.; Zhang, Y.; Yu, J.; Wu, S.; Zhang, Z.; Zheng, F.; Lou, X.; Guo, W. Tunable multi-mode diode laser absorption spectroscopy for methane detection. *Sens. Actuators A* **2013**, *199*, 106–110. [CrossRef]
95. Cai, T.; Gao, G.; Wang, M. Simultaneous detection of atmospheric CH_4 and CO using a single tunable multi-mode diode laser at 2.33 μm. *Opt. Express* **2016**, *24*, 859. [CrossRef]
96. Qi, R.B.; Du, Z.H.; Gao, D.Y.; Li, J.Y.; Xu, K.X. Wavelength modulation spectroscopy based on quasi-continuous-wave diode lasers. *Chin. Opt. Lett.* **2012**, *25*, 77–79.
97. Li, J.; Yu, B.; Zhao, W.; Chen, W. A review of signal enhancement and noise reduction techniques for tunable diode laser absorption spectroscopy. *Appl. Spectrosc. Rev.* **2014**, *49*, 666–691. [CrossRef]
98. Yang, R.; Dong, X.; Bi, Y.; Lv, T. A method of reducing background fluctuation in tunable diode laser absorption spectroscopy. *Opt. Commun.* **2018**, *410*, 782–786. [CrossRef]
99. Pilston, R.G.; White, J.U. A long path gas absorption cell. *J. Opt. Soc. Am.* **1954**, *44*, 572–573. [CrossRef]
100. Herriott, D.; Kogelnik, H.; Kompfner, R. Off-axis paths in spherical mirror interferometers. *Appl. Opt.* **1964**, *3*, 523–526. [CrossRef]
101. Chernin, S.M.; Barskaya, E.G. Optical multipass matrix systems. *Appl. Opt.* **1991**, *30*, 51–58. [CrossRef]
102. Xia, J.; Zhu, F.; Zhang, S.; Kolomenskii, A.; Schuessler, H. A ppb level sensitive sensor for atmospheric methane detection. *Infrared Phys. Technol.* **2017**, *86*, 194–201. [CrossRef]
103. Claude, R. Simple, stable, and compact multiple-reflection optical cell for very long optical paths. *Appl.Opt.* **2007**, *46*, 5408–5418.
104. Kasyutich, V.L. Laser beam patterns of an optical cavity formed by two twisted cylindrical mirrors. *Appl. Phys. B: Lasers Opt.* **2009**, *96*, 141–148. [CrossRef]
105. Ofner, J.; Heinz-Ulrich, K.; Zetzsch, C. Circular multireflection cell for optical spectroscopy. *Appl. Opt.* **2010**, *49*, 5001. [CrossRef]
106. Bjorklund, G.C. Frequency-modulation spectroscopy: A new method for measuring weak absorptions and dispersions. *Opt. Lett.* **1980**, *5*, 15. [CrossRef] [PubMed]
107. Zhang, L.; Tian, G.; Li, J.; Yu, B. Applications of absorption spectroscopy using quantum cascade lasers. *Appl. Spectrosc.* **2014**, *68*, 1095–1107. [CrossRef] [PubMed]

108. Silver, J.A. Frequency-modulation spectroscopy for trace species detection: Theory and comparison among experimental methods: Errata. *Appl. Opt.* **1992**, *31*, 4927. [CrossRef] [PubMed]
109. Osterwalder, J.M.; Rickett, B.J. Frequency modulation of GaAlAs injection lasers at microwave frequency rates. *IEEE J. Quantum Electron.* **1980**, *16*, 250–252. [CrossRef]
110. Güllük, T.; Wagner, H.E.; Slemr, F. A high-frequency modulated tunable diode laser absorption spectrometer for measurements of CO_2, CH_4, N_2O, and CO in air samples of a few cm^3. *Rev. Sci. Instrum.* **1997**, *68*, 230–239. [CrossRef]
111. Van Well, B.; Murray, S.; Hodgkinson, J.; Pride, R.; Strzoda, R.; Gibson, G.; Padgett, M. An open-path, hand-held laser system for the detection of methane gas. *J. Opt. A: Pure Appl. Opt.* **2005**, *7*, S420–S424. [CrossRef]
112. Uehara, K.; Tai, H. Remote detection of methane with a 1.66-microm diode laser. *Appl. Opt.* **1992**, *31*, 809–814. [CrossRef]
113. Iseki, T.; Tai, H.; Kimura, K. A portable remote methane sensor using a tunable diode laser. *Meas. Sci. Technol.* **2000**, *11*, 594. [CrossRef]
114. Gagliardi, G.; Borri, S.; Tamassia, F.; Capasso, F.; Gmachl, C.; Sivco, D.L.; Baillargeon, J.N.; Hutchinson, A.L.; Cho, A.Y. A frequency-modulated quantum-cascade laser for spectroscopy of CH_4 and N_2O isotopomers. *Isotopes. Environ. Health. Stud.* **2005**, *41*, 313–321. [CrossRef]
115. Chen, G.; Martini, R.; Park, S.W.; Bethea, C.G.; Chen, I.C.A.; Grant, P.D.; Dudek, R.; Liu, H.C. Optically induced fast wavelength modulation in a quantum cascade laser. *Appl. Phys. Lett.* **2010**, *97*. [CrossRef]
116. Eichholz, R.; Richter, H.; Wienold, M.; Schrottke, L.; Hey, R.; Grahn, H.T.; Hübers, H.W. Frequency modulation spectroscopy with a THz quantum-cascade laser. *Opt. Express* **2013**, *21*, 32199–32206. [CrossRef]
117. Hübers, H.W.; Pavlov, S.G.; Richter, H.; Semenov, A.D.; Mahler, L.; Tredicucci, A.; Beere, H.E.; Ritchie, D.A. High-resolution gas phase spectroscopy with a distributed feedback terahertz quantum cascade laser. *Appl. Phys. Lett.* **2006**, *89*. [CrossRef]
118. Cassidy, D.T.; Reid, J. Harmonic detection with tunable diode lasers—Two-tone modulation. *Appl. Phys. B* **1982**, *29*, 279–285. [CrossRef]
119. Janik, G.R.; Carlisle, C.B.; Gallagher, T.F. Two-tone frequency-modulation spectroscopy. *J. Opt. Soc. Am. B* **1986**, *3*, 1070–1074. [CrossRef]
120. Cooper, D.E.; Warren, R.E. Frequency modulation spectroscopy with lead-salt diode lasers: A comparison of single-tone and two-tone techniques. *Appl. Opt.* **1987**, *26*, 3726–3732. [CrossRef]
121. Modugno, G.; Corsi, C.; Gabrysch, M.; Marin, F.; Inguscio, M. Fundamental noise sources in a high-sensitivity two-tone frequency modulation spectrometer and detection of CO_2 at 1.6 µm and 2 µm. *Appl. Phys. B* **1998**, *67*, 289–296. [CrossRef]
122. Maddaloni, P.; Malara, P.; Gagliardi, G.; Natale, P.D. Two-tone frequency modulation spectroscopy for ambient-air trace gas detection using a portable difference-frequency source around 3 µm. *Appl. Phys. B* **2006**, *85*, 219–222. [CrossRef]

© 2019 by the authors. Licensee MDPI, Basel, Switzerland. This article is an open access article distributed under the terms and conditions of the Creative Commons Attribution (CC BY) license (http://creativecommons.org/licenses/by/4.0/).

Article

Near-Infrared C₂H₂ Detection System Based on Single Optical Path Time Division Multiplexing Differential Modulation Technique and Multi-Reflection Chamber

Biao Wang [1,*], Hongfei Lu [1,2], Chen Chen [3], Lei Chen [1,4], Houquan Lian [1,4], Tongxin Dai [1,4] and Yue Chen [1,2]

1. State Key Laboratory of Luminescence and Applications, Changchun Institute of Optics, Fine Mechanics and Physics, Chinese Academy of Sciences, Changchun 130033, China
2. University of Chinese Academy of Sciences, Beijing 100049, China
3. Key Laboratory of Geophysical Exploration Equipment, Ministry of Education, College of Instrumentation & Electrical Engineering, Jilin University, Changchun 130026, China
4. University of Science and Technology of China, Hefei 230026, China
* Correspondence: wb5996@163.com; Tel.: +86-130-7433-7472

Received: 30 May 2019; Accepted: 26 June 2019; Published: 28 June 2019

Featured Application: Near-infrared ppm-level C_2H_2 detection with time division multiplexing differential modulation, suitable for the precision improvement of other near-infrared single optical path gas detection systems.

Abstract: A time division multiplexing differential modulation technique is proposed to address the interference problem caused by the fluctuation of laser light intensity in the single optical path detection system. Simultaneously, a multi-reflection chamber is designed and manufactured to further improve the system's precision with an optical path length of 80 m. A near-infrared C_2H_2 detection system was developed. The absorption peak of the acetylene (C_2H_2) molecule near 1520 nm was selected as the absorption line. A laser driver is developed, and a lock-in amplifier is used to extract the second harmonic (2f) signal. A good linear relationship existed between C_2H_2 concentration and the 2f signal, and the correlation coefficient was 0.9997. In the detection range of 10–100 ppmv, the minimum detection limit was 0.3 ppmv, and the precision was 2%. At 50 ppmv, C_2H_2 and continuous detection for 10 h, the data average was 50.03 ppmv, and the fluctuation was less than ±1.2%. The Allan variance method was adopted to evaluate the long-term characteristic of the system. At 1 s of integration time, the Allan deviation was 0.3 ppmv. When the integration time reached 362 s, the Allan deviation was 0.0018 ppmv, which indicates the good stability of the detection system.

Keywords: near-infrared; C_2H_2 detection; TDLAS; time division multiplexing differential modulation; a multi-reflection chamber

1. Introduction

Acetylene (C_2H_2) is one of the most basic raw materials for industrial production. C_2H_2 is burned at high temperature to weld metals. C_2H_2 is easily decomposed, burned, and exploded, because its chemical character is very active. The explosion limit of C_2H_2 in air is between 2.3% and 72.3%. C_2H_2 has a lower explosion limit and a wider explosion range than other flammable and explosive gases. Therefore, highly sensitive and real-time detection of C_2H_2 concentration is especially necessary [1–4].

In recent years, tunable diode laser absorption spectrometry (TDLAS) has been widely used as a real-time and efficient online detection technique [5–9]. C_2H_2 detection is mostly performed in the near-infrared band because the laser process in the near-infrared band is mature. The reported TDLAS-based C_2H_2 detection system uses a wavelength modulation spectroscopy (WMS) technique to improve the signal-to-noise ratio (SNR) [10–12]. In 2018, Ming D et al. [13] developed a dual-optical C_2H_2 detection system near 1533 nm by using WMS, with minimum detection limit (MDL) of 7.9 ppmv at 6 m and 4 ppmv at 20 m. In the actual detection, to further eliminate the influence of the fluctuation of laser light intensity, the dual optical path differential method is used for detection [14,15]. For example, in 2018, the C_2H_2 detection system near 1534 nm, which was developed by He Q X et al. [15], has low MDL of 3.97 ppmv at 0.3 m and a precision of 3%. However, this dual optical path differential structure reduces the final detected optical power and increases system complexity. This study proposes a time division multiplexing differential modulation technique without changing the overall structure of the single optical path detection system. The DFB laser is driven and modulated by superimposing a high-frequency sine signal by a stair-stepping segmented low-frequency signal. Thus, it can eliminate the background noise caused by the fluctuation of laser light intensity, which further improves the detection precision. In addition, this study proposes the design and manufacture of a multi-reflection chamber that increases the absorption optical path and further improves the detection precision. The precision of the single-optical TDLAS gas detection system can be greatly improved by adopting a time division multiplexing differential modulation technique and the multi-reflection chamber method. It can better satisfy the high-precision detection requirements for C_2H_2 gas in practical applications without increasing the complexity of the system.

A near-infrared DFB laser is used to detect concentrations of C_2H_2 based on the TDLAS technique. The emitting peak wavelength of the laser is 1520 nm. A laser driver, which uses a time division multiplexing differential modulation technique to eliminate interference caused by the fluctuation of laser light intensity in the single optical path detection system, is developed. A laser temperature controller based on ARM was developed with a precision of 0.001 °C. The 2f signal's amplitude (SA) is extracted by the lock-in amplifier, and the experimental data can be processed by the algorithm to obtain the detection result. Gas detection experiments are carried out to study the response performance of the system. The system measures different concentrations of C_2H_2 with good linearity, and the MDL is 0.3 ppmv. In addition, the detection precision and stability of the system were analyzed with a precision of 2% and ranged from 10 to 100 ppmv, and the detection data fluctuation was less than ±1.2%. Conclusions were drawn based on the experimental results. The stair-stepping segmented a low-frequency signal that can be used to achieve high-precision C_2H_2 detection in a single optical path system, and there is no requirement for phase consistency compared with the 2f/1f normalization [16,17].

2. TDLAS-WMS Detection Theory

2.1. Absorption Line of C_2H_2

The transition between the ro-vibrational energy levels of a molecule need to absorb the energy of the infrared region, and it will form an infrared spectrum [18]. From the point of view of quantum mechanics, the transition between molecular energy levels satisfies the quantization conditions. Since the energy of the ro-vibrational level is discontinuous, when a level transition occurs, a specific molecule will only absorb a specific wavelength infrared light, which ensures the selectivity of the infrared absorption spectrum gas detection technique. The absorption peak of the C_2H_2 molecule in the mid-infrared band is evidently stronger than in the near-infrared band [19]. However, the process of the near-infrared laser is mature, widely used, and relatively inexpensive. In the actual detection process, the optical path coupling of the mid-infrared laser is more difficult. Therefore, this study selects the characteristic absorption line at the near-infrared band of the C_2H_2 molecule to detect its concentration.

The absorption spectrum of the C_2H_2 molecule in the range of 1520.05 to 1520.15 nm is shown in Figure 1, according to the high-resolution transmission molecular absorption database 2016 (HITRAN 2016). Evidently, the line intensity of the C_2H_2 molecule is the largest near 1520 nm at 10^{-20}. The line intensity of common gas molecules, such as CO_2, H_2O, and CH_4, is less than 10^{-23} at 1520 nm, which is far less than that of the C₂H₂ molecule. The interference generated can be ignored in the actual detection. Therefore, these disturbances can be ignored in the actual detection. The absorption line at 1520 nm of the C_2H_2 molecule is used comprehensively. The line intensity is set to 1.340×10^{-20} to obtain accurate detection results and avoid interference of other gas molecules as much as possible, considering these factors.

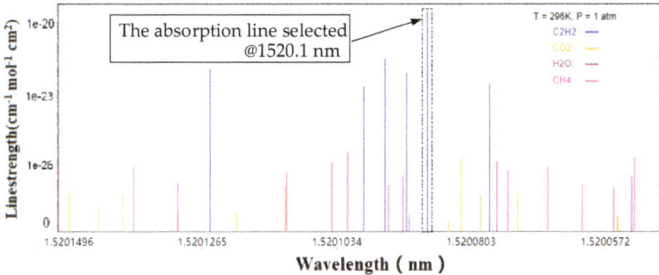

Figure 1. C_2H_2 (blue) absorption line in the range of 1520.05–1520.15 nm. The absorption line strength of CO_2, H_2O, and CH_4 is much smaller than that of C_2H_2.

2.2. Direct Absorption Spectroscopy

The Beer–Lambert law indicates that, when a beam of infrared light is transmitted through the C_2H_2 molecule, a specific wavelength of light is absorbed by the C_2H_2 molecule. The selective absorption of C_2H_2 molecules causes the attenuation of light energy, and the energy of light attenuation is proportional to the number of molecules of C_2H_2 [20,21]. The light intensity after absorption is obtained as follows.

$$I_t = I_0 e^{[-\alpha(v)CL]} \tag{1}$$

where I_0 is the initial emitting light intensity, L is the optical path length, C is the measured gas concentration, and $\alpha(v)$ is the absorption line shape function. At a standard atmospheric pressure, we use the Lorentz line shape function to describe the gas absorption coefficient, as follows.

$$\alpha(v) = \frac{\alpha_0}{\left(\frac{v-v_{cm}}{\Delta v}\right)^2 + 1} \tag{2}$$

where α_0 is the absorption coefficient at the center of the gas absorption line, v_{cm} is the center frequency of the gas absorption peak, and Δv is the half-width at half-peak of the absorption line.

2.3. Single Optical Path Time Division Multiplexing Differential Modulation

We can tune the laser output wavelength by modulating the drive current based on WMS [22]. A stair-stepping segmented low-frequency signal (10 Hz) was used to modulate the laser's injection current to further eliminate the fluctuation of laser light intensity on the single-light path detection system. As shown in Figure 2, the stair-stepping segmented low-frequency signal (10 Hz) is divided into L, M, and H segments. The M segment is a sawtooth signal. This segment is used for C_2H_2 concentration detection. The laser injection current is modulated by superimposing a high-frequency sine signal (5 kHz). Thus, the laser's output wavelength sweeps the absorption peak of C_2H_2. The currents of L and H segments are constant, and the laser's output wavelength driven by the currents is in the non-absorption region of the C_2H_2 molecule. The laser's mean value distribution can be considered in the short single period detection time of 100 ms at 10 Hz because its intensity fluctuation is mainly derived from its own background noise. Therefore, the average light intensity corresponding to the L and H segment waveforms is obtained, and the data processing is performed as the inversion basis of the light intensity I_{0M} of the M segment signal, which further eliminates the interference caused by the laser light intensity fluctuation in the single optical path detection system.

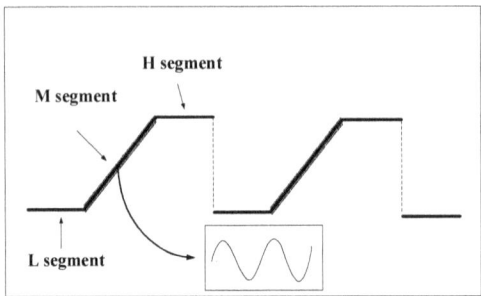

Figure 2. Waveform of stair-stepping segmented low-frequency signal.

After modulation, the output frequency v and the output intensity I_0' are obtained as follows.

$$v = v_0 + v_m \cos \omega t \tag{3}$$

$$I_0' = I_0(1 + \eta \cos \omega t) \tag{4}$$

where v_0 is the center frequency of the laser's output, v_m is the modulated SA, η is the light intensity modulation index, and $f = \omega/2\pi$ is the modulation frequency.

According to Equation (4) and Equation (1), the light intensity with gas absorption is the following.

$$I(v,t) = I_0(1 + \eta \cos \omega t)e^{[-\alpha(v)CL]} \tag{5}$$

The actual gas absorption and light intensity modulation are extremely small, i.e., $\alpha(v) << 1$ and $\eta << 1$. The Taylor series expansion is performed, according to Equation (5). The light intensity is obtained, ignoring the high-order terms, as follows.

$$I(v,t) = I_0[1 + \eta \cos \omega t - \alpha(v_0 + v_m \cos \omega t)]CL \tag{6}$$

From Equation (2), the center frequency v_0 of the laser is adjusted to coincide with the gas absorption peak, i.e., $v_0 = v_{cm}$. Then, the light intensity with gas absorption is obtained as follows.

$$I(v,t) = I_0(1 + \eta \cos \omega t - \frac{\alpha_0 CL}{1 + m^2 \cos^2 \omega t}) \tag{7}$$

$$m = \frac{v_m}{\Delta v} \tag{8}$$

The detection signals $H_L(t)$, $H_M(t)$, and $H_H(t)$ of the photodetector can be expressed as follows.

$$H_L(t) = spI_{0L}(1 + \eta \cos \omega t) \tag{9}$$

$$H_M(t) = spI_{0M}(1 + \eta \cos \omega t - \frac{\alpha_0 CL}{1 + m^2 \cos^2 \omega t}) \tag{10}$$

$$H_H(t) = spI_{0H}(1 + \eta \cos \omega t) \tag{11}$$

where s is the conversion factor of the photodetector, p is the amplification factor of the detection circuit, and I_{0L}, I_{0M}, I_{0H}, and are the incident light intensity corresponding to the L, M, and H segments of the low-frequency signal that satisfy $I_{0M} = \frac{I_{0L}+I_{0H}}{2}$. Differential signal $H(t) = \frac{H_L(t)+H_H(t)}{2} - H_M(t)$ is obtained as follows.

$$H(t) = spI_{0M}\left(\frac{\alpha_0 CL}{1+m^2\cos^2\omega t}\right) \tag{12}$$

The Fourier series expansion is performed on Equation (12) to obtain the second harmonic component coefficient H_{2f} as follows.

$$H_{2f} = \frac{kspI_{0M}CL\alpha_0}{2}$$

$$k = \frac{2(-2-m^2+2\sqrt{1+m^2})}{m^2\sqrt{1+m^2}} \tag{14}$$

From Equation (13), given that m, L, and I_{0M} are constant, the gas concentration (C), and the amplitude of the 2f signal (H_{2f}) are proportional. In the experiment, the 2f signal is extracted, and the concentration of C_2H_2 can be calculated. The relationship between H_{2f} and C can be obtained through calibration experiments.

3. System Configuration

The C_2H_2 detection system is mainly divided into two components, which includes optical and electrical modules. The system structure is shown in Figure 3a. The optical module includes a DFB laser, a self-developed multi-reflection gas chamber, and a photodetector. The electrical module includes a laser driver, a high-precision temperature controller, a lock-in amplifier, and a data processing module. The design of the C_2H_2 detection system is shown in Figure 3b.

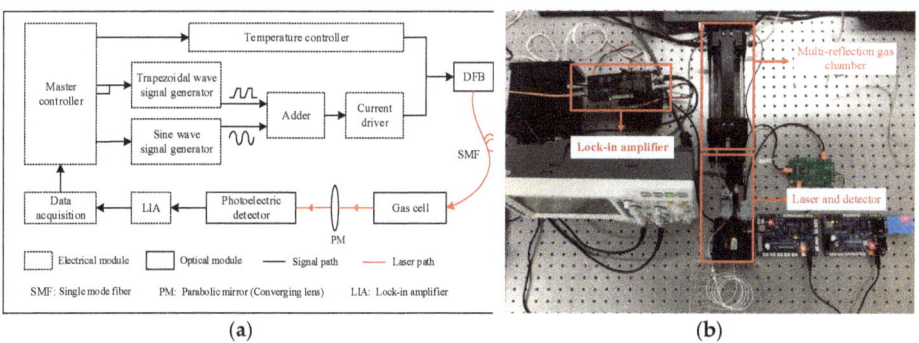

Figure 3. (a) Structure of the C_2H_2 detection system. (b) C_2H_2 concentration detection experimental platform.

3.1. Testing of Laser Performance

The near-infrared DFB laser (1520 nm, Nanoplus) is driven by a self-developed driver. The output wavelength of the laser is tuned by controlling the drive current and temperature, and the laser's temperature was set to 20 °C, 25 °C, and 30 °C by the self-developed temperature controller. As the driving current increases (20 to 150 mA), the emitting peak wavelength gradually increases, and the current tuning factor is approximately 0.011 nm/mA. At a stable driving current, the emitting peak wavelength increases with temperature, and the temperature tuning factor is approximately 0.113 nm/°C. The tuning characteristics of the laser are shown in Figure 4a.

The driving current is set to 80 mA, and the temperature is set to 25 °C with a monitoring experiment for 4 h (8:30–12:30). The laser's optical emission spectrometry is shown in Figure 4b. The emitting peak wavelength varies by approximately 0.001 nm (between 1520.011 and 1520.012 nm).

The monitoring experiment results show that the laser's emission peak wavelength slightly changes, and the working state is stable during a long-term continuous operation.

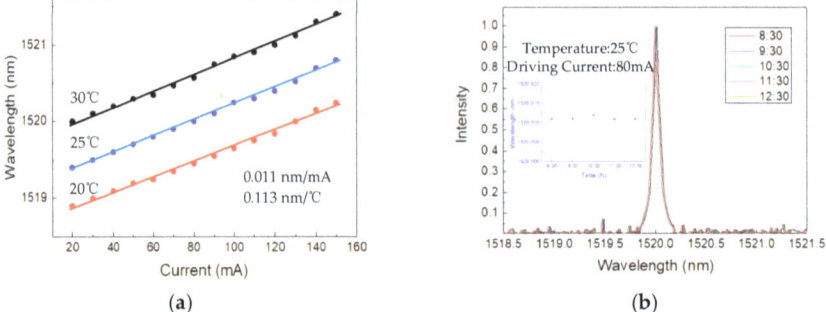

Figure 4. (a) Curve of the laser's emitting peak wavelength versus the driving current at 20 °C, 25 °C, and 30 °C. (b) Curve of the laser's emitting peak wavelength during four hours of continuous operation.

3.2. Multi-Reflection Chamber

The spots of the traditional optical chamber are roughly distributed in a circular shape on the spherical mirror, as shown in Figure 5a. When the distance between the mirrors is extremely close (i.e., the physical size is small), the optical path length cannot easily satisfy the detection requirements. To improve the detection sensitivity, the proposed gas chamber uses three mirrors with a mirror diameter of 60 mm. There are two semi-circular mirrors on the front and a circular mirror on the back. The mirror surface is coated with a silver film by a vacuum coating process. The effective reflection spectrum ranges from 450 nm to 20 µm and the reflectance is greater than 95%. The optical path length can be changed by manually adjusting the incident optical path. As shown in Figure 5b, the reflected spots on the mirror surface of the gas chamber are distributed in 13 rows, which increases the utilization of the mirror surface.

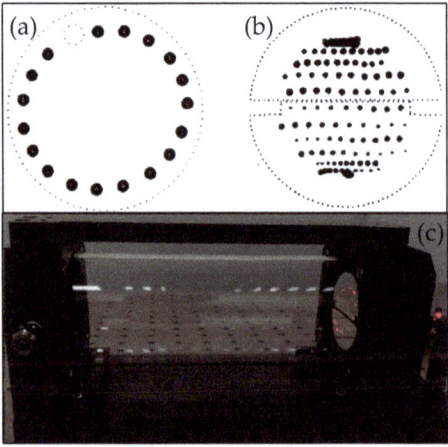

Figure 5. (a) Mirror spots of the conventional optical chamber distributed in a circular shape. (b) Mirror spots of a multi-reflection chamber distributed in 13 rows. (c) Photo of the multi-reflection chamber.

Since the mirror is made of fused silica (JGS1) material, and the main body of the gas chamber is made of 316 steel. Therefore, the material is easy to obtain, and the cost is relatively low compared with the products of the same market. The chamber adopts a ferrule connector/ asperity polishing connector (FC/APC) fiber coupling interface, the light is coupled into the gas chamber by the fiber, and the docking is convenient. The chamber's actual structure is shown in Figure 5c, and the optical path length is set to 80 m with 250 reflections by adjusting the incident optical path.

4. Experimental Process and Discussion

In the experiment, the laser temperature is set to 25 °C, and the emission peak wavelength is near 1520 nm. The light is coupled through a fiber into a gas chamber. We used the Gas Dilution System 4040 gas distribution system. The precision and repeatability of the instrument are ±1.0% and ±0.05%. The temperature and humidity of the laboratory are constant, and the precision of gas preparation can be controlled at ±1.2%.

4.1. Calibration Results and Response Time

To measure the relationship between the 2f SA and C_2H_2 concentration, 10 C_2H_2 samples with concentrations of 10, 20, 30, 40, 50, 60, 70, 80, 90, and 100 ppmv are prepared by dynamic gas preparation. The prepared C_2H_2 sample was sequentially introduced into the gas chamber at intervals of 5 min. As shown in Figure 6a, the 2f SA obtained in the experiment was expressed as max (2f). The 2f SA is linear with the C_2H_2 concentration, and Equation (15) is obtained by linear fitting (where C_{C2H2} is the measured C_2H_2 concentration).

$$\max(2f) = 0.0241 C_{C2H2} - 0.0015 \tag{15}$$

Then, Equation (16) is derived as follows.

$$C_{C2H2} = 41.4938 \max(2f) + 0.0622 \tag{16}$$

The 2f SA can be converted to C_2H_2 concentration by using Equation (16) with a linear correlation of 0.9997, and the fitted curve is shown in Figure 6b.

The response time of the system depends on the time of data processing and the speed of gas diffusion. The detection system can obtain the concentration value in one stair-stepping segmented signal cycle, and 10 concentration values are displayed after average processing. In the case of static gas preparation (i.e., using an air bag to inject C_2H_2 into the gas chamber), the response time is approximately 2 s.

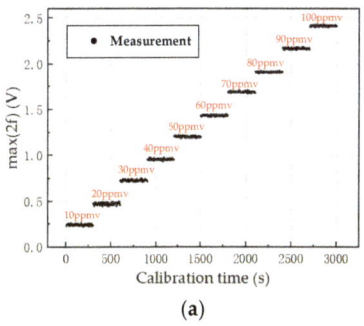
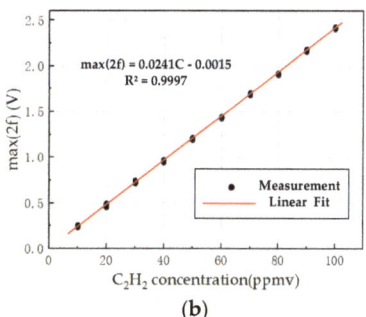

Figure 6. (a) Relationship between 2f signal's amplitude (SA) and concentration C_2H_2 (60 samples per group), and detection data recorded every 5 s. (b) Fitted curve of the 2f SA and concentration C_2H_2.

4.2. Detection Limit and Precision

To test the minimum detection limit (MDL), 50 ppmv C_2H_2 is introduced into the gas chamber. The 2f signal detected by the lock-in amplifier is shown in Figure 7a. When the C_2H_2 concentration is 0 ppmv, the 2f signal is as shown in Figure 7b. The 2f SA is 1.206 V when the C_2H_2 concentration is

50 ppmv, and the system noise's amplitude (NA) of 0 ppmv is 0.007 V. The SNR is 44.725 dB, and the MDL is as follows: 50 ppmv × NA / SA ≈ 0.3 ppmv.

He, Q.X. et al. used a 1.534-μm tunable diode laser and a miniature gas chamber to achieve C_2H_2 detection in Reference [19], and the minimum detectable absorption is 2.66×10^{-3}. In this paper, different concentrations of C_2H_2 were introduced into the gas chamber, and the minimum detectable concentration is 0.3 ppmv. Therefore, the minimum detectable absorption is 3.01×10^{-3}.

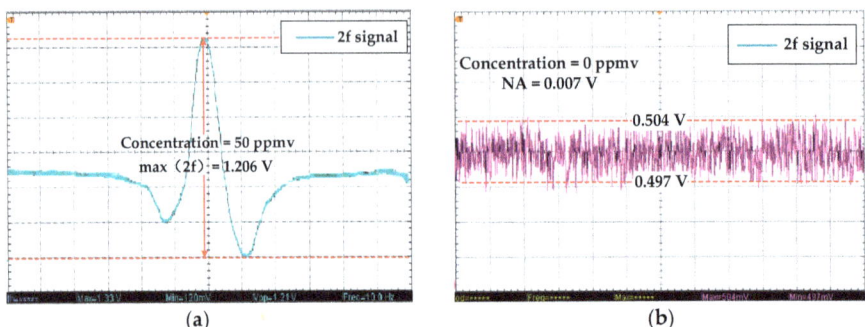

Figure 7. (a) 2f SA of 50 ppmv C_2H_2. (b) 2f SA of 0 ppmv C_2H_2.

To test the precision of the detection system, nine C_2H_2 samples at 10, 15, 25, 35, 45, 55, 75, 95, and 100 ppmv were prepared. The detection system calculated the concentration values according to the fitting formula. The error analysis of the detection results is shown in Figure 8. The blue line indicates the absolute error of the experimental results, and the red line reflects the relative error. The maximum absolute error is 0.3 ppmv, and the maximum relative error is within ±2%. Therefore, the precision is 2% in the detection range from 10 to 100 ppmv.

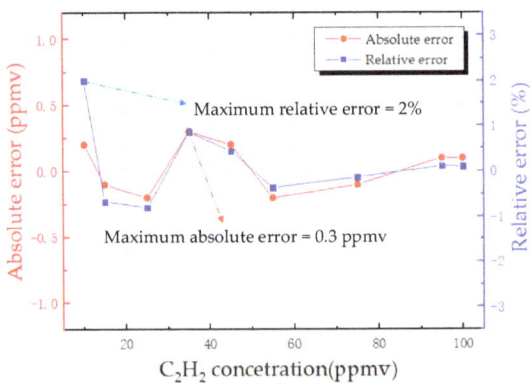

Figure 8. Detection error of the system on nine C_2H_2 samples with concentrations of 10, 15, 25, 35, 45, 55, 75, 95, and 100 ppmv.

4.3. Stability Analysis of the System

To test the stability of the detection system, the C_2H_2 sample at 50 ppmv was introduced into the gas chamber, and the concentration value was recorded every second. The experiment was carried out for 10 h. The result is shown in Figure 9a. The 2f SA ranged from 1.194 to 1.218 V, and the C_2H_2 concentration was between 49.6 and 50.6 ppmv. Therefore, the fluctuation was less than ±1.2%, and the average value of the data was 50.03 ppmv.

Allan deviation was adopted to analyze the long-term characteristics of the detection system [23]. The result shown in Figure 9b indicates that the Allan deviation at the initial integration time of 1 s is 0.3 ppmv. If the integration time continues to increase, then Allan deviation is greatly reduced. Moreover, the Allan deviation can be reduced to 0.0018 ppmv when the integration time is increased to 362 s. The results indicate that the detection system has good stability.

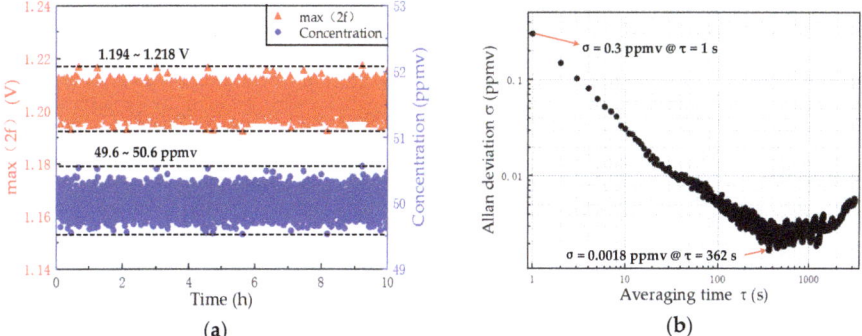

Figure 9. (a) Results of the stability test of 50 ppmv C_2H_2. (b) Analysis of Allan deviation for long-term characteristics of the detection system.

4.4. Comparison

We conducted a large number of experiments using a 15-cm single-pass cell and an 80-m multi-pass cell, which clarifies the increase in detection sensitivity caused by modulation and the increase in sensitivity attributable to the multi-reflection cell. Many researchers have recently conducted in-depth research on C_2H_2 detection technology in the near-infrared band. The performance comparison between the proposed C_2H_2 detection system and the reported near-infrared C_2H_2 detection system is shown in Table 1. The C_2H_2 detection system in References [13,15,19,24] uses the conventional driving method of the TDLAS-WMS technique. The laser's driving signal is generated by super-imposing low-frequency sawtooth waves and high-frequency sine waves. Precision and error are greater than 3%. Time division multiplexing differential modulation is used to eliminate light intensity fluctuations, and system stability has been further improved, with a lower detection limit and higher precision.

Table 1. Performance comparison between the proposed C_2H_2 detection system and the reported C_2H_2 detection system.

Reference/Type	Detection Wavelength	Detection Technique	Detection Limit (ppmv)	Detection Error (%)	System Stability (%)
[13]	1533 nm	TDLAS-WMS	4	–	–
[15]	1530 nm	TDLAS-WMS	3.97	3	3
[19]	1534 nm	TDLAS-WMS	200	5	2
[24]	1530 nm	TDLAS-WMS	1.46	–	–
Single-pass un-modulation	1520 nm	DAS	710	–	–
Single-pass modulation	1520 nm	TDLAS-WMS	142	–	–
Multi-pass un-modulation	1520 nm	DAS	1.5	–	–
Multi-pass modulation	1520 nm	TDLAS-WMS	0.3	±2	±1.2

5. Conclusions

A time division multiplexing differential modulation technique is proposed to eliminate the interference caused by the fluctuation of laser source in a single optical path detection system. A new multi-reflection chamber was designed to further improve the system detection precision. A near-infrared C_2H_2 detection system was developed based on TDLAS using a 1520 nm DFB laser. A laser driver and a high-precision temperature controller were developed, and a lock-in amplifier is used to extract the 2f signal. A number of experiments were performed using the detection system to study its performance. The experimental results indicated that the detection system has good performance indicators. The MDL is 0.3 ppmv, and the precision was 2% in the detection range of 10 to 100 ppmv and contains errors caused by dynamic gas preparation. A 10-h stability observation experiment was performed on 50 ppmv C_2H_2, and the fluctuation was less than ±1.2%. The Allan deviation value was 0.3 ppmv when the initial integration time was 1 s, according to the Allan deviation analysis experimental data. When the integration time increased to 362 s, Allan deviation value was 0.0018 ppmv, which indicated the system's good stability. In addition, the time division multiplexing differential modulation technique and the chamber structure adopted by the system were suitable for the precision improvement of other near-infrared single optical path gas detection systems. Therefore, the proposed approach was flexible and practical.

Author Contributions: Conceptualization, B.W. Methodology, B.W. and H.L. (Hongfei Lu). Software, B.W. Validation, B.W., H.L. (Hongfei Lu), and C.C. Formal analysis, L.C. Investigation, H.L. (Hongfei Lu). Writing—original draft preparation, H.L. (Hongfei Lu). Writing—review and editing, H.L. (Hongfei Lu), B.W., and C.C. Visualization, H.L. (Houquan Lian), T.D., and Y.C. Project administration, L.C. Funding acquisition, B.W.

Funding: The National Key R&D Program of China, grant number 2017YFB0405303, funded this research.

Acknowledgments: The author wishes to express his gratitude to the National Key R&D Program of China (2017YFB0405303) for their generous support toward this work.

Conflicts of Interest: The authors declare no conflicts of interest.

References

1. Utsav, K.C.; Nasir, F.E.; Aamir, F. A mid-infrared absorption diagnostic for acetylene detection. *Appl. Phys. B* **2015**, *120*, 223–232.
2. Cao, Y.; Jin, W.; Ho, H.L.; Qi, L.; Yang, Y.H. Acetylene detection based on diode laser QEPAS: combined wavelength and residual amplitude modulation. *Appl. Phys. B Lasers Opt.* **2012**, *109*, 359–366. [CrossRef]
3. D'Amico, A.; De Marcellis, A.; Di Carlo, C.; Di Natale, C.; Ferri, G.; Martinelli, E.; Paolesse, R.; Stornelli, V. Low-voltage low-power integrated analog lock-in amplifier for gas sensor applications. *Sens. B Chem.* **2010**, *144*, 400–406.
4. Kluczynski, P.; Jahjah, M.; Nahle, L.; Axner, O.; Belahsene, S. Detection of acetylene impurities in ethylene and polyethylene manufacturing processes using tunable diode laser spectroscopy in the 3-lm range. *Appl. Phys. B* **2011**, *105*, 427–434. [CrossRef]
5. Zheng, C.T.; Ye, W.L.; Li, G.L.; Yu, X.; Zhao, C.X.; Song, Z.W.; Wang, Y.D. Performance enhancement of mid-infrared CH_4 detection sensor by optimizing an asymmetric ellipsoid gas-cell and reducing voltage-fluctuation: Theory, design and experiment. *Sens. Actuators B Chem.* **2011**, *160*, 389–398. [CrossRef]
6. Xia, H.; Dong, F.Z.; Wu, B.; Zhang, Z.R.; Pang, T.; Sun, P.S.; Cui, X.J. Sensitive absorption measurements of hydrogen sulfide at 1.578 µm using wavelength modulation spectroscopy. *Chin. Phys. B* **2015**, *24*, 034204. [CrossRef]
7. Ye, W.L.; Zheng, C.T.; Yu, X.; Zhao, C.X.; Song, Z.W.; Wang, Y.D. Design and performances of a mid-infrared CH_4 detection device with novel three channel-based LS-FTF self-adaptive denoising structure. *Sens. Actuators B Chem.* **2011**, *155*, 37–45. [CrossRef]
8. Yue, J.X.; Xie, P.A.; Lu, Y.Z.; Liu, F.H.; Wang, F.P. Concentration monitoring of carbon dioxide in supersonic flow-fields by TDLAS. *Laser J.* **2018**, *39*, 30–33.
9. Lackner, M. Tunable diode laser absorption spectroscopy (TDLAS) in the process industries—A review. *Rev. Chem. Eng.* **2007**, *23*, 65–147. [CrossRef]
10. Hancock, G.; Van Helden, J.H.; Peverall, R.; Ritchie, G.A.D.; Walker, R.J. Direct and wavelength modulation spectroscopy using a cw external cavity quantum cascade laser. *Appl. Phys. Lett.* **2009**, *94*, 201110. [CrossRef]
11. Sur, R.; Sun, K.; Jeffries, J.B.; Socha, J.G.; Hanson, R.K. Scanned-wavelength-modulation-spectroscopy sensor for CO, CO_2, CH_4 and H_2O in a high-pressure engineering-scale transport-reactor coal gasifier. *Fuel* **2015**, *150*, 102–111. [CrossRef]

12. Sur, R.; Sun, K.; Jeffries, J.B.; Hanson, R.K.; Pummill, R.J.; Waind, T.; Wagner, D.R.; Whitty, K.J. TDLAS-based sensors for in situ measurement of syngas composition in a pressurized, oxygen-blown, entrained flow coal gasifier. *Appl. Phys. B* **2014**, *116*, 33–42. [CrossRef]
13. Dong, M.; Zheng, C.T.; Yao, D.; Zhong, G.Q.; Miao, S.Z.; Ye, W.L.; Wang, Y.D.; Tittel, F.K. Double-range near-infrared acetylene detection using a dual spot-ring Herriott cell (DSR-HC). *Opt. Express* **2018**, *26*, 12081–12091. [CrossRef] [PubMed]
14. He, Q.X.; Liu, H.F.; Li, B.; Pan, J.Q.; Wang, L.J.; Zheng, C.T.; Wang, Y.D. Online detection system on acetylene with tunable diode laser absorption spectroscopy method. *Spectrosc. Spectr. Anal.* **2016**, *36*, 3501–3505.
15. He, Q.X. Research on Gas Detection System Based on Infrared Laser Absorption Spectroscopy Technique. Ph.D. Thesis, Jilin University, Changchun, China, 2018.
16. Li, C.G.; Dong, L.; Zheng, C.T.; Tittel, F.K. Compact TDLAS based optical sensor for ppb-level ethane detection by use of a 3.34 μm room-temperature CW interband cascade laser. *Sens. Actuators B* **2016**, *232*, 188–194. [CrossRef]
17. Li, J.Y.; Luo, G.; Du, Z.H.; Ma, Y.W. Hollow waveguide enhanced dimethyl sulfide sensor based on a 3.3 μm interband cascade laser. *Sens. Actuators B* **2018**, *255*, 3550–3557. [CrossRef]
18. Yan, M.; Luo, P.L.; Iwakuni, K.; Millot, G.; Hänsch, T.W.; Picqué, N. Mid-infrared dual-comb spectroscopy with electro-optic modulators. *Light Sci. Appl.* **2016**, arXiv:1608.08013. [CrossRef] [PubMed]
19. He, Q.X.; Zheng, C.T.; Liu, H.F.; Li, B.; Wang, L.J.; Tittel, F.K. A near-infrared acetylene detection system based on a 1.534 μm tunable diode laser and a miniature gas chamber. *Infrared Phys. Technol.* **2016**, *75*, 93–99. [CrossRef]
20. Li, J.S.; Yu, B.L.; Zhao, W.X.; Chen, W.D. A Review of Signal Enhancement and Noise Reduction Techniques for Tunable Diode Laser Absorption Spectroscopy. *Appl. Spectrosc. Rev.* **2014**, *49*, 666–691. [CrossRef]
21. Zhu, X.R.; Lu, W.Y.; Rao, Y.Z.; Li, Y.S.; Lu, Z.M.; Yao, S.C. Selection of baseline method in TDLAS direct absorption CO_2 measurement. *Chin. Opt.* **2017**, *10*, 455–461.
22. Hosseinzadeh, S.S.; Khorsandi, A. Apodized 2f/1f wavelength modulation spectroscopy method for calibration-free trace detection of carbon monoxide in the near-infrared region: Theory and experiment. *Appl. Phys. B* **2014**, *116*, 521–531. [CrossRef]
23. Allan, D.W. Statistics of Atomic Frequency Standards. *Proc. IEEE* **1966**, *54*, 221–230. [CrossRef]
24. He, Y.; Zhang, Y.J.; Kan, R.F.; Xia, H.; Wang, M.; Cui, X.J.; Chen, J.Y.; Chen, D.; Liu, W.Q.; Liu, J.G. The development of acetylene online monitoring technology based on laser absorption spectrum. *Spectrosc. Spectr. Anal.* **2008**, *10*, 2228–2231.

© 2019 by the authors. Licensee MDPI, Basel, Switzerland. This article is an open access article distributed under the terms and conditions of the Creative Commons Attribution (CC BY) license (http://creativecommons.org/licenses/by/4.0/).

Article

Methane Detection Based on Improved Chicken Algorithm Optimization Support Vector Machine

Zhifang Wang, Shutao Wang *, Deming Kong and Shiyu Liu

Institute of Electrical Engineering, Yanshan University, Qinhuangdao 066004, China; wangzhifang0119@163.com (Z.W.); demingkong@ysu.edu.cn (D.K.); liushiyu0204@163.com (S.L.)
* Correspondence: wangshutao@ysu.edu.cn; Tel.: +86-138-3355-1640

Received: 26 March 2019; Accepted: 22 April 2019; Published: 28 April 2019

Abstract: Methane, known as a flammable and explosion hazard gas, is the main component of marsh gas, firedamp, and rock gas. Therefore, it is important to be able to detect methane concentration safely and effectively. At present, many models have been proposed to enhance the performance of methane predictions. However, the traditional models displayed inevitable shortcomings in parameter optimization in our experiment, which resulted in their having poor prediction performance. Accordingly, the improved chicken swarm algorithm optimized support vector machine (ICSO-SVM) was proposed to predict the concentration of methane precisely. The traditional chicken swarm optimization algorithm (CSO) easily falls into a local optimum due to its characteristics, so the ICSO algorithm was developed. The formula for position updating of the chicks of the ICSO is not only about the rooster of the same subgroup, but also about the roosters of other subgroups. Therefore, the ICSO algorithm more easily avoids falling into the local extremum. In this paper, the following work has been done. The sample data were obtained by using the methane detection system designed by us; In order to verify the validity of the ICSO algorithm, the ICSO, CSO, genetic algorithm (GA), and particle swarm optimization algorithm (PSO) algorithms were tested, and the four models were applied for methane concentration prediction. The results showed that he ICSO algorithm had the best convergence effect, relative error percentage, and average mean squared error, when the four models were applied to predict methane concentration. The results showed that the average mean squared error values of ICSO-SVM model were smaller than other three models, and that the ICSO-SVM model has better stability, and the average recovery rate of the ICSO-SVM is much closer to 100%. Therefore, the ICSO-SVM model can efficiently predict methane concentration.

Keywords: methane detection; support vector machine; chicken swarm optimization; algorithm; concentration prediction

1. Introduction

Air pollution is a serious environmental issue that has attracted more and more attention globally in recent years [1–3]. Methane is the main greenhouse pollutant, and also the main component of mine gas, biogas, and various liquid fuels [4,5]. It is stipulated that the lowest limit of explosion in air is 5.0%, the highest limit is 15.0%, and the explosive capacity is the strongest when the volume fraction is 9.5% [6]. Methane in the atmosphere can also cause a greenhouse effect and accelerate global warming [7,8]. It is for these reasons that methane detection is an indispensable field of research. Traditional detection generally adopts chemical methods, which require chemical reagents. These reagents have many disadvantages, such as danger, need to replace, and short life, and so these methods are not conducive to online real-time detection. In addition, absorption spectroscopy is a detection method that offers a rapid, direct, and selective technique to measure the concentration of methane [9], and it has become the dominant detecting method [10,11]. Methane concentrations are obtained through infrared spectroscopy with an appropriate forecasting model.

Currently, it is urgent that an accurate and effective methane prediction model be developed, such as least squares fit, multi-element linear regression, back propagation neural network, or support vector machine (SVM). SVM, based on the principle of structural risk minimization, has higher efficiency when the number of training samples is small [12,13]. The performance of SVM is highly related to its kernel parameters and penalty factor, so choosing appropriate parameters is the key to improving the prediction accuracy. At present, there are a lot of parameter optimization algorithms. For example, Zhou and Lu used the genetic algorithm (GA) to select features and optimize the SVM parameters to improve the prediction accuracy of a hospitalization expense model [14]. Wang and Guan used the particle swarm optimization algorithm optimized support vector machine (PSO-SVM) classifier to classify the maximum tensile shear strength of spot-welded joints; the results showed that the PSO-SVM classifier had a good accuracy [15]. The PSO-SVM based on adaptive mutation was used to classify the increased volume and complexity of flow cytometry (FCM) data by Wang [16]. Liu proposed a short-term wind speed forecasting method, which consists of ensemble empirical mode decomposition (EEMD) for data preprocessing, and an SVM optimized by the cuckoo search algorithm (CS). The experimental results indicated that the proposed model can not only improve the forecasting accuracy, but also can be an effective tool in assisting the management of wind power plants [17]. A new cuckoo search algorithm based on a chaotic catfish effect optimization of the SVM was proposed by He and Xia, who applied it to oil layer recognition [18]. Dai and Niu proposed a SVM optimization based on differential evolution and the grey wolf optimization (DE-GWO-SVM) algorithm to predict power grid investment, which proved that the DE-GWO-SVM model had strong generalization capacity and had a good prediction effect on power grid investment forecasting in China [19].

The CSO algorithm, known as a novel nature-inspired algorithm, was proposed by Meng Xianbing in 2014 [20]. CSO is a stochastic optimization method based on the search behavior of the chicken, which simulates the hierarchy order and the behaviors of a chicken swarm. The chicken swarm is divided into subgroups, and each subgroup consists of chicks, some hens, and a rooster. There is competition between subgroups, that is to say, there is a global optimization result. However, in the paper [18], it was indicated that CSO easily falls into a local optimum, and its progress and speed are greatly influenced by initial values. To this end, an ICSO is proposed. In the ICSO algorithm, a position update equation was added the chicks' learning from the rooster, and learning factors of the chicks were introduced. The convergence accuracy of ICSO is improved, and the ICSO algorithm can easily jump out of the local optimal [21]. The ICSO algorithm has been applied variously in many fields. In Reference [22], the ICSO algorithm was applied in tracking control of the maximum power point of a photovoltaic system. Liang and Wang employed an ICSO algorithm to improve the efficiency of synthetic aperture radar (SAR) image segmentation [23] and so on. However, no paper has yet reported on the application of ICSO algorithm in gas detection.

Based on the above research, this paper proposes a prediction model for methane based on ICSO and SVM. The methane concentration is predicted by SVM, and the ICSO is used to optimize the penalty factor and kernel parameters of SVM. To validate the performance of ICSO-SVM model, the parameters of SVM were also optimized by CSO, GA, and PSO. This paper is organized as follows: Section 2 introduces the methodology, including the SVM, CSO, and ICSO algorithms, and proposes the ICSO-SVM forecasting model. Section 3 gives a brief introduction to the evaluation criteria for forecasting performance. The experiment device and its performance analysis are introduced in Section 4. In Section 5, the results and the superiority of the ICSO-optimized SVM are discussed, using comparisons with the CSO-, GA-, and PSO-optimized SVM. Finally, Section 6 summarizes this paper.

2. Methodology

2.1. SVM

The SVM was proposed by Vapnik in 1995 [24,25]. SVM, as a machine learning method, is effective for small samples, nonlinear, high dimensional, etc. SVM is developed from solving linear problems; it can construct an optimal hyperplane under the condition of linear and divisible. However, in practical applications, most problems are nonlinear. Therefore, the nonlinear input data map to a high-dimensional feature space. For example, given a set of array lengths n, which belong to $\mathbf{R}^d$:

$$(x_1, y_1), (x_2, y_2), \ldots, (x_n, y_n) x_i \in \mathbf{R}^d, y_i \in \{-1, +1\}, i = 1, \ldots, n \tag{1}$$

where $x_i \in \mathbf{R}^d$ are input training samples, y_i are training samples, and d is dimension.

If the samples are separable, there is a classification hyperplane that separates the two types of samples. Crosses and open circles represent two types of samples in Figure 1, respectively. The nearest points to the classified hyperplane are named the support vector. H is a classification hyperplane. Hyperplanes H_1 and H_2 are linked to two types of support vectors and are parallel to H. The distance between H_1 and H is equal to the distance between H_2 and H. The distance between H_1 and H_2 is called the classification interval.

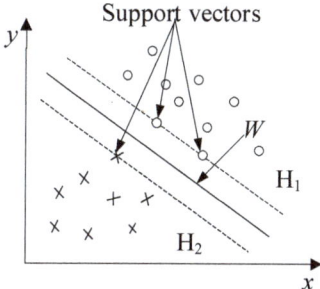

Figure 1. Optimization classification of a hyperplane under a linear condition.

A hyperplane divides the data into two categories, as follows:

$$\begin{cases} (w \cdot x_i) + b \geq 0, & y_i = +1 \\ (w \cdot x_i) + b \leq 0, & y_i = -1 \end{cases} \tag{2}$$

where w is the vector of the hyperplane, x is the input vector of the training set, and b is the constant term of the hyperplane. A hyperplane over two types of sample support vector is defined as:

$$\begin{cases} (w \cdot x_i) + b = +1 \\ (w \cdot x_i) + b = -1 \end{cases} \tag{3}$$

The interval d between hyperplanes H_1 and H_2 can be obtained from Equation (3):

$$d = \frac{2}{|w|} \tag{4}$$

The regression function of classification of the hyperplane is defined as:

$$f(x) = (w \cdot x) + b, w \in \mathbf{R}^d, b \in \mathbf{R} \tag{5}$$

The optimal hyperplane has a maximum margin between two classes. The optimal hyperplane problem is transformed into solving the quadratic optimization, and the slack variable is introduced [26]. The quadratic form can be represented as:

$$\min \tfrac{1}{2}\|w\|^2 + C\sum_{i=1}^{n} \xi_i + \xi_i^*$$
$$s.t. \begin{cases} y_i - (w \cdot x_i) - b \leq \varepsilon + \xi_i \\ (w \cdot x_i) + b - y_i \leq \varepsilon + \xi_i^* \\ \xi_i, \xi_i^* \geq 0 \end{cases} \quad (6)$$

where ξ_i, ξ_i^* are relaxation factors, C is the penalty factor, ε is the insensitivity coefficient, and s.t. is constraint.

Due to the complexity of the calculations of the quadratic optimization, the Equation (6) is transformed into a dual problem with Lagrange duality theory, as Equation (7) and (8)

$$L(w, \xi_i, \xi_i^*, \alpha, \alpha^*, C, \beta, \beta^*) = \tfrac{1}{2}\|w\|^2 + C\sum_{i=1}^{n} \xi_i + \xi_i^*$$
$$- \sum_{i=1}^{n} \alpha_i[(wx_i) + b - y_i + \varepsilon + \xi_i]$$
$$- \sum_{i=1}^{n} \alpha_i^*[y_i - (wx_i) - b + \varepsilon + \xi_i^*]$$
$$- \sum_{i=1}^{n} (\beta_i \xi_i + \beta_i^* \xi_i^*) \quad (7)$$

where $\alpha_i, \alpha_i^*, \beta_i, \beta_i^*$ are all Lagrange multipliers.

$$\max_{\alpha_i, \alpha_i^*} -\varepsilon \sum_{i=1}^{l}(\alpha_i + \alpha_i^*) + \sum_{i=1}^{l} y_i(\alpha_i - \alpha_i^*)$$
$$-\tfrac{1}{2} \sum_{i,j=1}^{l}(\alpha_i - \alpha_i^*)(\alpha_j - \alpha_j^*)(x_i \cdot x_j)$$
$$s.t. \begin{cases} \sum_{i=1}^{l}(\alpha_i - \alpha_i^*) = 0 \\ \alpha_i, \alpha_i^* \in [0, C] \end{cases} \quad (8)$$

The expression of $f(x)$ in Equation (5) is expressed as:

$$f(x) = \sum_{i=1}^{l} \beta_i(x_i \cdot x) + b \quad (9)$$

β_i is non-support vector. When the data set is certain, $\beta_i = 0$. The $f(x)$ is represented by the remaining support vectors as:

$$f(x) = \sum_{i \in N} \beta_i(x_i x) + b \quad (10)$$

where N is a subset of the input data set. For a particular problem, a model for this problem can be determined by a subset of given data. When the problem is nonlinear, Equation (10) cannot accurately represent $f(x)$. The nonlinear input data map to a high-dimensional feature space using nonlinear mapping $\Phi(x)$. In order to reduce the amount of calculation involved, the inner product operation of the high-dimensional feature space is converted into a function transport of the input space by using the kernel function $K(x_i, x)$.

$$K(x_i, x_j) = (\Phi(x_i) \cdot \Phi(x_j)) \quad (11)$$

The expression of $f(x)$ of Equation (10) can be expressed as:

$$f(x) = \sum_{i \in N} \beta_i K(x_i, x) + b \tag{12}$$

Common kernel functions include the linear kernel function, polynomial kernel function, sigmoid kernel function, and Gaussian radical kernel function. The Gaussian radical kernel function is better for problems with less a priori information [20]. The expression of the Gaussian radical kernel function is expressed as:

$$K(x_i, x_j) = \exp\left(-\frac{\|x_i - x_j\|^2}{2r^2}\right) \tag{13}$$

where r is the radius of radial basis kernel function, and $g = 2r^2$ is the kernel parameter.

The values of g and penalty factor C heavily affect the performance of SVM. The ICSO algorithm is intended to optimize the parameters of SVM rather than relying on random selection.

2.2. Chicken Swarm Optimization Algorithm (CSO)

The CSO algorithm is a kind of bionic random search algorithm, which imitates the foraging behaviors of a chicken swarm. The CSO algorithm consists of several subgroups. Each subgroup consists of a rooster, some hens, and several chicks. The roosters have the best fitness value, and the chicks have the worst fitness value. The position of each individual (roosters, hens, and chicks) represents a solution of the problem. The rooster has the best search ability compared to hens and chicks in each subgroup.

The rooster is a leader of the subgroup, with its position update equation defined by the following:

$$x_{i,j}(t+1) = x_{i,j}(t) \cdot \left(1 + \Phi(0, \sigma^2)\right) \tag{14}$$

$$\sigma^2 = \begin{cases} 1, f_{ir} \leq f_{kr} \\ \exp\left(\frac{f_{kr} - f_{ir}}{|f_{ir}| + \varepsilon}\right) \end{cases} \tag{15}$$

where $x_{i,j}(t+1)$ is the position of the rooster at the time $t+1$. $x_{i,j}(t)$ is the position of at the time t. i is the subgroup number, j is the rooster index, $\Phi(0, \sigma^2)$ is the Gaussian distribution with zero mean and standard deviation σ. f_{ir} and f_{kr} are the fitness value of rooster, which is randomly selected ($k \neq i$). ε is the smallest constant that is not equal to 0.

The hens follow the rooster when foraging, so their position is affected by roosters in both the same subgroup and other subgroups. Hens' position update equation is as follows:

$$x_{i,j}(t+1) = x_{i,j}(t) + C_1 \cdot rand \cdot \left(x_{r_1,j}(t) - x_{i,j}(t)\right) + C_2 \cdot rand \cdot \left(x_{r_2,j}(t) - x_{i,j}(t)\right) \tag{16}$$

$$C_1 = \exp((f_{ih} - f_{r_1})/(abs(f_i + \varepsilon))) \tag{17}$$

$$C_2 = \exp((f_{r_2} - f_i)) \tag{18}$$

where $rand$ is a random number over [0, 1]. f_{r1} is the fitness value of the r_1th rooster, which belongs to the same subgroup as the ith hen. r_2 is an index of chicken (rooster or hen), which is randomly selected and not equal to r_1. C_1 and C_2 are the weight of the same subgroup and a different subgroup to the hen, respectively.

The chicks follow their mothers when foraging; their position update equation is as follows:

$$x_{i,j}(t+1) = x_{i,j}(t) + F \cdot \left(x_{m,j}(t) - x_{i,j}(t)\right) \tag{19}$$

where $x_{m,j}$ is the position of the ith chick's mother. F (F ∈ [0, 2]) is a following coefficient, which means that the chick will follow its mother to go foraging.

2.3. Improved Chicken Swarm Optimization (ICSO)

In the swarm, the chicks will follow their mother hen when foraging. The chicks have the worst foraging, and have the smallest foraging range—that is to say, the chicks have the worst global search ability. In the CSO algorithm, the number of hens is the largest. Therefore, the search ability of hens has a great influence on the convergence of the CSO algorithm. From Equation (16), we can see that the position of hens is affected by roosters in the same subgroup and other subgroups, and hens have no self-learning ability. The roosters fall into the local optimum, which results in the hens and chicks falling into the local optimum and affecting the convergence of the whole algorithm. In the Improve Chicken Swarm Optimization (ICSO) algorithm, learning factors C_3 and C_4 are introduced to the chicks' position equation to solve the above problem. The chicks' position is not just about the rooster of same subgroup, but the roosters of other subgroups. The chicks' position update equation is modified as follows:

$$x_{i,j}(t+1) = x_{i,j}(t) + F \cdot \left(x_{m,j}(t) - x_{i,j}(t)\right) + C_3 \cdot \left(x_{r3,j}(t) - x_{i,j}(t)\right) + C_4 \cdot \left(x_{r4,j}(t) - x_{i,j}(t)\right) \quad (20)$$

where C_3 and C_4 are constants, which are learning factors by which the chicks follow roosters of the same subgroup and other subgroups, respectively. $x_{r3,j}$ is position of the rooster in same subgroup as the chicks. $x_{r4,j}$ is position of the rooster in other subgroups.

2.4. ICSO Optimized SVM Model

The steps of ICSO optimized SVM are as follows:

(1) Parameter setting. The population size *pop*: namely, the number of chickens (roosters, hens, and chicks). The maximum number of iterations M: the chickens finish their forage after repeating their search procedure M times. Reconstruction coefficient G: the role assignment of chickens and the subgroup divisions will be done every G times. The numbers of roosters is denoted as RP, hens are HP, mother hens are MP, and chicks are CP. The values of the learning factors are denoted as C_3 and C_4. The penalty factor C and the kernel parameter g are set within a range.

(2) Calculate the best fitness of the individuals, and find the optimum position according to the value of their fitness. Initialize the personal best position p best and the global best position g best. Initialize the current iteration number $t = 1$.

(3) If $t \% G = 1$, rank the fitness of chickens and sort chickens according to their fitness values in descending order. Select the chickens with the best fitness values as roosters. Those chickens with the worst fitness values are chicks, and the other chickens are hens. The chickens are divided into subgroups, the number of subgroups equals to the number of roosters. The hens and chicks are randomly assigned. The hens are assigned randomly as the chicks' mothers, and chicks are in the same subgroup as their mothers.

(4) Update the position of each chicken with Equations (14), (16), and (20), and recalculate the fitness values of the chickens. Update the value of p best and g best.

(5) Repeat steps (3) and (4) until the iteration stop condition is reached, and output the optimum value.

3. Performance Evaluation Criterion

For the spectral data of methane with large variations, pretreatment should be done before training. Experimental data were given normalized treatment as follows:

$$y = \frac{x - x_{min}}{x_{max} - x_{min}} \quad (21)$$

where x is the raw data, y is the processed data, and $x, y \in \mathbf{R}^m$. $x_{min} = \min(x)$. $x_{max} = \max(x)$. The fluctuation range of processed data is 0–1, and $y_i \in [0, 1]$, $i = 1, 2, \ldots n$.

The mean squared error (MSE), relative error (RE), and the recovery rate (r) were used to evaluate the predictive effect of the model. Their values can be computed as follows:

$$MSE = \frac{\sum_{i}^{N}(y_i' - y_i)^2}{N} \quad (22)$$

$$r = \frac{y_i'}{y_i} \times 100\% \quad (23)$$

$$RE = \frac{y_i' - y_i}{y_i} \times 100\% \quad (24)$$

where y_i is the true concentration value, y_i' is the predicted concentration value, and N is the numbers of the sample set.

4. Introduction of Datasets

The experiment was carried out using the methane detection system shown in Figure 2. Based on the infrared spectrum absorption characterization of methane gas, the long optical distance differential absorption method for methane detection was studied. The system is mainly composed of light source, filter system, double-chamber, signal collector, and a processing part.

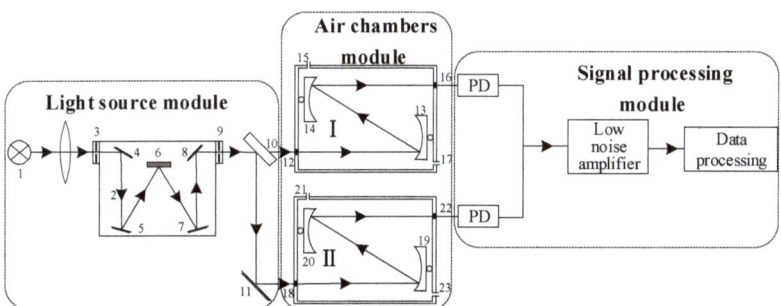

Figure 2. Structural diagram of methane gas detection system. 1—super light emitting diode light source; 2—condensing mirror; 3, 9—slit; 4, 8, 11—plane mirror; 5—collimator; 6—grating; 7—focus lens; 10—beam splitter; 12, 16, 18, 22—gradient index lens; 13, 14, 19, 20—spherical mirror; 15, 21—air inlet; 17, 23—air outlet.

The light source uses a super light emitting diode (SLED). The power spectrum of the SLED was obtained using the steady-state spectrograph (AQ6317C, YOKOGAWA, Tokyo, Japan), as shown in Figure 3. The filter system uses slits, a collimator, a grating, a focus lens, and plane mirrors to obtain the necessary experimental monochrome. The chamber consists of two parts: reference chamber I and test chamber II. The length of chambers is 0.9 m. Reference chamber I is filled with nitrogen, and the test chamber II. Is filled with the target gas (methane). As shown in Figure 2, the effective optical path can be extended to 2.7 m because the light was reflected twice in chamber. Light from the light source is scattered in the air inlet of the chamber, therefore, a graduated refractive index (GRIN) rod lens is placed at the inlet and outlet of the chamber. The pigtail of the GRIN rod lens is fused with transmission optical fiber.

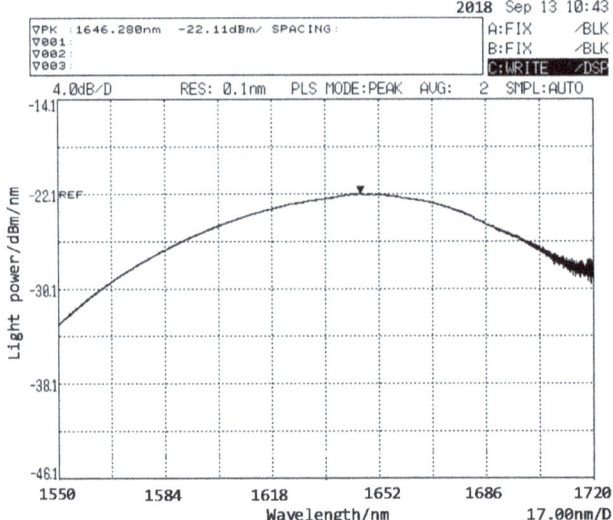

Figure 3. The power spectrum of SLED.

Nitrogen was used as the diluting gas to create concentration standards of methane gas. Concentrations of methane at 2000 ppm, 3000 ppm, 4000 ppm, 5000 ppm, 6000 ppm, 7000 ppm, 8000 ppm, 9000 ppm, 10,000 ppm, 11,000 ppm, 12,000 ppm, 13,000 ppm, 14,000 ppm, 15,000 ppm, 16,000 ppm, 17,000 ppm, 18,000 ppm, 19,000 ppm, and 20,000 ppm were prepared. For each concentration of gas, we made three repeated measurements. The measurement results are shown in Table 1. Table 1 reveals that the maximum measuring error was 0.045, and the average error was 0.0075. The four-concentration absorption spectra of methane are shown in Figure 4. The linear relationship between optical power and methane concentration is shown in Figure 5. The linear correlation coefficient is 0.9888, and the linear equation is $y = -0.2344 x - 41.41$.

From the experimental results, we can see that the methane detection system shown in Figure 2 can be used to detect methane. The data sets in this paper were obtained from the above detection system.

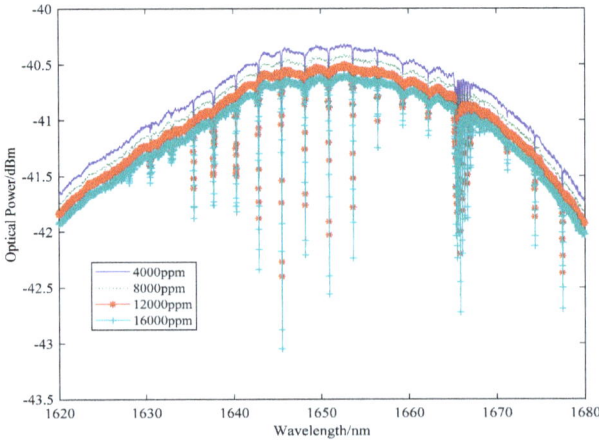

Figure 4. Spectra of four concentrations of methane gas.

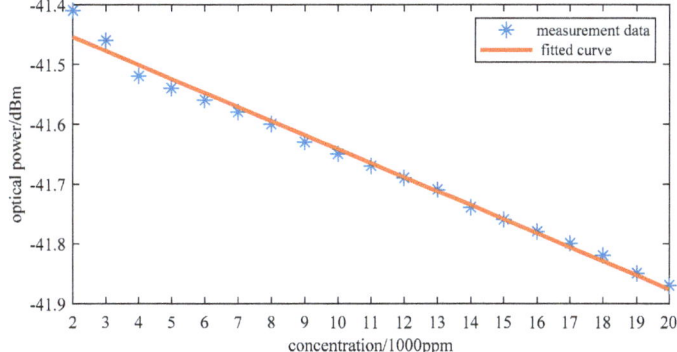

Figure 5. Linear relationship between optical power and methane concentration.

Table 1. Methane gas test results and relative errors.

Standard Concentration of Methane/ppm	Detectable Concentration/ppm			Average Concentrations/ppm	Relative Error
	1	2	3		
2000	2100	2010	2150	2090	0.0450
4000	3900	4070	3870	3970	−0.0075
6000	6110	6030	5900	6010	0.0017
8000	8060	8270	8100	8140	0.0175
10,000	10,110	9770	9960	9950	−0.0050
12,000	12,120	11,970	12,010	12,030	0.0025
14,000	14,240	14,170	14,050	14,150	0.0107
16,000	16,300	16,110	16,250	16,220	0.0138
18,000	17,950	17,870	18,130	17,980	−0.0011
20,000	19,860	19,930	20,020	19,940	−0.0030

5. Results and Analysis

The Windows 7 Ultimate operating system was used to perform the experiments. The specific version of the software used to conduct the proposed model was Matlab2014a. The details of the hardware are as follows: Intel(R) Core (TM) i3-4160 CPU (Fourth Generation Standard Edition, Intel Corporation, Santa Clara, CA, USA and 2014), and 4 GB RAM. The effectiveness and superiority of our method were verified through the following aspects.

The results of the ICSO algorithm were compared with the CSO, PSO, and GA algorithms.

5.1. Parameter Setting and Analysis

In this subsection, we give all parameter settings used in this paper and focus on analyzing some parameters used in our method.

The parameters settings and analysis of ICSO, CSO, PSO, and GA are given after experimental verification, as follows:

First, considering that the population size pop and the iterations M were small, it was difficult to converge to a global optimum. If their values are too large, it will take much time. We set their values to be 100 and 100, respectively, after experimental verification, and set the cross-validation value to be 3. Other parameters of the four algorithms are listed in Table 2. The four algorithms ran independently, and the average convergence curve obtained is shown in Figures 6–9.

As can be seen from Figures 6–9, the ICSO algorithm found the optimal fitness value after the 4th iteration, CSO after the 9th iteration, GA after about the 10th iteration, and PSO after about the 13th iteration. The average fitness value of the ICSO algorithm began to converge after the 21st iteration, the CSO after about the 22th iteration, and the GA after the 9th iteration, but it did not converge to the optimal fitness. The average fitness value of PSO stabilized at the 3rd iteration, but it has a large gap between the average fitness curve and optimal fitness curve.

According to the above results, the ICSO algorithm is the fastest algorithm that is convergent to a global optimum solution to solve optimization problems. The CSO algorithm is also convergent to a global optimum solution, but the convergence speed is slower. The GA and PSO algorithms cannot converge to the global optimum. Comprehensive comparison shows that the convergence effect of the ICSO algorithm is the best.

Table 2. The parameters of the four algorithms.

The Algorithms	Parameters
GA [1]	$C \in [0.1, 1000], g \in [0.001, 100]$
PSO [2]	$C_1 = 1.5, C_2 = 1.7, w = 0.7, C \in [0.1, 1000], g \in [0.001, 100]$
CSO [3]	$RP = 0.15 * pop, HP = 0.7 * pop, MP = 0.5 * HP, CP = pop - RP - HP - MP, G = 10, C \in [0.1, 1000], g \in [0.001, 100]$
ICSO [4]	$RP = 0.15 * pop, HP = 0.7 * pop, MP = 0.5 * HP, CP = pop - RP - HP - MP, G = 10, C \in [0.1, 1000], g \in [0.001, 100]$

[1] genetic algorithm; [2] particle swarm optimization algorithm; [3] chicken swarm optimization algorithm; [4] improved chicken swarm optimization algorithm.

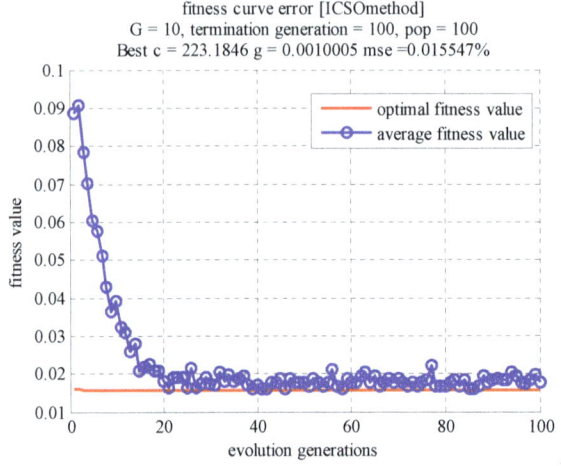

Figure 6. The fitness curve of improved chicken swarm optimization algorithm.

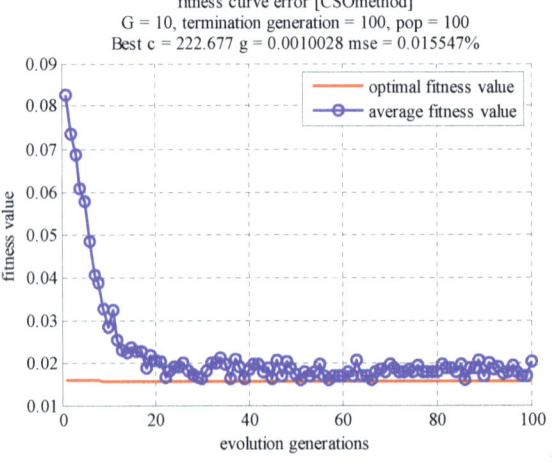

Figure 7. The fitness curve of chicken swarm optimization algorithm.

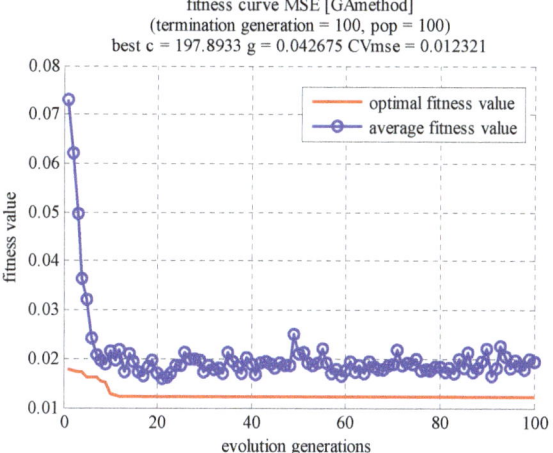

Figure 8. The fitness curve of genetic algorithm.

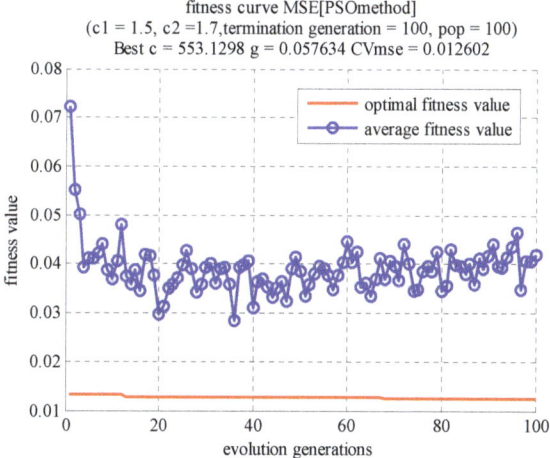

Figure 9. The fitness curve of particle swarm optimization algorithm.

5.2. Prediction Results

In our experiment, there were 40 concentrations (1000 ppm–40,000 ppm) of methane. We randomly split the dataset into 80% training and 20% test sets. In other words, 32 samples were selected for training the classifiers, while the rest of the samples were used to test the model. The training set and testing set were randomly selected from the whole dataset. We repeated the train–test procedure five times with four models (ICSO-SVM, CSO-SVM, GA-SVM, and PSO-SVM), and calculated the mean value. The predicted results of the four models are shown in Table 3.

In order to analyze the performances of four models clearly, we calculated the relative error percentages of the four models, as shown in Figure 10.

As shown in Figure 10, the fluctuations of the ICSO-SVM and CSO-SVM relative error lines are stable, while the GA-SVM and PSO-SVM relative error lines are volatile. The maximum relative error percentage of the ICSO-SVM model was 4%, which is obviously lower than the other three models.

Appl. Sci. **2019**, *9*, 1761

Table 3. The predicted results of the four models.

Samples Number	Ture Value/ppm	ICSO-SVM/ppm	CSO-SVM/ppm	GA-SVM/ppm	PSO-SVM/ppm
1	2000	2300	2300	2600	2800
2	7000	6900	7200	7400	7700
3	11,000	11,300	11,500	11,800	11,800
4	14,000	14,100	14,200	13,700	13,600
5	19,000	18,800	18,900	18,600	18,700
6	26,000	26,200	26,200	26,400	26,700
7	31,000	31,200	31,300	30,700	30,700
8	38,000	37,900	37,900	38,300	38,200

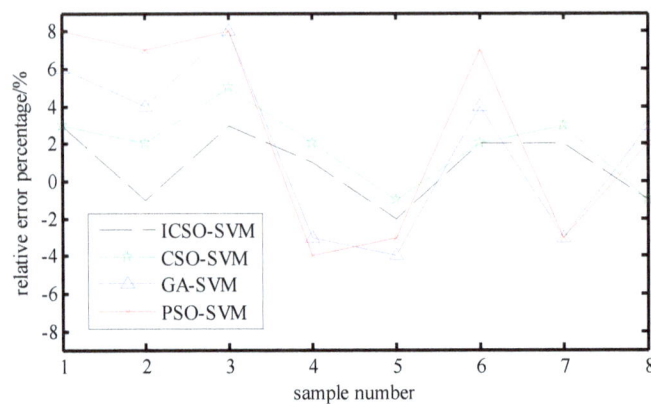

Figure 10. The relative error percentage for 8test samples.

To eliminate bias in the test results, we repeated this train–test procedure 50 times with different random splits. We then averaged the recovery of each test to get the recovery rate and the mean squared error for each model.

The recovery rate can be calculated with Equation (23), and the mean squared error with Equation (22). The recovery rates for 50 repetitions of the four models are shown in Figures 11–14. The recovery rates and the mean squared errors of ICSO-SVM, CSO-SVM, GA-SVM, and PSO-SVM models are shown in Table 4.

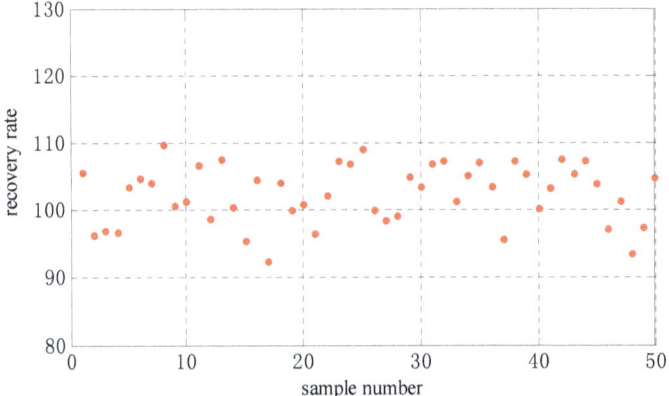

Figure 11. The recovery rate of ICSO-SVM.

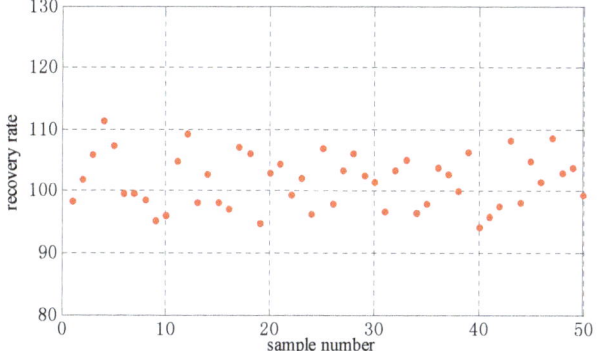

Figure 12. The recovery rate of CSO-SVM.

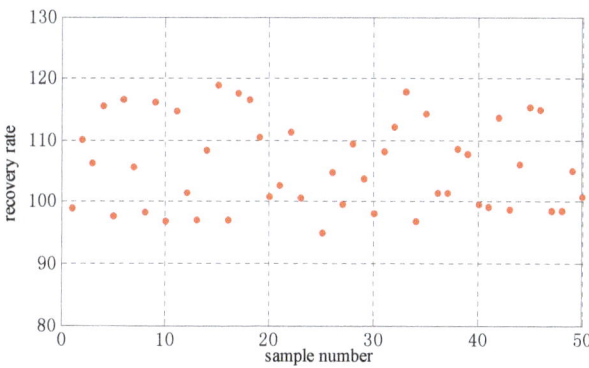

Figure 13. The recovery rate of GA-SVM.

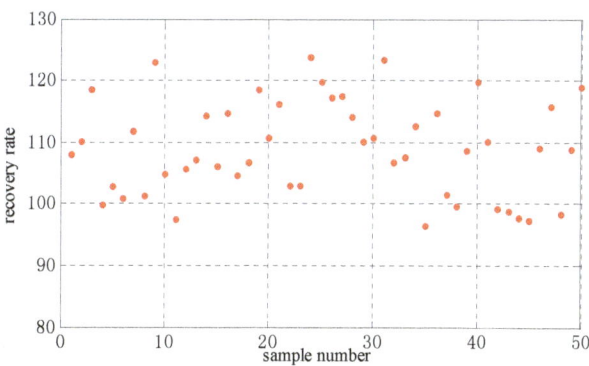

Figure 14. The recovery rate of PSO-SVM.

Table 4. The recovery rates and the mean squared errors of four models.

Models	ICSO-SVM	CSO-SVM	GA-SVM	PSO-SVM
Average recovery rate/%	101.23	103.15	113.58	125.61
Average mean squared error	1.12×10^{-5}	1.23×10^{-5}	3.56×10^{-5}	3.22×10^{-5}

It can be seen from Figures 11–14 that the recovery rate of the ICSO-SVM model remained stable within the values of [90, 110], the CSO-SVM and GA-SVM models within [80, 120], and the PSO-SVM model within [75, 120]. The results of the stability study showed that the ICSO-SVM model has better stability. From Table 4, the four models could be indexed on their average recovery rate, as follows: ICSO-SVM > CSO-SVM > GA-SVM > PSO-SVM. The four models could also be indexed on their average mean squared error, as follows: ICSO-SVM > CSO-SVM > PSO-SVM > GA-SVM. The results from the experiments indicate that the ICSO-SVM has the best prediction performance.

6. Conclusions

In order to detect the concentration of methane accurately, the support vector machine optimized by improved chicken swarm optimization (ICSO-SVM) was used in this paper. First, the data were obtained by the methane detecting system. Next, in order to verify the validity of the ICSO-SVM model for predicting methane, CSO-SVM, GA-SVM, and PSO-SVM were used for comparison.

This study draws the following conclusions:

(1) The mean squared error was adopted as the fitness function of the models. The experimental results show that the ICSO algorithm more easily finds a global optimum, and can converge more stably than the other three algorithms. The results also show that the ICSO algorithm has satisfactory convergence, and that it is effective for the improvement of the CSO algorithm.

(2) The samples were randomly selected from the whole dataset. The train–test procedure was repeated five times with four models. Compared with the other three optimization algorithms, the prediction values and predicted average relative error percentage of the ICSO-SVM model are obviously superior.

(3) From the 50 train–test repeats experiment, we can see that the recovery rate of ICSO-SVM model shows better stability than other three models. The average recovery rates of ICSO-SVM, CSO-SVM, GA-SVM, and PSO-SVM were 101.23, 103.15, 113.58, and 125.61, respectively. The average mean squared errors of the four models were 1.12×10^{-5}, 1.23×10^{-5}, 3.56×10^{-5}, and 3.22×10^{-5}, respectively. These experimental results verify the feasibility and validity of ICSO-SVM for predicting the concentration of methane.

These are initial steps. Further research should focus on integrating the detection system and algorithm into methane detection equipment, meaning it would be possible to detect methane concentration and obtain a concentration value quickly. Finally, the equipment should be tested at in civil, ambient, and industrial spaces.

Author Contributions: Z.W. and S.W. conceived, designed, and performed the experiment; Z.W. and S.L. proposed the methodology; Z.W. wrote the original draft preparation, S.W. and S.L. reviewed and edited the paper; S.W. and D.K. supervised the whole work and acquired funding.

Funding: This research was funded by National Natural Science Foundation of China (Nos. 61771419, 61501394), and Natural Science Foundation of Hebei Province of China (No. F2017203220).

Conflicts of Interest: The authors declare no conflict of interest.

References

1. Apte, J.S.; Marshall, J.D.; Cohen, A.J.; Brauer, M. Addressing global mortality from ambient PM2.5. *Environ. Sci. Technol.* **2015**, *49*, 8057–8066. [CrossRef] [PubMed]
2. Di, Q.; Kloog, I.; Koutrakis, P.; Lyapustin, A.; Wang, Y.; Schwartz, J. Assessing PM2.5 exposures with high spatiotemporal resolution across the continental United States. *Environ. Sci. Technol.* **2016**, *50*, 4712–4721. [CrossRef] [PubMed]
3. Li, C.; Zhu, Z. Research and application of a novel hybrid air quality early-warning system: A case study in China. *Sci. Total Environ.* **2018**, *626*, 1421–1438. [CrossRef]
4. Liu, K.; Wang, L.; Tan, T.; Wang, G.; Zhang, W.; Chen, W.; Gao, X. Highly Sensitive Detection of Methane by Near-infrared Laser Absorption Spectroscopy Using A Compact Dense-pattern Multipass Cell. *Sens. Actuators B-Chem.* **2015**, *220*, 1000–1005. [CrossRef]

5. Cao, Y.; Sanchez, N.P.; Jiang, W.; Griffin, R.J.; Xie, F.; Hughes, L.C.; Zah, C.-E.; Tittel, F.K. Simultaneous Atmospheric Nitrous Oxide, Methane and Water Vapor Detection with A Single Continuous Wave Quantum Cascade Laser. *Opt. Express* **2015**, *23*, 2121–2132. [CrossRef]
6. Tan, W.; Yu, H.; Huang, C. Discrepant responses of methane emissions to additions with different organic compound classes of rice straw in paddy soil. *Sci. Total Environ.* **2018**, *630*, 141–145. [CrossRef]
7. Zhou, G. Research prospect on impact of climate change on agricultural production in China. *Meteorol. Environ. Sci.* **2015**, *38*, 80–94.
8. Contribution of Working Group I to the 5th Assessment Report of the Intergovernmental Panel on Climate Change. *The Physical Science Basis*; IPCC: Geneva, Switzerland, 2013.
9. Shemshad, J.; Aminossadati, S.M.; Kizil, M.S. A review of developments in near infrared methane detection based on tunable diode laser. *Sens. Actuators B Chem.* **2012**, *171*, 77–92. [CrossRef]
10. Ma, Y.; Lewicki, R.; Razeghi, M.; Tittel, F.K. QEPAS based ppb-level detection of CO and N2O using a high power CW DFB-QCL. *Opt. Express* **2013**, *21*, 1008–1019. [CrossRef] [PubMed]
11. He, Y.; Ma, Y.; Tong, Y.; Yu, X.; Tittel, F.K. Ultra-high sensitive light-induced thermoelastic spectroscopy sensor with a high Q-factor quartz tuning fork and a multipass cell. *Opt. Lett.* **2019**, *44*, 1904–1907. [CrossRef] [PubMed]
12. Hao, Z.; Zhao, H.; Zhang, C.; Wang, H.; Jiang, Y.; Yi, Z. Estimating winter wheat area based on an SVM and the variable fuzzy set method. *Remote Sens. Lett.* **2019**, *10*, 343–352. [CrossRef]
13. Haobin, P.E.N.G.; Guohua, C.H.E.N.; Xiaoxuan, C.H.E.N.; Zhimin, L.U.; Shunchun, Y.A.O. Hybrid classification of coal and biomass by laser-induced breakdown spectroscopy combined with K-means and SVM. *Plasma Sci. Technol.* **2019**, *21*, 034008. [CrossRef]
14. Tao, Z.; Huiling, L.; Wenwen, W.; Xia, Y. GA-SVM based feature selection and parameter optimization in hospitalization expense modeling. *Appl. Soft Comput. J.* **2019**, *75*, 323–332. [CrossRef]
15. Wang, X.; Guan, S.; Hua, L.; Wang, B.; He, X. Classification of spot-welded joint strength using ultrasonic signal time-frequency features and PSO-SVM method. *Ultrasonics* **2019**, *91*, 161–169. [CrossRef]
16. Wang, Y.; Meng, X.; Zhu, L. Cell Group Recognition Method Based on Adaptive Mutation PSO-SVM. *Cells* **2018**, *7*, 135. [CrossRef]
17. Liu, T.; Liu, S.; Heng, J.; Gao, Y. A New Hybrid Approach for Wind Speed Forecasting Applying Support Vector Machine with Ensemble Empirical Mode Decomposition and Cuckoo Search Algorithm. *Appl. Sci.* **2018**, *8*, 1754. [CrossRef]
18. He, Z.; Xia, K.; Niu, W.; Aslam, N.; Hou, J. Semi supervised SVM Based on Cuckoo Search Algorithm and Its Application. *Mathe. Probl. Eng.* **2018**, *2018*, 8243764.
19. Dai, S.; Niu, D.; Han, Y. Forecasting of Power Grid Investment in China Based on Support Vector Machine Optimized by Differential Evolution Algorithm and Grey Wolf Optimization Algorithm. *Appl. Sci.* **2018**, *8*, 636. [CrossRef]
20. Meng, X.; Liu, Y.; Gao, X.; Zhang, H. *A New Bio-Inspired Algorithm: Chicken Swarm Optimization*; Advances in Swarm Intelligence; Springer: Berlin, Germany, 2014; pp. 86–94.
21. Wu, D.; Kong, F.; Gao, W.; Shen, Y.; Ji, Z. Improved Chicken Swarm Optimization. In Proceedings of the 5th Annual IEEE International Conference on Cyber Technology in Automation, Shenyang, China, 8–12 June 2015.
22. Wu, Z.; Yu, D.; Kang, X. Application of improved chicken swarm optimization for MPPT in photovoltaic system. *Optim. Control Appl. Meth.* **2018**, *39*, 1029–1042. [CrossRef]
23. Liang, J.; Wang, L.; Ma, M.; Zhang, J. A fast SAR image segmentation method based on improved chicken swarm optimization algorithm. *Multimed. Tools Appl.* **2018**, *77*, 31787–31805. [CrossRef]
24. VAPNIK, V. *The Nature of Statistical Learning Theory*; Springer: New York, NY, USA, 1995.
25. Cherkassky, V. The Nature of Statistical Learning Theory. *IEEE Trans. Neural Netw.* **1997**, *8*, 1564. [CrossRef] [PubMed]
26. Smola, A.J.; Scholkopf, B. A tutorial on support vector regression. *Stat. Comput.* **2004**, *14*, 199–222. [CrossRef]

© 2019 by the authors. Licensee MDPI, Basel, Switzerland. This article is an open access article distributed under the terms and conditions of the Creative Commons Attribution (CC BY) license (http://creativecommons.org/licenses/by/4.0/).

Article

High-Precision $^{13}CO_2/^{12}CO_2$ Isotopic Ratio Measurement Using Tunable Diode Laser Absorption Spectroscopy at 4.3 um for Deep-Sea Natural Gas Hydrate Exploration

Hanquan Zhang [1], Mingming Wen [1,*], Yonghang Li [1], Peng Wan [1] and Chen Chen [2,*]

[1] Key Laboratory of Marine Mineral Resources, Ministry of Natural Resources, Guangzhou Marine Geological Survey, Guangzhou 510760, China
[2] College of Instrumentation & Electrical Engineering, Key Laboratory of Geophysical Exploration Equipment, Ministry of Education of China, Jilin University, Changchun 130026, China
* Correspondence: wmingming@mail.cgs.gov.cn (M.W.); cchen@jlu.edu.cn (C.C.); Tel.: +86-135-4439-9196 (M.W.); +86-137-5606-4009 (C.C.)

Received: 12 July 2019; Accepted: 17 August 2019; Published: 21 August 2019

Abstract: For the detection of deep-sea natural gas hydrates, it is very important to accurately detect the $^{13}CO_2/^{12}CO_2$ isotope ratio of dissolved gas in seawater. In this paper, a $^{13}CO_2/^{12}CO_2$ isotope ratio sensor is investigated, which uses a tunable diode laser absorption spectroscopy (TDLAS) technique at 4.3 μm. The proposed sensor consists of a mid-infrared interband cascade laser (ICL) operating in continuous wave mode, a long optical path multi-pass gas cell (MPGC) of 24 m, and a mid-infrared mercury cadmium telluride (MCT) detector. Aiming at the problem of the strong absorption intensity of the two absorption lines of $^{13}CO_2$ and $^{12}CO_2$ being affected by temperature, a high-precision temperature control system for the MPGC was fabricated. Five different concentrations of CO_2 gas were configured to calibrate the sensor, and the response linearity could reach 0.9992 for $^{12}CO_2$ and 0.9996 for $^{13}CO_2$. The data show that the carbon isotope measurement precision was assessed to be 0.0139‰ when the integration time was 92 s and the optical path length was 24 m. The sensor is combined with a gas–liquid separator to detect the $^{13}CO_2/^{12}CO_2$ isotope ratio of CO_2 gas extracted from water. Results validate the reported sensor system's potential application in deep-sea natural gas hydrate exploration.

Keywords: deep-sea natural gas hydrate exploration; $^{13}CO_2/^{12}CO_2$ isotope ratio detection; TDLAS technique; mid-infrared ICL

1. Introduction

Natural gas hydrates (NGHs) are formed on the seabed under low-temperature and high-pressure conditions [1]. Compared to traditional energy, such as coal and petroleum, NGHs as an alternative energy source have the characteristics of clean, high-efficiency, and abundant reserves. To accelerate the utilization rate of NGHs, many efforts were made in the exploration of such energy sources, and the most important thing is to accurately identify the distribution area. Stable isotope compositions of volatiles bound in the formation of NGHs are due to physical processes and chemical reactions [2,3]. Therefore, the isotope compositions can be used to identify and even quantitatively analyze the source or settlement position of NGHs.

At present, a very effective method for detecting deep-sea natural gas hydrate is measuring the gas dissolved in the seawater. The type of gas measured mainly includes carbon dioxide (CO_2) and its carbon isotope ratio. So far, high-precision chemical element measured methods mainly include mass spectrometry [4], chromatography [5], flame ionization [6], etc. However, these detection methods are complex in structure and poor in long-term measurement stability, which are not suitable for detection in a complex seabed environment. Recently, the tunable diode laser absorption spectroscopy (TDLAS) [7–11] technique was applied in many non-contact gas detection applications, such as environmental, industrial, biology, and safety applications, as a consequence of its high sensitivity and high resolution. In 2006, Lau et al. measured the CO_2 isotopic ratio under the conditions of 45 Torr and a light path length of 100.9 m using a distributed feedback laser in the absorption spectrum of 1.6 µm, achieving the precision of ±1.0‰ at 3000 ppm [12]. In 2008, AERODYNE RESEARCH, Inc. proposed a CO_2 isotope monitor using a mid-infrared absorption spectroscope with optical path lengths up to 76 m to achieve 0.10‰ precision for $\delta^{13}C$ at 1 s [13]. In the same year, Tuzson described in situ, continuous, and high-precision isotope ratio measurements of atmospheric CO_2 using a quantum cascade laser-based absorption spectrometer, and achieved isotope ratios of ambient CO_2 with precision of 0.03 at 100 s of integration time [14]. In 2012, Kasyutich illustrated a laser spectrometer based on a continuous-wave thermoelectrically cooled distributed feedback quantum cascade laser at 2308 cm^{-1} for the measurement of $^{13}CO_2/^{12}CO_2$ isotopic ratio changes in exhaled breath samples. Typical short-term $\delta^{13}C$ precisions of 1.1‰ (1 s integration time) and 0.5‰ (8 to 12 s integration time) were estimated from the Allan variance plots of recorded data [15]. In 2018, a vertical cavity surface emitting laser was utilized by Ghetti et al. in the detection of CO_2 isotopic ratio collected from exhaled small samples of CO_2 (0.1 L) with a resolution of about 0.2% [16]. Although these instruments achieved good carbon isotope detection performance, they cannot meet the requirement of deep-sea natural gas hydrate exploration ($\delta^{13}C$ of ~0.01‰). Moreover, these sensing systems are bulky, which makes them unsuitable for CO_2 gas detection on the seabed with compact requirements.

Compared to state-of-the-art studies, this paper presents the following: (a) a high-precision detection method for CO_2 isotopes applied for deep-sea NGHs, (b) a linear optical structure combined with a multi-pass gas cell (MPGC) to overcome the challenges of a physically compact requirement for NGH exploration, (c) a high-precision temperature control system for the MPGC to improve the measurement precision of carbon isotopes, (d) a compact cylinder cell placed between the interband cascade laser (ICL) and MPGC to remove the interference from CO_2 in the environment and to improve the measurement accuracy, and (e) a prototype using a mid-infrared absorption spectroscope capable of CO_2 isotope detection in indoor experiments. In this study, a sensing system capable of highly precise measurements of the $^{13}CO_2/^{12}CO_2$ isotope ratio of CO_2 gas separated from seawater was developed. The TDLAS technique is utilized to detect the carbon isotope absorption line which is located in the strong spectrum band at 4.2 µm. The described sensor utilizes a mid-infrared ICL operating in continuous wave mode with low-temperature, long-path multi-reflection gas absorption chambers, and a mid-infrared MCT detector with high responsivity. An indoor experiment was conducted to verify the work performance and application feasibility of the proposed sensing system.

The remainder of this paper is organized as follows: Section 2 introduces the carbon isotope detection principle and absorption line selection. Section 3 presents the system design architecture that includes an optical part and electrical part. In addition, the sensor's calibration and its response combined with a gas–liquid separator were determined in an indoor experiment, which is described in Section 4. Finally, the conclusions and future research prospects of carbon isotopic ratio measurement are illustrated in Section 5.

2. Carbon Isotope Detection Principle and Absorption Line Selection

2.1. Carbon Isotope Detection Principle

The basic principle of the TDLAS technique for carbon isotope detection is the Beer–Lambert law [17]. After passing through the MPGC filled with the measured gas, the laser measures the absorption of specific gas molecules with a given optical path length to obtain the concentration characteristics of the gas molecules. The Beer–Lambert law can be described by the following formula:

$$I_1(v) = I_0(v)e^{-\alpha(v)PCL},\tag{1}$$

where $I_0(v)$ is the emitting light intensity of the laser, $I_1(v)$ is the light intensity after passing the measured gas, L is the effective length of the absorption optical path, P is the pressure in the cell, C is the gas concentration, and $\alpha(v)$ is the molecular absorption coefficient. Then, $\alpha(v)$ can be expressed as follows:

$$\alpha(v) = S(T) \cdot g(v - v') \cdot C,\tag{2}$$

where $S(T)$ is the absorption intensity of gas at temperature T, $g(v - v')$ is the absorption line function of measured gas, which relates to the concentration C and pressure P of the measured gas, and v' is the initial frequency of the energy level transition of gas molecule. To improve the minimum detection limit (MDL) performance of the sensor, tunable diode laser absorption spectroscopy-wavelength modulation spectroscopy (TDLAS-WMS) was adopted to eliminate the 1/f noise caused by the ICL or external environmental disturbances [18]. The time-dependent wavelength of the ICL can be described as follows:

$$v = v_0 + A\cos(\omega t),\tag{3}$$

where v_0 is the central frequency of the emitting light, which is determined by the low-frequency component of the driving signal, and A and ω are the amplitude and frequency of the high-frequency component of the driving signal, respectively. By substituting Equation (3) into Equation (1) and expanding it in the form of a cosine Fourier series, we get

$$I_1(v) = \sum_{n=0}^{\infty} A_n(v_0) \cos(n\omega t),\tag{4}$$

where A_n is the amplitude of each harmonic component, which can be expressed as follows [19]:

$$A_n(v_0) = \frac{I_0(v) \cdot 2_{n-1} \cdot C \cdot L}{n!} \cdot A^n \cdot \left.\frac{d^n \alpha(v)}{dv^n}\right|_{v=v_0}.\tag{5}$$

According to Equation (5), the amplitude of the second harmonic component is

$$A_2(v_0) = \frac{I_0(v) \cdot C \cdot L}{4} \cdot A^2 \cdot \left.\frac{d^2\alpha}{dv^2}\right|_{v=v_0}.\tag{6}$$

Based on the above formulas, the second harmonic components at the center frequency reach maximum values, which are positively proportional to the gas concentration. In summary, TDLAS-WMS is the optimum choice to analyze the measured gas concentration, which can effectively reduce the 1/f noise, increase the signal-to-noise ratio, and improve the MDL performance of the sensor [18].

For carbon isotope detection, $\delta^{13}C$ is defined to represent the carbon isotope ratio, shown as follows [20]:

$$\delta^{13}C = 1000 \times \left(\frac{[^{13}CO_2]/[^{12}CO_2]}{RPDB} - 1\right), ‰,\tag{7}$$

where RPDB is the $^{13}C/^{12}C$ ratio of Pee Dee Belemnite (PDB), assumed to 0.01124 [21].

The parameters of spectral line absorption intensity are a function of temperature; therefore, when the temperature of the measured gas changes, it will bring uncertainty to the measurement results. The reason is that the temperature has different influences on the absorption intensity of the two pair lines of the carbon isotope. In the detection of isotope $\delta^{13}C$ values, the effect of temperature on abundance values can be described by

$$\Delta\delta \approx \frac{\Delta T \cdot \Delta E}{k \cdot B \cdot T^2} \times 1000‰,\tag{8}$$

where $\Delta\delta$ is the isotope ratio change, ΔT is the gas temperature change, B is the Boltzmann constant, k is a constant, T is the absolute temperature, and ΔE is the low energy level difference between the two absorption lines.

To reduce the effect of temperature change on carbon isotope detection, it is necessary to select absorption line pairs with low energy levels as close as possible to each other. In addition, high-stability temperature control is required for the MPGC.

2.2. Absorption Line Selection

The high-resolution transmission (HITRAN) molecular absorption database is a worldwide standard for calculating or simulating atmospheric molecular transmission and radiation [22,23]. It covers a wide spectral region from microwave to ultraviolet radiation. In the infrared band, the absorption spectra mainly comprise vibrational and rotational spectra, and each gas has multiple absorption bands. The HITRAN absorption spectra of 10 ppmv CO_2 and 1% H_2O at a gas pressure of 20 Torr with a 24-m effective optical path length are depicted in Figure 1.

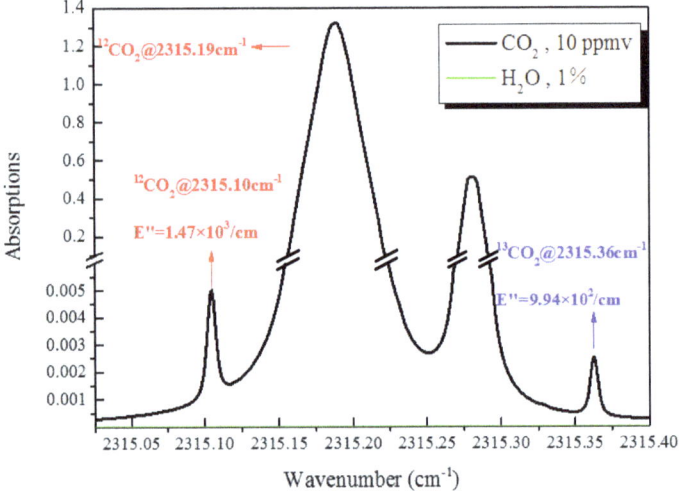

Figure 1. The high-resolution transmission (HITRAN) absorption spectra of 10 ppmv CO_2 and 1% H_2O at a gas pressure of 20 Torr with a 24-m effective optical path length.

The selection of the absorption line pair of $^{13}CO_2/^{12}CO_2$ is especially critical for obtaining high sensitivity and precision. The absorption line pair was chosen at 2315.10 cm^{-1} and 2315.36 cm^{-1} with appropriate spacing, as it does not overlap with other chemical substances, such as methane or sulfuretted hydrogen. The absorption lines are strong, resulting in a better signal-to-noise ratio and avoiding the need for a larger absorption cell, which is beneficial for the miniaturization of the sensing system. Due to the similar low energy levels of absorption line pairs, the detected isotopic stability is less dependent on temperature. In particular, there is no water vapor absorption line between the line pairs; thus, the sensing system does not need to over-consider the influence of ubiquitous water vapor.

Using the selected absorption line pair, the parameters shown in Equation (8) can be determined as follows: Boltzmann constant $B = 1.328 \times 10^{-23}$ J·K^{-1}, $\Delta E = 4.76 \times 10^2$ cm^{-1}, $k = 1.9865 \times 10^{-23}$ J·cm, and the target temperature $T = 300$ K. For the target CO_2 isotope $\delta^{13}C$ value of 0.6‰, the temperature change amount ΔT should be less than 0.127 K. If the calculation of the MDL according to Allan variance can reach lower than ~0.01‰, it will meet the requirement of deep-sea NGH exploration.

3. System Configuration

Using the above detection principle, the schematic block diagram and physical image of the sensing system are shown in Figure 2a,b, respectively.

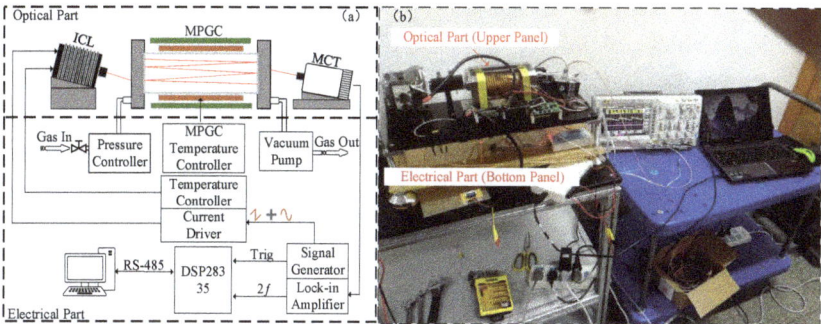

Figure 2. (**a**) Schematic block diagram of the sensing system including an optical part and an electrical part. (**b**) Gas sensor physical image including an optical part and an electrical part. MPGC: multi-pass gas cell; ICL: interband cascade laser; MCT: mercury cadmium telluride.

The sensing system mainly includes an optical part and an electrical part. In the optical part, a continuous-wave ICL with a thermoelectric cooling function produced by Nanoplus was used as a luminous source. In order to meet the compact requirements of the sensing system, the physical size of the MPGC was $20 \times 7.6 \times 10.5$ cm^3, and the effective optical path was 24 m. The emitting light of the ICL enters into the MPGC (custom design multi-pass White type absorption cell) with correct position and angle, and exits to the mid-infrared MCT detector after 215 reflections in the interior of the MPGC. The mid-infrared detector (PVI-4TE-5, VIGO, Ożarów Mazowiecki, Poland) with a thermoelectric cooling function was manufactured by VIGO Systems.

In the electrical part, a self-fabricated ICL driver and temperature controller were used to replace commercial instruments, thereby reducing the size and cost of the proposed sensor. A low-power, high-performance, floating-point digital signal processor (DSP, Texas Instruments, Dallas, TX, USA) was used as the controller for the sensing system. Under the control of the DSP processor, a triangular wave signal and a sine wave signal are superimposed by an adder and supplied to a current source for scanning and modulating the ICL. The signal output from the mid-infrared MCT is demodulated using a self-developed lock-in amplifier (LIA). The DSP processor uses an analog-to-digital converter (ADC) to acquire the peak of secondary harmonic signal pairs, and then obtains the $^{13}CO_2/^{12}CO_2$ isotope ratio according to the calibration curve. A pressure controller (model IQ + Flow, $4 \times 3 \times 1$ cm^3) manufactured by Bronkhorst was used to control the pressure of the MPGC at 20 Torr, while the micro-DC pump (TOPSFLO, Changsha, China) manufactured by Knf Neuberger was used to extract the separated gas into the MPGC.

As the absorption coefficient of the selected absorption line pairs to be measured can be affected by the temperature of CO_2, the temperature of CO_2 directly affects the accuracy and precision of the sensing system. A high-precision temperature control system for the MPGC was designed and developed in this paper. The schematic block diagram and physical image are shown in Figure 3a,b.

In terms of hardware in the circuit, a polyimide electrothermal film as a heating device and a PT1000 platinum resistor with ±0.03 °C accuracy (1/10 B Class) as a temperature sensor were used to form a closed-loop temperature control system. The precision of the temperature measurement could reach 10 mK. For software, the Ziegier–Nichols method was used to set the values of the three proportional, integral, and difference parameters (P, I, and D) [24]. Aiming to address the temperature overshoot caused by the complex structure and slow response of the controlled object, the integral separation PID control algorithm was used to rapidly control the temperature and avoid overshoot, resulting in reliable performance for the $^{13}CO_2/^{12}CO_2$ isotope sensing system.

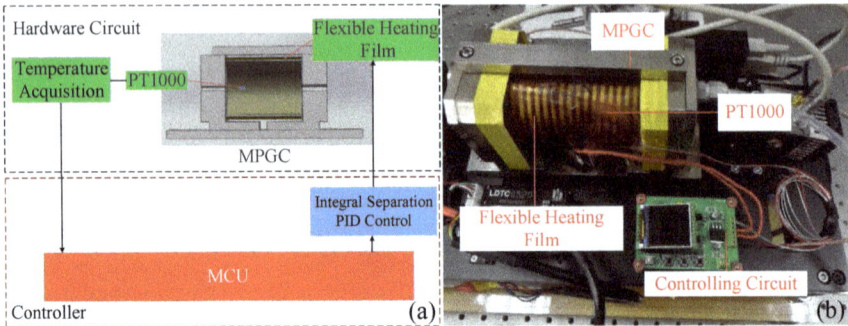

Figure 3. (**a**) Schematic block diagram of the temperature control system for the multi-pass gas cell (MPGC), which includes a controller and hardware circuit. (**b**) Physical image of the temperature control system including a PT1000 temperature sensor, flexible heating film, and controlling circuit.

4. Indoor Experiment

4.1. System Operating Temperature Control

The experiment for the MPGC working temperature was conducted to verify the performance of the investigated temperature control system, and the experimental results are shown in Figure 4.

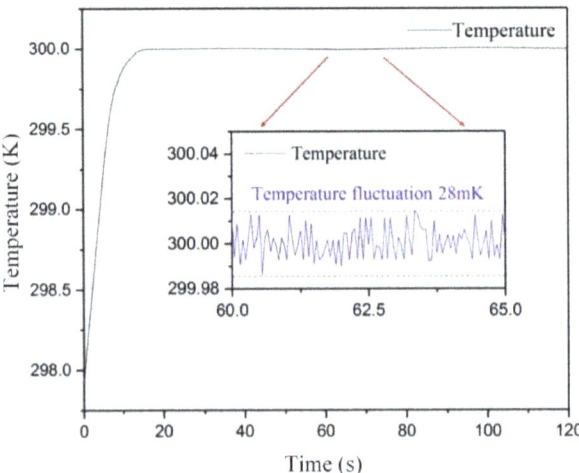

Figure 4. The experimental results of the MPGC working temperature with 120 s of control time; the temperature fluctuation results are inserted.

It can be seen from the above figure that the working temperature of the MPGC reached a stable state after 15 s with no overshoot. In the stable state, the temperature fluctuation of 0.028 K was less than 0.127 K. Before the set threshold of working temperature was reached, the PD algorithm was utilized. Then, the PID algorithm was used during the stable state. As the mentioned integral separation PID control algorithm was adopted, the working temperature of the MPGC increased quickly with no overshoot. Thus, the problem of slow recovery following heating overshoot could be avoided. As we used a polyimide electrothermal film wrapping with a cylindrical MPGC for heating, there was still a temperature gradient of 0.17 °C/cm in the radial direction of the MPGC. This is the main limitation of the sensing system for carbon isotope detection accuracy.

4.2. System Response

Because the sensor uses the TDLAS technique, the signal-to-noise ratio of the output signal is related to the modulation depth of the laser and the line width of the target gas molecular absorption line. In general, the optimal modulation coefficient is 2.2, that is, the modulation depth is 2.2 times the half-width at half-maximum of the absorption line peak. In the measurement, because the widths of the absorption spectral lines of $^{13}CO_2$ and $^{12}CO_2$ s were not consistent, the tradeoff modulation depth was 0.3 mA.

In the sensing system response experiment, the waveform of the second harmonic signal could be obtained by subtracting the background signal represented by the non-absorbent wing. Then, the relationship between the peak value of second harmonic signal and the carbon line pairs could be obtained. The second harmonic signals obtained from five different concentrations (20, 30, 40, 50, and 60 ppm) are shown in Figure 5.

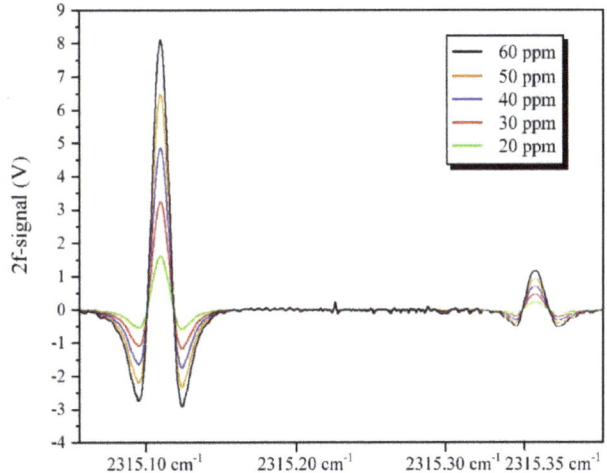

Figure 5. Second harmonic signals with the background absorption removed, obtained from five different concentrations (20, 30, 40, 50, and 60 ppm).

As shown in the above figure, five different concentrations of CO_2 gas were separately arranged using a gas dilution system with a concentration accuracy of ±1% (Environics, 4040); they were sequentially detected by the proposed carbon isotope sensing system. The peak value of the harmonic signal corresponding to different concentrations of $^{13}CO_2$ and $^{12}CO_2$ can be acquired to calculate the carbon isotope ratio R^{13}.

4.3. System Calibration

In order to accurately measure carbon isotopes, it is necessary to calibrate the sensing system using known concentrations of gas. The five different concentrations of CO_2 (20 ppm, 30 ppm, 40 ppm, 50 ppm, and 60 ppm) configured above were pumped into the MPGC with 5 min of measurement and a constant pressure of 20 Torr. The peak values of the second harmonic signal of $^{13}CO_2$ and $^{12}CO_2$ are shown in Figure 6.

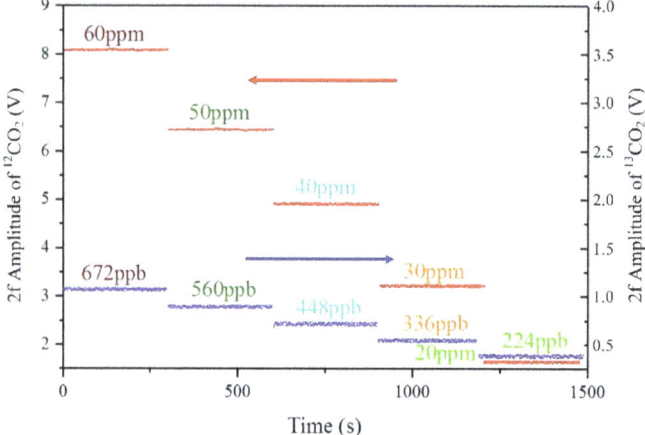

Figure 6. Peak values of the second harmonic signal of $^{13}CO_2$ and $^{12}CO_2$ for five different concentration levels of 20 ppm, 30 ppm, 40 ppm, 50 ppm, and 60 ppm.

The five datasets were averaged firstly, and then the relationship between the average voltage and calibration concentration was obtained, as shown in Figure 7.

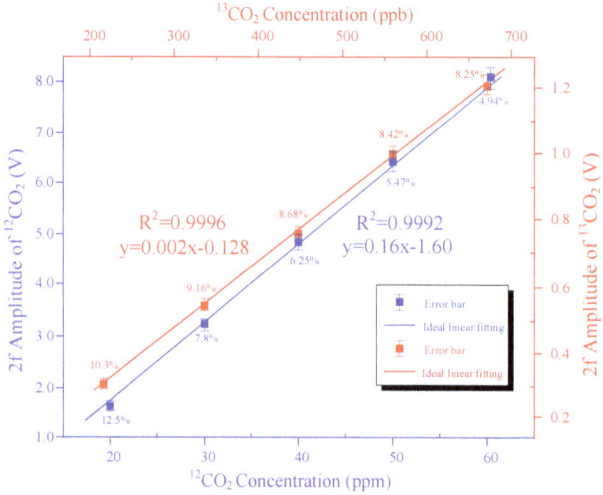

Figure 7. Measured concentration data points and linear fitting curves of the CO_2 concentration versus 2f amplitude value.

In Figure 7, the blue data points indicate the relationship between peak values of the second harmonic signal of $^{12}CO_2$ and the corresponding concentration. On the other hand, the red data points indicate the relationship between peak values of the second harmonic signal of $^{13}CO_2$ and the corresponding concentration. Using the acquired data, we can obtain the following formulas via zero-crossing linear fitting:

$$^{12}CO_2 = 0.16 \times \max 2f(^{12}CO_2) - 1.60, \tag{9}$$

$$^{13}CO_2 = 0.002 \times \max 2f(^{13}CO_2) - 0.128, \tag{10}$$

where $max2f(^{12}C)$ and $max2f(^{13}C)$ are the peak values of the second harmonic signals of $^{13}CO_2$ and $^{12}CO_2$, respectively. The values of zero crossing shown in Equations (9) and (10) were −1.60 for $^{12}CO_2$ and −0.128 for $^{13}CO_2$, respectively. As the output response of the sensor was calibrated by linear fitting, the response linearity could reach 0.9992 for $^{12}CO_2$ and 0.9996 for $^{13}CO_2$, as shown in Figure 7. Calibrated errors were inevitably introduced to the equations. For $^{12}CO_2$, the range of calibrated error was between 12.5% and 4.94%. The calibration error decreased with the increase in concentration. For $^{13}CO_2$, the range of calibrated error was between 10.3% and 8.25%. In addition, with the increase in concentration, the calibrated error decreased, and the trend was consistent with that of $^{12}CO_2$. The above two formulas can calculate the $^{13}CO_2$ and $^{12}CO_2$ concentration by deriving the peak values of the second harmonic signal; thus, the carbon isotope ratio can be obtained accurately.

The stability of the pressure control could reach ~0.2%, and the precision could reach ±0.04 Torr. The stability and precision of the pressure control affects the measurement performance of the sensing system. The static gas distribution mode, which closes the valves at the front and back ends of the MPGC (sealing the MPGC) when the gas pressure in the MPGC reaches the target value, was adopted to stabilize the gas pressure and reduce calibration error.

4.4. Carbon Isotope Measurement Precision

Based on the selected absorption spectrum lines of $^{13}CO_2/^{12}CO_2$, the MDL concentration of the carbon isotope could reach sub-ppmv levels. Thus, the obtained performance can fully meet the requirements of NGH exploration, where the background value of CO_2 concentration in seawater is several hundred ppmv. When measuring the isotope ratio of $^{13}CO_2/^{12}CO_2$, the measured data drift over time. To test the measurement precision and long-term stability of the sensing system, a configured concentration of 500 ppmv CO_2 was pumped into the MPGC. The Allen variance $\delta^{13}C$ calculated from the measured data is shown in Figure 8.

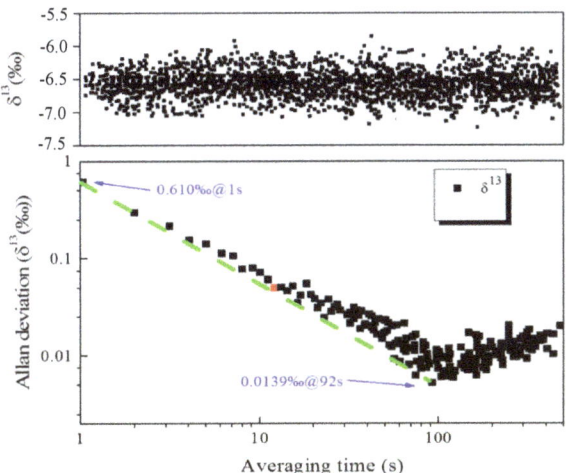

Figure 8. Long-term measurement of a configured concentration of 500 ppmv CO_2 and the measured Allan deviation analysis $^{13}\delta$ based on the data.

The experiment results show that the precision was 0.610‰ with an integral time of 1 s. Moreover, when the integral time was increased to 92 s, the corresponding precision could be significantly reduced to 0.0139‰. Since white noise was the main component before 92 s, the precision decreased along with the increase in integral time. After 92 s, drift became the dominant source of noise, and the precision began increasing. The green dashed line in the figure describes the theoretical expectation of the system response when white noise plays a dominant role.

4.5. Working Performance with Gas–Liquid Separator

In order to test the measurement performance of the $^{13}CO_2/^{12}CO_2$ isotope ratio of CO_2 gas extracted from water, an experiment using the proposed sensing system connected with a gas–liquid separator was conducted. The schematic block diagram of the system configuration and the physical image are shown in Figure 9a,b.

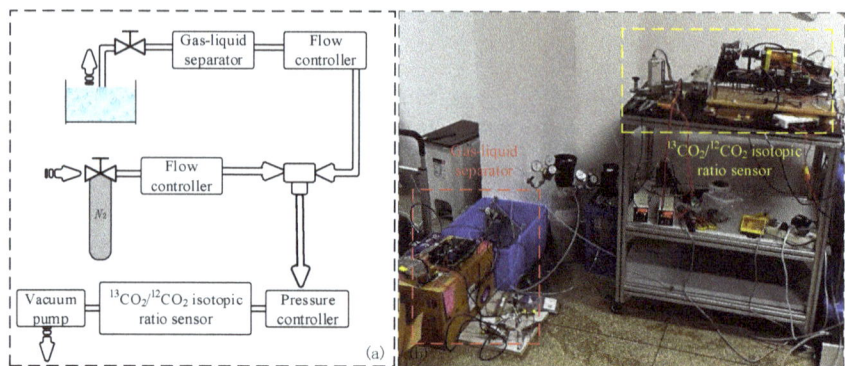

Figure 9. (a) Schematic block diagram of system configuration, and (b) physical image of the sensing system for deployment in the detection of dissolved CO_2 in water assisted by a gas–liquid separator and carrier gas.

Using the gas–liquid separator, the dissolved CO_2 gas in the water was separated but had limited degassing quality. In order to meet the gas pressure requirement of 20 Torr in the MPGC, N_2 was used as a supplementary gas. In the experiment, two flow controllers were used to control the flow rate of the dissolved CO_2 gas and N_2, at 2.5 sccm and 177.5 sccm, respectively. The separated CO_2 and the N_2 carrier gas formed the mixed gas that was pumped into the MPGC with two hours of measurement. The measured data are described in Figure 10.

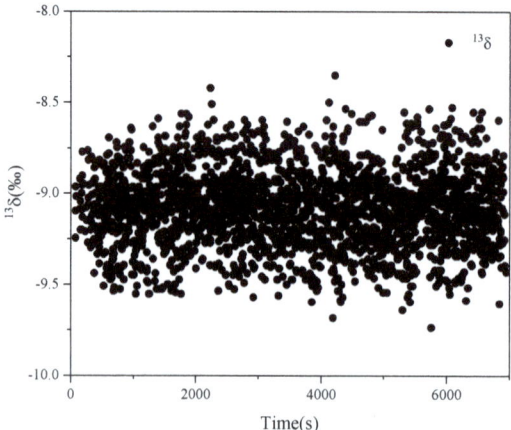

Figure 10. Carbon isotope measurement of mixed gas including the separated CO_2 and the N_2 carrier gas with two hours of measurement.

In two hours of testing, the carbon isotope ratio of the CO_2 gas separated from water was between −8.351‰ and −9.736‰, and the average value was −9.081‰; the largest fluctuation was 0.73‰, and the standard deviation was as low as 0.2‰ [25]. Since N_2 carrier gas with a purity of 99.99%

contained CO_2 in the order of ppm, the carbon isotope ratio was constant. During the long-term experiment, there was a flow velocity fluctuation between the gas separated by the gas–liquid separator and the pure N_2 carrier gas, which resulted in fluctuations of the carbon isotope ratio. The drift noise of the sensing system was not shown here.

However, these measurements were conducted using a continuous-wave mid-infrared ICL, a compact MPGC, a mid-infrared MCT detector, and optical and electrical components being exposed to air. Here, it should be noted that there is no sequestration of CO_2 in exhaled air. The performances of the accuracy and precision of carbon isotope measurements were worse than the results achieved in the experimental measurement using a configured concentration of CO_2. We expect that the accuracy and precision of the $\delta^{13}C$ measurements performed in this work can be improved by adopting a cylinder cavity to seal the entire sensing system with pure nitrogen. In addition, as the temperature gradient exists in the radial direction and axial direction of the MPGC, the accuracy and precision of carbon isotope measurement deteriorated. A smaller MPGC with the same effective optical path length will be investigated, and the effect of the temperature gradient will be reduced. It is, therefore, expected that an improvement of the accuracy and precision can be achieved.

5. Discussion and Conclusions

In this paper, we reported the design and performance of a $^{13}CO_2/^{12}CO_2$ isotope ratio sensor based on the TDLAS technique. The sensing system consisted of a mid-infrared ICL, a mid-infrared MCT detector, and an MPGC with a physical size of 20 × 7.6 × 10.5 cm^3. The ICL with a center wavelength of 4.3 μm covered the carbon isotope absorption line pairs located at 2314.36 cm^{-1} and 2315.10 cm^{-1}. In order to achieve high-precision measurements of carbon isotope ratio, a high-precision temperature control system for MPGC was fabricated. In the indoor experiment, the response linearity of the proposed system reached 0.9996%, and the MDL was as low as 0.0139‰ when the integration time was 92 s. Finally, the sensor was combined with a gas–liquid separator to measure the $^{13}CO_2/^{12}CO_2$ isotope ratio of CO_2 gas extracted from water. In conclusion, although the performance of the proposed sensor can meet the requirements of NGH exploration, this was only achieved in laboratory conditions, which suggests the proposed prototype of the $^{13}CO_2/^{12}CO_2$ isotope sensor has potential application in deep-sea natural gas hydrate exploration. Further research is needed before the sensors can really be applied in NGH exploration. In terms of measurement technique, as 2f/1f WMS technique had a negligible absorption-independent background and was immune to the absorption-independent systematic losses. Thus, the 2f/1f WMS technique is a possible approach to reduce noise from mechanical vibration and to make the sensor more robust when applied in NGH exploration. In terms of sensors, cooling devices, antistatic devices, and shock absorption/isolation devices should be investigated. In terms of loading equipment, a cylindrical deep-sea hull that can withstand tens of mega-pascals on the sea floor is needed, as well as a mechanical fastening device and electrical interface cable.

Compared with the high-precision carbon isotope measured methods of mass spectrometry, although the mass spectrometer can achieve MDL of ppt level for single chemical elements and ppm level for isotopes, the requirements of a high vacuum and high stability make it more suitable for ultra-high-performance detection in a laboratory environment, whereas it is not suitable for application in complex environments. For chromatographic analysis, its advantages lie in the detection of a variety of chemical components, and the MDL can reach to ppm level; however, it seems inadequate to meet the MDL requirement of ppb level. As far as flame ionization is concerned, it is responsive to almost all organic materials, and the response is proportional to the number of carbon atoms; however, it is insensitive to inorganic substances such as CO_2, H_2O, CS_2, etc. Based on the above discussion, the mid-infrared TDLAS technique can realize the MDL of ppb level for single gas components, as well as tens of ppm for isotopes, and it can be applied in complex environments.

In many practical applications, the ICL is subjected to low-frequency environmental perturbations such as thermal fluctuations and mechanical vibrations, resulting in the laser wavelength fluctuating slightly. A laser wavelength locking (LWL) technique is required to keep the operating point of the ICL constant, thereby reducing the fluctuation. Thus, the precision of the proposed sensing system can be further improved by using the LWL technique, which requires a redesign of the system structure. The stability of the system can be further improved by adding a damping and waterproof device to strengthen the antijamming capability.

Author Contributions: H.Z. performed the research. M.W., Y.L., P.W., and C.C. conceptualized the research and wrote the paper.

Funding: This research was supported by the National Key R&D Program of China (No. 2016YFC0303902, No. 2018YFC1503802), the National Natural Science Foundation of China (No. 61871199), and the Science and Technology Department of Jilin Province of China (No. 20180201022GX).

Conflicts of Interest: The authors declare no conflicts of interest.

References

1. Edwards, N.; Mir, R.; Willoughby, E.; Schwalenberg, K.; Scholl, C. The assessment and evolution of offshore gas hydrate deposits using seafloor controlled source electromagnetic methodology. In Proceedings of the OCEANS'10 IEEE SYDNEY, Sydney, Australia, 24–27 May 2010; pp. 1–10.
2. Hachikubo, A.; Yanagawa, K.; Tomaru, H.; Lu, H.; Matsumoto, R. Molecular and Isotopic Composition of Volatiles in Gas Hydrates and in Sediment from the Joetsu Basin, Eastern Margin of the Japan Sea. *Energies* **2015**, *8*, 4647–4666. [CrossRef]
3. Kennett, J.P.; Cannariato, K.G.; Hendy, I.L.; Behl, R.J. Carbon isotopic evidence for methane hydrate instability during Quaternary interstadials. *Science* **2000**, *288*, 128–133. [CrossRef] [PubMed]
4. Aebersold, R.; Mann, M. Mass spectrometry-based proteomics. *Nature* **2003**, *422*, 198. [CrossRef] [PubMed]
5. Giddings, J. *Dynamics of Chromatography: Principles and Theory*; CRC Press: Boca Raton, FL, USA, 2017.
6. Farajzadeh, M.; Mogaddam, M.; Aghdam, S.; Nouri, N.; Bamorrowat, M. Application of elevated temperature-dispersive liquid-liquid microextraction for determination of organophosphorus pesticides residues in aqueous samples followed by gas chromatography-flame ionization detection. *Food Chem.* **2016**, *212*, 198–204. [CrossRef] [PubMed]
7. Wienhold, F.; Fischer, H.; Hoor, P. TRISTAR-a tracer in situ TDLAS for atmospheric research. *Appl. Phys. B Lasers Opt.* **1998**, *67*, 411–417. [CrossRef]
8. Werle, P. A review of recent advances in semiconductor laser based gas monitors. *Spectrochim. Acta Part A* **1998**, *54*, 197–236. [CrossRef]
9. Lins, B.; Zinn, P.; Engelbrecht, R.; Schmauss, B. Simulation-based comparison of FENBIE noise effects in wavelength modulation spectroscopy and direct absorption TDLAS. *Appl. Phys. B Lasers Opt.* **2010**, *100*, 367–376. [CrossRef]
10. Liu, Y.; Biao, W.; Kai, Y.; Lei, C.; Yue, C.; Ning, Y. The simulation research of gas absorption process based on TDLAS. *Laser J.* **2017**, *38*, 37–40.
11. Min, L.; Bo, F. Design of High Sensitivity Infrared Methane Detector Based on TDLAS-WMS. *Laser J.* **2015**, *36*, 75–79.
12. Lau, S.; Salffner, K.; Löhmannsröben, H. Isotopic resolution of carbon monoxide and carbon dioxide by NIR diode laser spectroscopy. In Proceedings of the SPIE Photonics Europe, Strasbourg, France, 3–7 April 2006; pp. 538–543.
13. Nelson, D.D.; McManus, J.B.; Herndon, S.C.; Zahniser, M.S.; Tuzson, B.; Emmenegger, L. New Method for Isotopic Ratio Measurements of Atmospheric Carbon Dioxide Using a 4.3 µm Pulsed Quantum Cascade Laser. *Appl. Phys. B* **2008**, *90*, 301–309. [CrossRef]
14. Tuzson, B.; Mohn, J.; Zeeman, M.J.; Werner, R.A.; Eugster, W.; Zahniser, M.S.; Nelson, D.D.; McManus, J.B.; Emmenegger, L. High precision and continuous field measurements of δ^{13}C and δ^{18}O in carbon dioxide with a cryogen-free QCLAS. *Appl. Phys. B* **2008**, *92*, 451. [CrossRef]
15. Kasyutich, V.; Philip, A.M. $^{13}CO_2/^{12}CO_2$ isotopic ratio measurements with a continuous-wave quantum cascade laser in exhaled breath. *Infrared Phys. Technol.* **2012**, *55*, 60–66. [CrossRef]
16. Ghetti, A.; Cocola, L.; Tondello, G.; Galzerano, G.; Poletto, L. Performance evaluation of a TDLAS system for carbon dioxide isotopic ratio measurement in human breath. In Proceedings of the SPIE Photonics Europe, Strasbourg, France, 22–26 April 2018; p. 106800U.

17. Chen, C.; Biao, W. Review of infrared gas technology. *Laser J.* **2019**, *40*, 1–5.
18. Klein, A.; Witzel, O.; Ebert, V. Rapid, Time-Division Multiplexed, Direct Absorption- and Wavelength Modulation-Spectroscopy. *Sensors* **2014**, *14*, 21497–21513. [CrossRef] [PubMed]
19. Wang, M.; Zhang, Y.; Liu, W.; Liu, J.; Wang, T.; Tu, X.; Gao, S.; Kan, R. Second-harmonic Detection Research with Tunable Diode Laser Absorption Spectroscope. *Opt. Tech.* **2005**, *31*, 279–285.
20. Schaeffer, S.; Miller, J.; Vaughn, B.; White, J.; Bowling, D. Long-term field performance of a tunable diode laser absorption spectrometer for analysis of carbon isotopes of CO_2 in forest air. *Atmos. Chem. Phys.* **2008**, *8*, 5263–5277. [CrossRef]
21. Srivastava, A.; Verkouteren, R. Metrology for stable isotope reference materials: $^{13}C/^{12}C$ and $^{18}O/^{16}O$ isotope ratio value assignment of pure carbon dioxide gas samples on the Vienna PeeDee Belemnite-CO_2 scale using dual-inlet mass spectrometry. *Anal. Bioanal. Chem.* **2018**, *410*, 4153–4163. [CrossRef] [PubMed]
22. Gordon, I.; Rothman, L.; Hill, C. The HITRAN 2016 molecular spectroscopic database. *J. Quant. Spectrosc. Radiat. Transf.* **2017**, *203*, 3–69. [CrossRef]
23. Šimečková, M.; Jacquemart, D.; Rothman, L.; Gamache, R.; Goldman, A. Einstein A-coefficients and statistical weights for molecular absorption transitions in the HITRAN database. *J. Quant. Spectrosc. Radiat. Transf.* **2006**, *98*, 130–155. [CrossRef]
24. Åström, K.J.; Hägglund, T. Revisiting the Ziegler–Nichols step response method for PID control. *J. Process Control* **2004**, *14*, 635–650. [CrossRef]
25. Ahn, S.; Jeffrey, A.F. *Standard Errors of Mean, Variance, and Standard Deviation Estimators*; EECS Department, The University of Michigan: Ann Arbor, MI, USA, 2003; pp. 1–2.

 © 2019 by the authors. Licensee MDPI, Basel, Switzerland. This article is an open access article distributed under the terms and conditions of the Creative Commons Attribution (CC BY) license (http://creativecommons.org/licenses/by/4.0/).

Article

TDLAS Monitoring of Carbon Dioxide with Temperature Compensation in Power Plant Exhausts

Xiaorui Zhu [1,3], Shunchun Yao [1,3,*], Wei Ren [2], Zhimin Lu [1,3] and Zhenghui Li [1,3]

1. School of Electric Power, South China University of Technology, Guangzhou 510640, China; 201620110487@mail.scut.edu.cn (X.Z.); zhmlu@scut.edu.cn (Z.L.); epzhenghui@mail.scut.edu.cn (Z.L.)
2. Department of Mechanical and Automation Engineering, The Chinese University of Hong Kong, Shatin, Hong Kong SAR 2912, China; renwei@mae.cuhk.edu.hk
3. Guangdong Province Engineering Research Center of High Efficiency and Low Pollution Energy Conversion, Guangzhou 501640, China
* Correspondence: epscyao@scut.edu.cn; Tel.: +86-139-2515-0807

Received: 7 January 2019; Accepted: 24 January 2019; Published: 28 January 2019

Featured Application: Applying to complex industrial environment, TDLAS monitoring with temperature compensation can come handy.

Abstract: Temperature variations of flue gas have an effect on carbon dioxide (CO_2) emissions monitoring. This paper demonstrates accurate CO_2 concentration measurement using tunable diode laser absorption spectroscopy (TDLAS) with temperature compensation methods. A distributed feedback diode laser at 1579 nm was chosen as the laser source for CO_2 measurements. A modeled flue gas was made referring to CO_2 concentrations of 10–20% and temperatures of 298–338 K in the exhaust of a power plant. Two temperature compensation methods based on direct absorption (DA) and wavelength modulation (WMS) are presented to improve the accuracy of the concentration measurement. The relative standard deviations of DA and WMS measurements of concentration were reduced from 0.84% and 0.35% to 0.42% and 0.31%, respectively. Our experimental results have validated the rationality of temperature compensations and can be further applied for high-precision measurement of gas concentrations in industrial emission monitoring.

Keywords: carbon dioxide monitoring; absorption spectroscopy; temperature compensation; wavelength modulation

1. Introduction

Carbon dioxide (CO_2) is a major component of greenhouse gases. Measurements of CO_2 in large emission sources such as coal-fired power plants provide an effective support of carbon emission statistics. Outlet ducts in power plants provide a challenging detection environment for gas absorption detection due to the instability and poor timeliness. Compared with the current spectroscopic techniques such as Fourier transform infrared (FTIR) [1] and non-dispersive infrared (NDIR) spectroscopy [2], tunable diode laser absorption spectroscopy (TDLAS) with high sensitivity and short response time [3] offers significant advantages for in situ CO_2 measurements in the hot, humid, and dusty environments of power plants. With the rapid advancements in semiconductor laser industry, TDLAS has been widely applied in different fields such as combustion diagnostics [4–8], isotopic analysis [9–13], and atmospheric gas detection [14–18]. Because of their exquisite constituents and compact packaging, as well as low cost, tunable semiconductor diode lasers are nearly the ideal source for highly-sensitive and highly-spectral-resolved measurements in industrial fields [19]. However, Doppler-broadening at high temperatures alters a larger portion of the integrated absorbance for a particular transition to the far-wing region. For in-situ CO_2 measurements in outlet ducts, the temperature-varied absorption line intensity leads to measurement error in both direct absorption (DA) and wavelength modulation spectroscopy (WMS) signals.

Numerous studies have been performed to investigate the influence of temperature variation on TDLAS measurement. Many of them used spectroscopic parameters such as line intensity [20] from databases like HITRAN (High Resolution Transmission) or HITEMP (High Temperature Molecular Spectroscopic). Nevertheless, these databases, which mainly derive from theoretical calculation, may have some uncertainty while applied in high-temperature measurement. Several of previous studies [21–25] measured spectral parameters such as line intensity, line position, and broadening coefficients in a harsh experimental condition. Under a certain range of pressure and temperature, they employed these spectroscopic parameters for temperature compensation. Empirical corrections to the temperature dependence of the absorption for CO_2–CO_2, CO_2–N_2, and CO_2–O_2 at 193–300 K and the wavelength near 4.3 μm are discussed [26,27]. These corrections are applied via the corrective shape χ-function to the line shape function of individual absorption features. Perrin et al. [28] continued the studies at 4.3 μm to develop a temperature dependent χ-function for corrections of CO_2–CO_2 and CO_2–N_2 collisions at temperatures up to 800 K. Additionally, Zhang et al. [29] proposed a 1f ratio method to correct the harmonic signals affected by temperature fluctuations. They employed a DFB (Distributed Feedback) laser at 760.77 nm to monitor 21% oxygen in the range of 300–900 K and the results showed that the proposed method was effective for minimizing the influence of temperature changes. Shu et al. [30] introduced a temperature correction coefficient for HCl concentration using TDLAS at 1.7 μm. Zhang et al. [31] studied the temperature influence on the detection of ammonia slip based on TDLAS. The empirical formula for temperature compensation was proposed and the error was reduced to about 5.1%. Qi et al. [32] simulated the process of direct TDLAS measurement of H_2O based on the HITRAN spectra database and gave the temperature correction curve in atmosphere detection. A temperature correction function was optimized based on open-path TDLAS for determining ammonia emission rules in soil environments [33]. Liu et al. [34] combined regularization methods with TDLAS to measure non-uniform temperature distribution, relying on the measurements of 12 absorption transitions of water vapor from 1300 nm to 1350 nm. The results showed that regularization methods are less sensitive to the noise caused by temperature variation.

Despite some temperature compensation methods applied in industrial process monitoring, little work is found in practical emission measurements of CO_2 at 1579 nm in power plants. In this paper, TDLAS experiments were conducted in a temperature field simulating the outlet duct of a power plant. Both DA and WMS methods are used to measure CO_2 concentrations at different temperatures. Two temperature compensation methods are used to reduce the measurement errors.

2. Absorption Spectroscopy

2.1. Direct Absorption Spectroscopy

The fundamental theory that governs absorption spectroscopy for narrow-linewidth radiation sources is embodied in the Beer–Lambert law (Equation (1)). The ratio of the transmitted intensity I_t and initial (reference) intensity I_0 of laser radiation through an absorbing medium at a particular frequency v is exponentially related to the transition line intensity $S(T)$ ($cm^{-2} \cdot atm^{-1}$), line-shape function $\varphi(v)$ (cm), total pressure P (atm), volume concentration of the absorbing species X, and the path length L (cm):

$$\frac{I_t}{I_0} = \exp[-PS(T)\phi(v)XL] \tag{1}$$

If both sides of Equation (1) are computed logarithmically and integrated over the whole frequency domain, the volume concentration X is described by [34]:

$$X = \frac{\int_{-\infty}^{+\infty} -\ln\frac{I_t}{I_0} dv}{PS(T)L} = \frac{A}{PS(T)L} \tag{2}$$

If the minimum detectable integrated absorbance A_{min} is known, i.e., SNR (signal-to-noise ratio) = 1, the detection limit $X_{j,min}$ of species j can be expressed as:

$$X_{j,min} = \frac{A_{min}}{PS(T)L} \qquad (3)$$

The temperature-dependent line intensity for a particular transition is determined by its line intensity S at a reference temperature T_0 with the following equation:

$$S(T) = S(T_0)\left(\frac{Q(T_0)}{Q(T)}\right)_{T_0} \exp\left[-\frac{hcE_i''}{k}\left(\frac{1}{T}-\frac{1}{T_0}\right)\right] \cdot \left[\frac{1-\exp(-hcv_{0,i}/kT)}{1-\exp(-hcv_{0,i}/kT_0)}\right] \qquad (4)$$

where $Q(T)$ is the partition function of the absorbing molecule (CO_2), $v_{0,i}$ is the frequency of the transition, E'' is the lower-state energy of the transition, h is the Planck's constant (6.626 × 10^{-27} erg·s), k is the Boltzmann's constant (1.38 × 10^{-16} erg/K), and c is the speed of light in vacuum [35].

Temperature correction is required for TDLAS measurements in the environment with varying gas temperatures. According to Equation (2), in the case of the constant optical path and pressure, the ratio of integrated absorbance and line intensity, $A/S(T)$ is theoretically proportional to gas concentration X in the same system. Therefore, the linear correction between $A/S(T)$ and gas concentration can be applied to compensate the errors caused by temperature influence as shown in the Equation (5):

$$y_{A/S(T)} = a + bx_X \qquad (5)$$

In Equation (5), a and b are fitting coefficients. a is the background signal without absorption, and b is the gradient of the linear correction.

2.2. Wavelength Modulation Spectroscopy

Wavelength modulation spectroscopy could be used to improve the sensitivity of measurement. In WMS, the wavelength of the laser used is modulated by a combination of slow triangle wave and a fast-sinusoidal signal while the transmitted light is demodulated at a harmonic of the modulation frequency. For a sufficiently slow ramp, the frequency output of the laser can be expressed as [36]:

$$v(t) = \bar{v} + v_0 \cos(2\pi ft) \qquad (6)$$

where $\bar{v}$ is the laser center frequency, v_0 is the modulation amplitude, and f is the modulation frequency. The transmitted laser intensity can be expanded in a Fourier cosine series as:

$$I(t) = I_0(t)T[\bar{v} + v_0\cos(2\pi ft)] = I_0(t)\sum_{k=0}^{k=\infty} H_k(\bar{v}, v_0)\cos(2\pi kft) \qquad (7)$$

where $H_k(\bar{v}, v_0)$ is the kth harmonic Fourier component of the transmission coefficient. The second harmonic Fourier coefficient which is given by:

$$H_2(x,m) = \frac{2XLPS(T)}{\pi \Delta v_c}\left\{\frac{2}{m^2}\left[\frac{2+m^2}{(1+m^2)^{1/2}}-2\right]\right\} \qquad (8)$$

Two dimensionless numbers $x = 2(\bar{v} - v_0)/\Delta v_c$, $m = 2a/(\Delta v_c)$, Δv_c (cm^{-1}) is the collisional linewidth (half width at half maximum, HWHM) of the probed transition, respectively; m is the modulation index, a dimensionless parameter defined as the ratio of the wavelength-modulation amplitude to the linewidth Δv_c. CO_2 concentration X is inferred explicitly from [37]:

$$X \propto \frac{I_0 P_{2f} \pi \Delta v_c}{S(T)PL} \left\{ \frac{2}{m^2} \left[\frac{2+m^2}{(1+m^2)^{1/2}} - 2 \right] \right\}^{-1} \quad (9)$$

where P_{2f} is the measured peak 2f signal at the line center v_0; therefore, the measured column density depends on the line intensity and linewidth of the transition, both of which vary with temperature.

For the optical thin condition due to the weak absorption in the near-infrared and low-gas concentration in industrial emissions, the peak 2f signal is linearly proportional to concentration:

$$y_X = a + b x_{P_{2f}} \quad (10)$$

The variation of line intensity and collisional linewidth caused by temperature change introduces errors to the above linear fitting. According to Equation (6), parameters like line intensity $S(T)$ and collisional linewidth Δv_c are considered in the correction of temperature influence, which is given by:

$$y_X = a + b x_{\frac{P_{2f} \Delta v_c}{S(T)}} \quad (11)$$

3. Experiment Details

The exhaust of a power plant has a complicated composition, possibly including H_2O, SO_2, N_2, O_2, CO, and CO_2. According to the HITRAN2012 database [38], O_2 and SO_2 have no transitions near 1579 nm, and the strongest transition of H_2O is up to 10^{-26} in magnitude. Figure 1 shows the simulated absorbance of H_2O (0.8%), CO (0.001%), and CO_2 (10%) in the near-infrared at a typical condition of the power plant exhaust from 1578 to 1580 nm (6330–6335.5 cm^{-1}). The CO_2 absorbance near 1579.12 nm (6332.7 cm^{-1}) selected in this study is approximately 1000 times stronger than the transitions of other gases and isolated from CO interference transitions, while it is still accessible by distributed feedback diode lasers. Thus, this transition is sufficiently strong and ideal for in situ CO_2 measurement in the outlet ducts.

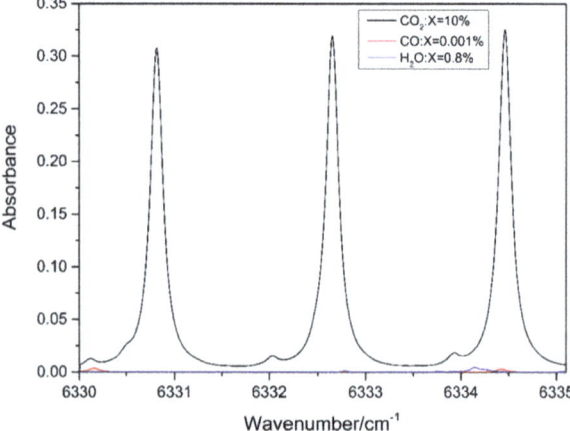

Figure 1. Simulated absorbance of CO_2 (10%), CO (0.001%), and H_2O (0.8%) from 6330–6335.5 cm^{-1} based on the HITRAN (High Resolution Transmission) database (298 K, 1 atm).

The experimental setup for CO_2 concentration measurement is schematically shown in Figure 2. The setup consists of two main parts: gas preparation and laser detection. For the gas preparation, the CO_2 mixtures were generated by mixing the pure N_2 and CO_2 (0%–20%, interval of 2%) continuous gas flows controlled by two mass flow controllers (Seven Stars, Beijing, China). A gas mixer was used to obtain the uniform gas mixture. For the laser detection, the 1.58 µm DFB laser with ~5 mW output power was used as the laser source to exploit the CO_2 absorption line. A photodetector (SM05PD5A InGaAs Detector, THORLABS) was used to detect the transmitted laser beam through the test gas. Laser intensity and wavelength were varied by a combination of temperature and injection current using a commercial controller (PCI-1DA) with the real time spectrum display and capture features. The electrical signals received by the controller were processed into real-time spectral images. The wavelength was scanned (laser temperature at 35 °C, injection current scan between 30 and 120 mA) with a linear ramp of current from a function generator. The temperature and current tuning coefficients were measured to be 0.413 cm^{-1}/K and 0.049 cm^{-1}/mA, respectively. For the second harmonic signals, the laser controller contains lock-in amplification and demodulator for 2f signal extraction.

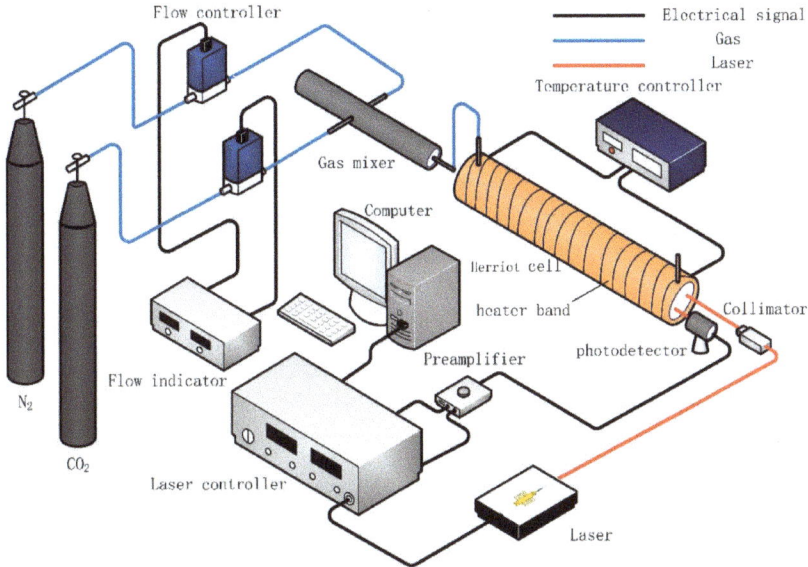

Figure 2. Schematic diagram of the experimental setup.

The laser beam was directed into a Herriot cell (29.5 cm long, 571 mL volume, effective optical path length of 20 m) by a collimator and detected by the InGaAs detector. The 20-meter optical path allowed the system to have sufficient detection sensitivity. The cell was mounted with two wedged (1.5°) sapphire windows to avoid unwanted interference fringes. Three type-K thermocouples were equally spaced along the center section of the Herriot cell to determine the temperature of CO_2; the maximum temperature difference observed among these thermocouples was <0.1 K from 298 K to 338 K. The gas pressure was continuously measured by a pressure gauge (YN60, full range 0.6 MPa) with an accuracy grade of ±0.015 MPa.

Before each experiment, the Herriot cell was purged with pure N_2 for 5 minutes to remove the residual gas. The cell was operated at 298–338 K (interval of 10 K) at an operational pressure of 1 atm. When a stabilized temperature was obtained, the test gases at six different CO_2 concentrations (10%–20%, interval of 2%) were measured with a single sweep of the laser. The continuous tuning range (at constant temperature) was within 5 cm^{-1}. The diode laser was scanned at a 5 Hz repetition rate continuously, and the data were sampled at a frequency of 24 kHz to provide 4800 data points/sweep. The response time of real-time online monitoring can be further reduced by decreasing the sampling frequency. Each measured spectrum was obtained with 20 averaged sweeps, yielding a 4 s measurement time that was fast enough to satisfy the industrial application. Experiments using both direct absorption and wavelength modulation were repeated for three times to evaluate the repeatability of the sensor system.

4. Direct Absorption Measurement and Correction

In this part, the direct absorption spectra with and without temperature corrections are compared to analyze the measurement accuracy by relative errors. The x-axis of measured spectral data were sampling points, which were time domain signals. The frequency calibration of the spectrum was required to convert the time domain to the frequency domain. The simulations based on HITRAN2012 were used to calibrate the frequency domain. As shown in Figure 3, three wavenumbers in HITRAN2012 correspond to three points in measured spectral. Therefore, the curvilinear equation of quadratic polynomial was obtained to convert the time domain to the frequency domain. Figure 4 plots the measured incident signal I_0 and the direct absorption signals I_t for 10% CO_2 at 318 K. The incident laser intensity I_0 was inferred by third-order polynomial fit in the far-wing parts without absorption. The Lorentz profile was chosen to fit the line shape for improving accuracy due to the domination of Lorentzian component at 1 atm. After fitting, the logarithmic ln (I_t/I_0) was integrated over the whole frequency domain to calculate the integrated absorbance A. Thus, the concentration X can be calculated from Equation (2), with line intensity $S(T)$ obtained from Equation (4), total pressure P, and path length L as known parameters.

Figure 5 shows the measured absorption spectra of 16% CO_2 in the 6331.8–6333.6 cm^{-1} spectral region at five different temperatures from 298 to 338K. The absorption signals in Figure 5 decrease with the increased temperature. Figure 6 plots the CO_2 line shape near 6332.7 cm^{-1} (P = 1atm, T = 298 K, 10% CO_2 in N_2) well overlaid with the best-fit Lorentz profile. The obtained Lorentzian width of 0.15412 ± 0.0042 cm^{-1} was broadened by pressure and temperature. The residuals of the curve to data is shown in the bottom panel of Figure 6, showing that the fits are similar. Table 1 shows the relative errors of direct absorption from 298 to 338 K. The error of inferred concentration obtained from the spectroscopic measurement generally rises with the increase of temperature. At 308 K, the biggest relative error was up to 5.66%, which cannot meet the requirement of industrial application.

To further explore the cause of measurement errors, line intensities from the HITRAN database were compared with measured line intensities which are calculated by Equation (4). With a certain $S(T)$, the obtained integrated absorbance A varies linearly with P·X·L where is the product of total pressure (P = 1 atm), absorption length (L = 2000 cm), and concentration (X = 10–20%, interval 2%). The relationship between A and P·X·L can be represented by a linear fitting and the slope of the fitting line is the intensity of the spectral line at the corresponding temperature from 298 K to 338 K, as depicted in Figure 7. Quality of the fit was very good in five temperatures, shown by R squared values from 0.99354 to 0.99875. As shown in Table 2, the calculated values show good agreement with the HITRAN values in variation tendency, but the former's gradient decreases more with the temperature rising. The discrepancy indicates that a 3–10% change of line intensity leads to 1–6% change in the measured CO_2 concentration.

Figure 8 depicts that both the integrated absorbance and the line intensity of 14% CO_2 decrease with the increase of temperature. The discrepancy leads to large errors in the concentration measurement, when the total pressure P and path length L are invariable. This information can be used to compensate these errors caused by change of temperature. According to the Equation (5), the least squares method is used for linear fitting between $A/S(T)$ and concentration X at different temperatures. Figure 9 shows the result of linear fit between $A/S(T)$ and concentration at T = 318 K. The fitting results show that the R squared value reaches 0.999, indicating a good linear relationship between $A/S(T)$ and concentration.

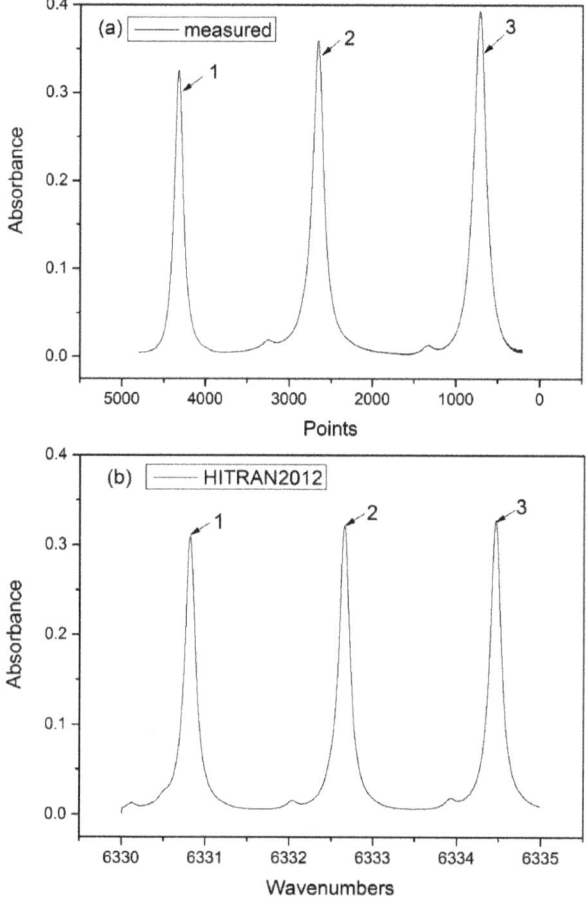

Figure 3. (a) Measured and (b) simulated absorbance of CO_2 at 296 K and 1 atm.

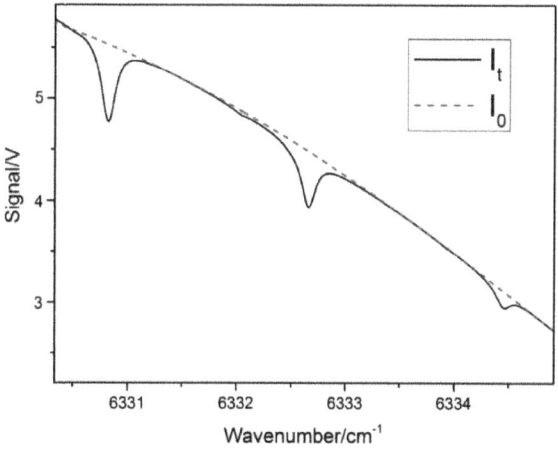

Figure 4. Direct absorption signals I_t and baseline signal I_0 (T = 318 K, 10% CO_2, L = 20 m).

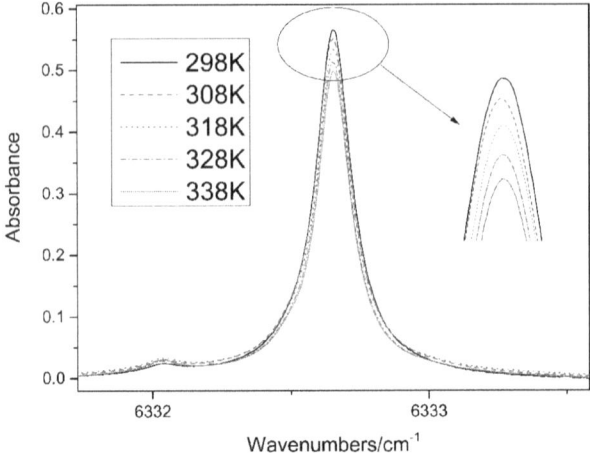

Figure 5. Absorption spectra of 16% CO_2 in the 6331.8 to 6333.6 cm^{-1} spectral region at T = 298–338 K.

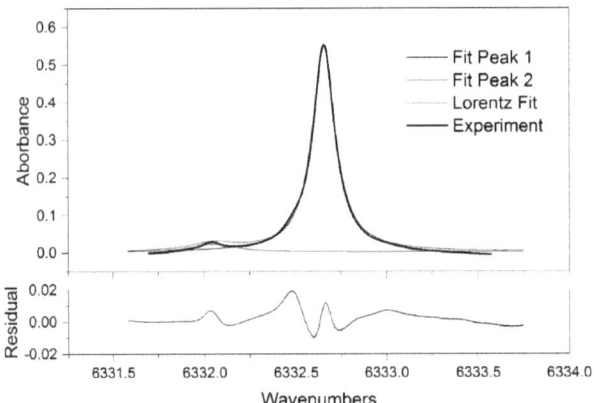

Figure 6. Measured CO_2 spectrum near 6332.7 cm^{-1} with Lorentz fitting (P = 1 atm, T = 338 K, 18% CO_2).

Table 1. The relative errors of direct absorption results from 298 to 338 K.

True Concentration	Inferred Concentration				
	298 K	308 K	318 K	328 K	338 K
10%	0.30%	5.66%	1.33%	2.51%	−2.25%
12%	−2.78%	−0.58%	1.35%	−1.24%	−2.62%
14%	0.28%	−0.60%	−0.38%	−2.88%	−3.35%
16%	−1.24%	0.09%	0.65%	−4.70%	−3.58%
18%	3.74%	1.52%	1.59%	1.56%	−3.14%
20%	−1.21%	3.54%	0.61%	−3.72%	−2.83%

Table 2. Line intensity (10^{-4} $cm^{-2} \cdot atm^{-1}$) values from database and calculated values.

S(T)	298 K	308 K	318 K	328 K	338 K
Database	3.8967	3.8147	3.7318	3.6483	3.5648
Calculated	4.0639	3.9329	3.8824	3.5011	3.2471

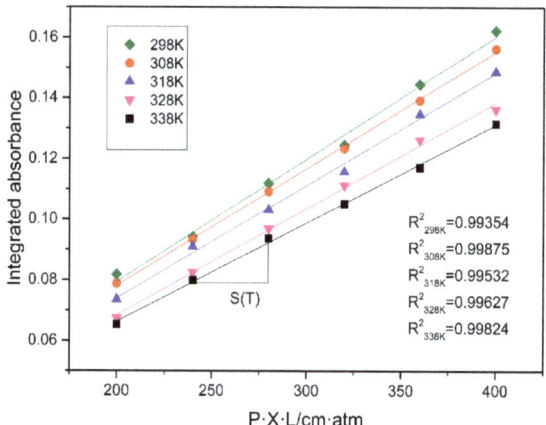

Figure 7. Integrated absorbance as a function of $P \cdot X \cdot L$ (T = 298–338 K, P = 1 atm).

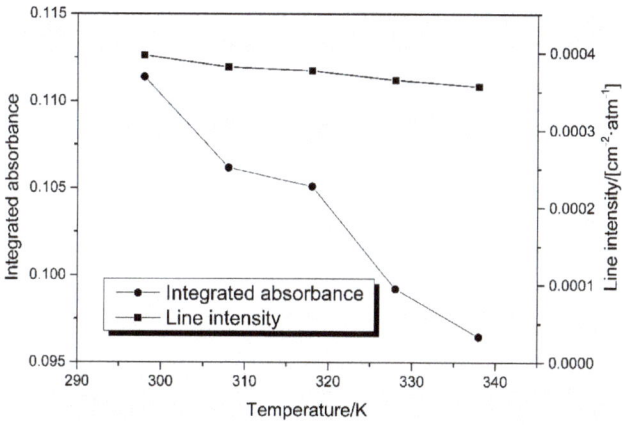

Figure 8. Integrated absorbance and spectral intensity with temperature change (14% CO_2).

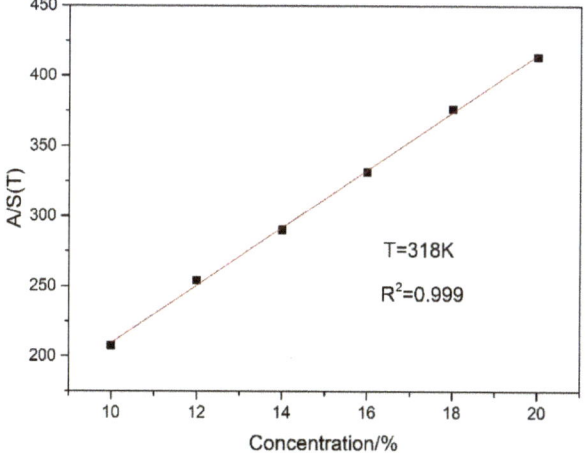

Figure 9. A linear fit between $A/S(T)$ and concentration at T = 318 K.

Figure 10 illustrates the absolute values of the relative error of the measured results before and after the correction. As shown in Figure 10, after temperature compensation, the maximum relative error is −2.57% corresponding to the measurement of 14% CO_2 at 318 K. However, the absolute error was only 0.36% and the relative error of the mean square value was less than 3.5%. All these measurements confirm that this temperature compensation enables the direct detection of CO_2 in outlet ducts temperature with a detection limit of 8.4×10^{-5} at SNR = 1. Measurements of CO_2 concentration can be made for a variety of carbon emission sources, and the data are helpful for developing more accurate detection mechanisms.

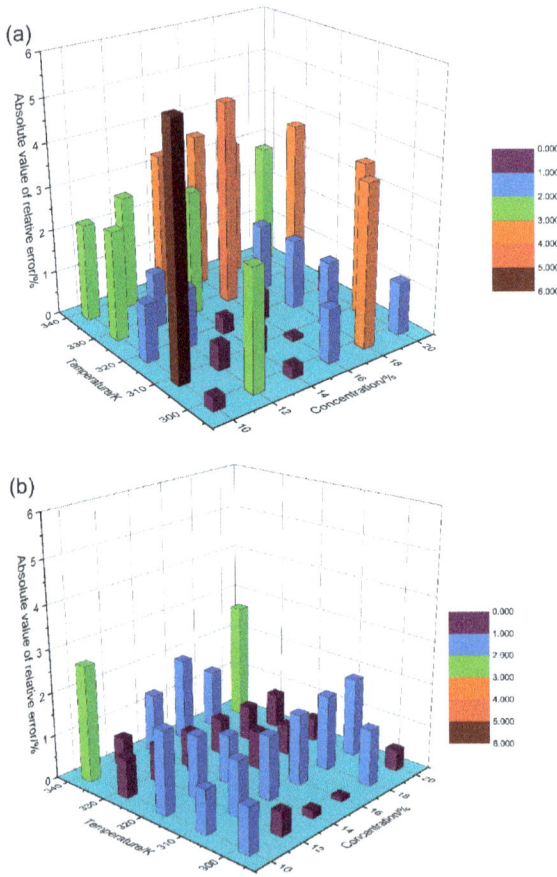

Figure 10. Absolute value of the relative error changes with concentration and temperature. (**a**) Before and (**b**) after compensation.

5. WMS-2f Measurement and Correction

Wavelength modulation spectroscopy (WMS) was applied to restrain a variety of background noise during measurement. Due to the low absorption of CO_2 in the near-infrared, the second harmonic (2f) signal was directly proportional to concentration [39]. When the optical path and pressure were constant, the WMS-2f peak amplitude was related to gas concentration and the wavelength modulation index m, which is dependent on the modulation current I_{mod}. Therefore, an appropriate choice of I_{mod} was required to improve the detection precision.

The WMS-2f peak signals of CO_2 measured at different concentrations (10%–20%, interval 5%) and modulation currents (0 mA–18 mA, interval 1mA) are plotted in Figure 11. These measurements were conducted by controlling the gas flow controller and laser controller to maintain CO_2 concentrations and modulation currents. According to the experimental results shown in Figure 11, with the choice of I_{mod} of ~15 mA, the WMS-2f peak signals are stable at all concentrations. Hence, I_{mod} = 15 mA was selected in this study.

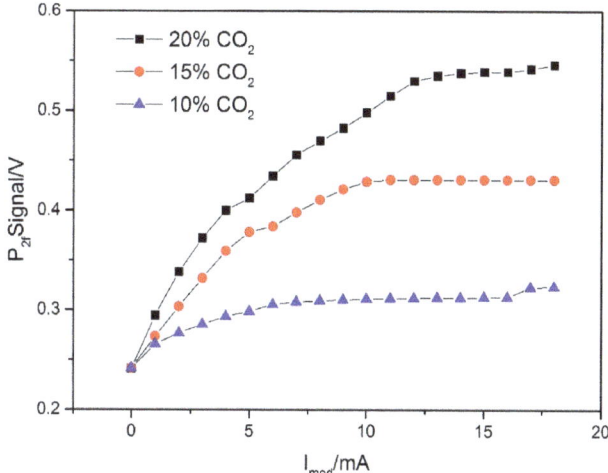

Figure 11. WMS (Wavelength modulation spectroscopy)-2f peak signals as a function of the laser modulation current I_{mod} at different concentrations of CO_2.

In this study, the experimental operation was the same as the direct absorption experiment except a scanning frequency of 50 Hz and modulation frequency of 31.2 KHz were used. There was sufficient time to stabilize the second harmonic peak before the second harmonic data were recorded. Because the shape of the harmonic signal was affected by optical fringe noise and beam noise, the multi-sampling average (40 times) were used to eliminate the negative influences. Gas concentrations and peaks of second harmonic signals show good linear relationships at 298–338 K by employing the mean filtering process. The relationships can be expressed as the fitting of $X = a \cdot P_{2f} + b$, as Equation (10), which was obtained by the linear least squares method, and P_{2f} was the peak of the second harmonic (V), and X was gas concentration (%). The R squared value for linear fitting was >0.99 and the system reproducibility was tested after 60 min of continuous monitoring. In consideration of the temperature effect on line intensity and linewidth, the compensation method can be used to improve the accuracy of measurements. Following Equation (11), the linear relations between signals of $P_{2f} \cdot \Delta v_c / S(T)$ and corresponding concentrations were worked out by the least squares method. The averaged fitting R squared was improved up to 0.997, as shown in Figure 12. The peaks of WMS-2f signals at 6332.7 cm^{-1} were plotted as a function of CO_2 concentration (linear fit equation: y = 0.4114 − 0.0167x, T = 318 K). The results are shown in Tables 3 and 4. Compared with the results without correction, the averaged relative error decreased by 0.24% and the total R squared value increased by 0.28%.

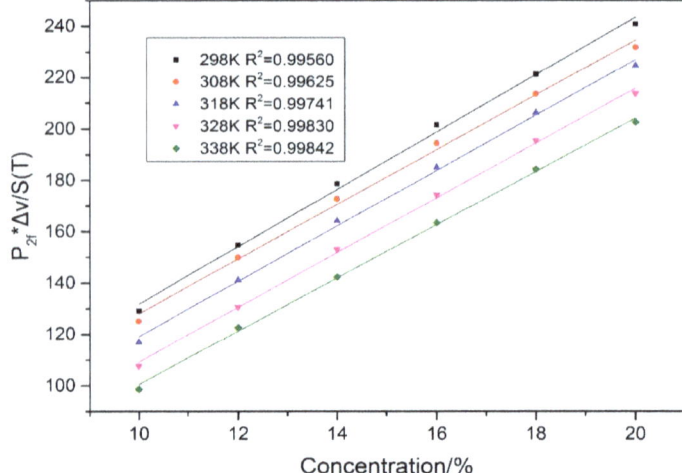

Figure 12. A linear fit of between $P_{2f} \cdot \Delta v_c / S(T)$ and concentration at different temperatures (298–338 K).

Table 3. The relative errors of wavelength modulation results before correction (298–338 K).

True Concentration	Inferred Concentration				
	298 K	308 K	318 K	328 K	338 K
10%	−2.86%	−3.02%	−2.60%	−2.62%	−2.67%
12%	1.49%	0.99%	0.43%	0.46%	1.04%
14%	0.68%	1.64%	1.84%	1.73%	1.31%
16%	2.89%	1.43%	1.15%	1.51%	1.06%
18%	−1.50%	0.31%	0.54%	0.10%	0.32%
20%	−0.89%	−1.64%	−1.65%	−1.48%	−1.35%

Table 4. The relative errors of wavelength modulation results after correction (298–338 K).

True Concentration	Inferred Concentration				
	298 K	308 K	318 K	328 K	338 K
10%	−2.30%	−2.58%	−1.97%	−1.94%	−2.25%
12%	1.52%	0.83%	0.11%	0.02%	0.66%
14%	−0.16%	1.43%	1.56%	1.27%	1.17%
16%	2.85%	1.13%	0.82%	1.48%	1.14%
18%	−1.34%	0.23%	0.66%	0.30%	0.25%
20%	−0.73%	−1.32%	−1.41%	−1.40%	−1.22%

Figure 13 merges the inferred concentration values of all measurements at temperature 298–338 K and concentration 10–20%. The error bars represent the repeatability of three repeated experiments by SD. After compensation, the normal repeatability of CO_2 values of DA and WMS were 5.55% and 1.79% at a SD (1σ). The mean relative error, which was caused by the uncertainties of temperature, pressure, and absorption length measurements, decreased from 3.06% to 1.17%. For the wavelength modulation, the mean relative errors were 1.44% and 1.20%, before and after compensation respectively. The relative standard deviation (RSD) for DA and WMS were 0.42% and 0.31%, while the maximum relative errors and detection limits (SNR = 1) for DA and WMS were 2.7% and 2.85%, 0.0084% and 0.0017%. It should be noted that the experiment in this study examined only high CO_2 concentrations of 10–20%. As can be seen from the data, the effect on the direct absorption method was more outstanding.

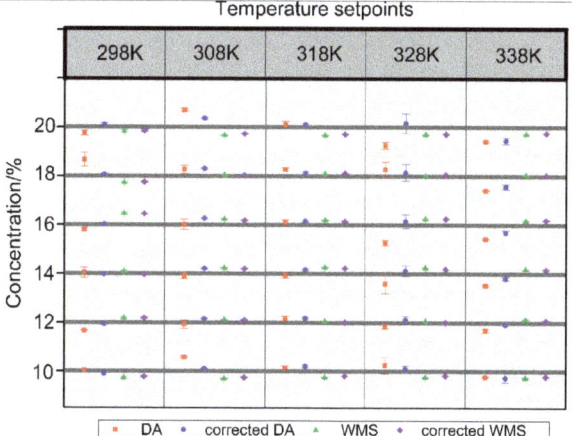

Figure 13. Measured concentration of all four measurements at temperatures 298–338 K and concentrations 10–20%.

6. Summary and Conclusions

Temperature compensations of CO_2 concentration measurements using TDLAS were reported. Absorption spectra at different temperatures were derived from HITRAN to find suitable transitions for in situ CO_2 measurements and recorded using DA and WMS. The temperature compensations based on temperature influence of line intensity were then applied to reduce measurement errors. Our experimental results show that there was an obvious decrease in the influence of temperature on signals after temperature compensation, and the detection limits of both methods meet the needs of the in situ CO_2 measurement. However, compared with WMS, DA is less sensitive to fluctuations of temperature and pressure, resulting in a superior long-term stability. The DA measurement based on temperature compensation has enough accuracy to be used for reliable determination of high concentration CO_2 in power plant exhausts.

Author Contributions: Conceptualization, X.Z. and S.Y.; Methodology, X.Z. and Z.L. (Zhenghui Li); Investigation, X.Z.; Writing—Original Draft Preparation, X.Z.; Writing—Review & Editing, W.R. and Z.L. (Zhimin Lu); Funding Acquisition, S.Y.

Funding: This research was funded by the Guangdong Province Train High-Level Personnel Special Support Program, grant number 2014TQ01N334, and the Guangdong Province Key Laboratory of Efficient and Clean Energy Utilization, grant number 2013A061401005.

Acknowledgments: The authors would like to thank the Guangdong Province Train High-Level Personnel Special Support Program (2014TQ01N334) and the Guangdong Province Key Laboratory of Efficient and Clean Energy Utilization (2013A061401005), South China University of Technology.

Conflicts of Interest: The authors declare no conflict of interest.

References

1. Eng, R.S.; Butler, J.F.; Linden, K.J. Tunable diode laser spectroscopy—An invited review. *Opt. Eng.* **1980**, *19*, 945–960. [CrossRef]
2. Werle, P. A review of recent advances in semiconductor laser based gas monitors. *Spectrochim. Acta Part A Mol. Biomol. Spectrosc.* **1998**, *54*, 197–236. [CrossRef]
3. Nwaboh, J.A.; Werhahn, O.; Ortwein, P.; Schiel, D.; Ebert, V. Laser-spectrometric gas analysis: CO_2-TDLAS at 2 μm. *Meas. Sci. Technol.* **2012**, *24*, 015202. [CrossRef]
4. Wang, F.; Cen, K.F.; Li, N.; Jeffries, J.; Huang, Q.X.; Yan, J.H.; Chi, Y. Two-dimensional tomography for gas concentration and temperature distributions based on tunable diode laser absorption spectroscopy. *Meas. Sci. Technol.* **2010**, *21*, 045301. [CrossRef]

5. Farooq, A.; Jeffries, J.B.; Hanson, R.K. CO_2 concentration and temperature sensor for combustion gases using diode-laser absorption near 2.7 µm. *Appl. Phys. B* **2008**, *90*, 619–628. [CrossRef]
6. Kamimoto, T.; Deguchi, Y.; Shisawa, Y.; Kitauchi, Y.; Eto, Y. Development of fuel composition measurement technology using laser diagnostics. *Appl. Therm. Eng.* **2016**, *102*, 596–603. [CrossRef]
7. Wang, F.; Wu, Q.; Huang, Q.; Zhang, H.; Yan, J.; Cen, K. Simultaneous measurement of 2-dimensional H_2O concentration and temperature distribution in premixed methane/air flame using TDLAS-based tomography technology. *Opt. Commun.* **2015**, *346*, 53–63. [CrossRef]
8. Liu, C.; Cao, Z.; Li, F.; Lin, Y.; Xu, L. Flame monitoring of a model swirl injector using 1D tunable diode laser absorption spectroscopy tomography. *Meas. Sci. Technol.* **2017**. [CrossRef]
9. Li, J.S.; Durry, G.; Cousin, J.; Joly, L.; Parvitte, B.; Zeninari, V. Self-broadening coefficients and positions of acetylene around 1.533 µm studied by high-resolution diode laser absorption spectrometry. *J. Quant. Spectrosc. Radiat. Transf.* **2010**, *111*, 2332–2340. [CrossRef]
10. Wen, X.F.; Lee, X.; Sun, X.M.; Wang, J.L.; Hu, Z.M.; Li, S.G.; Yu, G.R. Dew water isotopic ratios and their relationships to ecosystem water pools and fluxes in a cropland and a grassland in China. *Oecologia* **2012**, *168*, 549–561. [CrossRef]
11. Durry, G.; Li, J.S.; Vinogradov, I.; Titov, A.; Joly, L.; Cousin, J.; Decarpenterie, T.; Amarouche, N.; Liu, X.; Parvitte, B.; et al. Near infrared diode laser spectroscopy of C_2H_2, H_2O, CO_2, and their isotopologues and the application to TDLAS, a tunable diode laser spectrometer for the martian PHOBOS-GRUNT space mission. *Appl. Phys. B* **2010**, *99*, 339–351. [CrossRef]
12. Barbu, T.L.; Zéninari, V.; Parvitte, B.; Courtoisa, D.; Durrya, G. Line strengths and self-broadening coefficients of carbon dioxide isotopologues ($^{13}CO_2$, and $^{18}O^{12}C^{16}O$) near 2.04 µm for the in situ laser sensing of the Martian atmosphere. *J. Quant. Spectrosc. Radiat. Transf.* **2006**, *98*, 264–276. [CrossRef]
13. Barbu, T.L.; Vinogradov, I.; Durry, G.; Korablev, O.; Chassefière, E.; Bertaux, J.-L. TDLAS a laser diode sensor for the in situ monitoring of H_2O, CO_2, and their isotopes in the Martian atmosphere. *Adv. Space Res.* **2006**, *38*, 718–725. [CrossRef]
14. Li, J.; Parchatka, U.; Fischer, H. Applications of wavelet transform to quantum cascade laser spectrometer for atmospheric trace gas measurements. *Appl. Phys. B* **2012**, *108*, 951–963. [CrossRef]
15. Pogán, A.; Wagner, S.; Werhahn, O.; Eberta, V. Development and metrological characterization of a tunable diode laser absorption spectroscopy (TDLAS) spectrometer for simultaneous absolute measurement of carbon dioxide and water vapor. *Appl. Spectrosc.* **2015**, *69*, 257–268. [CrossRef] [PubMed]
16. Tadic, I.; Parchatka, U.; Königstedt, R.; Fischer, H. In-flight stability of quantum cascade laser-based infrared absorption spectroscopy measurements of atmospheric carbon monoxide. *Appl. Phys. B* **2017**, *123*, 146. [CrossRef]
17. Underwood, R.; Gardiner, T.; Finlayson, A.; Few, J.; Wilkinson, J.; Bell, S.; Merrison, J.; Iversonb, J.J.; de Podesta, M. A combined non-contact acoustic thermometer and infrared hygrometer for atmospheric measurements. *Meteorol. Appl.* **2016**, *22*, 830–835. [CrossRef]
18. Nwaboh, J.A.; Qu, Z.; Werhahn, O.; Ebert, V. Interband cascade laser-based optical transfer standard for atmospheric carbon monoxide measurements. *Appl. Opt.* **2017**, *56*, E84–E93. [CrossRef]
19. Zhou, X.; Liu, X.; Jeffries, J.B.; Hanson, R.K. Development of a sensor for temperature and water concentration in combustion gases using a single tunable diode laser. *Meas. Sci. Technol.* **2003**, *14*, 1459–1468. [CrossRef]
20. Phillips, W.J.; Plemmons, D.H.; Galyen, N.A. *HITRAN/HITEMP Spectral Databases and Uncertainty Propagation by Means of Monte Carlo Simulation with Application to Tunable Diode Laser Absorption Diagnostics*; No. AEDC-TR-11-T-2; Office of the Deputy Undersecretary of Defense (Science and Technology) Washington DC Joint Technology Office on Hypersonic: Washington, DC, USA, 2011; pp. 15–17.
21. Chen, J.; Li, C.; Zhou, M.; Liu, J.; Kan, R.; Xu, Z. Measurement of CO_2, concentration at high-temperature based on tunable diode laser absorption spectroscopy. *Infrared Phys. Technol.* **2017**, *80*, 131–137. [CrossRef]
22. Joly, L.; Gibert, F.; Grouiez, B.; Grossela, A.; Parvittea, B.; Durry, G.; Ze´ninari, V. A complete study of CO_2, line parameters around 4845 cm^{-1}, for Lidar applications. *J. Quant. Spectrosc. Radiat. Transf.* **2008**, *109*, 426–434. [CrossRef]
23. Pogány, A.; Ott, O.; Werhahn, O.; Ebert, V. Towards traceability in CO_2, line strength measurements by TDLAS at 2.7 µm. *J. Quant. Spectrosc. Radiat. Transf.* **2013**, *130*, 147–157.

24. Cai, T.; Gao, G.; Wang, M.; Wang, G.; Liu, Y.; Gao, X. Experimental study of carbon dioxide spectroscopic parameters around 2.0 μm region for combustion diagnostic applications. *J. Quant. Spectrosc. Radiat. Transf.* **2017**, *201*, 136–147. [CrossRef]
25. Sur, R.; Spearrin, R.M.; Peng, W.Y.; Strand, C.L.; Jeffries, J.B.; Enns, G.M.; Hanson, R.K. Line intensities and temperature-dependent line broadening coefficients of Q-branch transitions in the v_2 band of ammonia near 10.4 μm. *J. Quant. Spectrosc. Radiat. Transf.* **2016**, *175*, 90–99. [CrossRef] [PubMed]
26. Doucen, R.L.; Cousin, C.; Boulet, C.; Henry, A. Temperature dependence of the absorption in the region beyond the 4.3-micron band head of CO_2. I—Pure CO_2 case. *Appl. Opt.* **1985**, *24*, 3899–3907. [CrossRef] [PubMed]
27. Cousin, C.; Doucen, R.L.; Boulet, C.; Henry, A. Temperature dependence of the absorption in the region beyond the 4.3-μm band head of CO_2. 2: N_2 and O_2 broadening. *Appl. Opt.* **1985**, *24*, 3899–3907. [CrossRef] [PubMed]
28. Perrin, M.Y.; Hartmann, J.M. Temperature-dependent measurements and modeling of absorption by CO_2-N_2 mixtures in the far line-wings of the 4.3 um CO_2 band. *J. Quant. Spectrosc. Radiat. Transf.* **1989**, *42*, 311–317. [CrossRef]
29. Zhang, Z.R.; Wu, B.; Xia, H.; Pang, T.; Wang, G.X.; Sun, P.S.; Dong, F.Z.; Wang, Y. Study on the temperature modified method for monitoring gas concentrations with tunable diode laser absorption spectroscopy. *Acta Phys. Sin.* **2013**, *62*, 1706–1721.
30. Shu, X.W. An Investigation of Temperature Compensation of HCL Gas Online Monitoring Based on TDLAS Method. *Spectrosc. Spectr. Anal.* **2010**, *30*, 1352–1356.
31. Zhang, Z.; Zou, D.; Chen, W.; Zhao, H.J.; Xu, K.X. Temperature influence in the TDLAS detection of escaping ammonia. *Opto-Electron. Eng.* **2014**. [CrossRef]
32. Qi, R.B.; He, S.K.; Li, X.T.; Wang, X.Z. Simulation of TDLAS direct absorption based on HITRAN database. *Spectrosc. Spectr. Anal.* **2015**, *35*, 172–177.
33. He, Y.; Zhang, Y.J.; You, K.; Gao, Y.W.; Zhao, N.J.; Liu, W.Q. Volatilization characteristics analysis of ammonia from soil by straw returning to the field based on the infrared spectroscopy technology. *J. Infrared Millim. Waves* **2017**, *36*, 397–402.
34. Liu, C.; Xu, L.; Cao, Z. Measurement of no uniform temperature and concentration distributions by combining line-of-sight tunable diode laser absorption spectroscopy with regularization methods. *Appl. Opt.* **2013**, *52*, 4827–4842. [CrossRef] [PubMed]
35. You, K.; Zhang, Y.J.; Wang, L.M.; Li, H.B.; He, Y. Improving the Stability of Tunable Diode Laser Sensor for Natural Gas Leakage Monitoring. *Adv. Mater. Res.* **2013**, *760–762*, 84–87. [CrossRef]
36. Yao, C.; Wang, Z.; Wang, Q.; Bian, Y.; Chen, C.; Zhang, L.; Ren, W. Interband cascade laser absorption sensor for real-time monitoring of formaldehyde filtration by a nanofiber membrane. *Appl. Opt.* **2018**, *57*, 8005–8010. [CrossRef] [PubMed]
37. Gharavi, M. Quantification of near-IR tunable diode laser measurements in flames. In Proceedings of the 2nd Joint Meeting of the US Sections of the Combustion Institute, Oakland, CA, USA, 25–28 March 2001.
38. Gordon, I.E.; Rothmana, L.S.; Hill, C.; Kochanovac, R.V.; Tan, Y.; Bernath, P.F.; Birk, M.; Boudon, V.; Campargue, A.; Chance, K.V.; et al. The HITRAN2016 molecular spectroscopic database. *J. Quant. Spectrosc. Radiat. Transf.* **2017**, *203*, 3–69. [CrossRef]
39. Liu, J.T.; Jeffries, J.B.; Hanson, R.K. Large-modulation-depth 2f spectroscopy with diode lasers for rapid temperature and species measurements in gases with blended and broadened spectra. *Appl. Opt.* **2004**, *43*, 6500–6509. [CrossRef]

© 2019 by the authors. Licensee MDPI, Basel, Switzerland. This article is an open access article distributed under the terms and conditions of the Creative Commons Attribution (CC BY) license (http://creativecommons.org/licenses/by/4.0/).

Article

Investigating the Relation between Absorption and Gas Concentration in Gas Detection Using a Diffuse Integrating Cavity

Xue Zhou [1], Jia Yu [2], Lin Wang [1] and Zhiguo Zhang [1,*]

[1] Condensed-Matter Science and Technology Institute, Harbin Institute of Technology, Harbin 150001, China; zhouxuehit@163.com (X.Z.); wl0106050234@163.com (L.W.)
[2] Physical Science and Technology, Heilongjiang University, Harbin 150001, China; yujia19134@163.com
* Correspondence: zhangzhiguo@hit.edu.cn

Received: 2 August 2018; Accepted: 7 September 2018; Published: 12 September 2018

Featured Application: The current work provides twofold benefits on using a diffuse integrating cavity as a gas cell: (1) Understand the non-linear relation between absorption and gas concentration; (2) Provide the experimental method for determining the transition point.

Abstract: The relationship between absorption and gas concentration was studied using a diffuse integrating cavity as a gas cell. The light transmission process in an arbitrary diffuse cavity was theoretically derived based on a beam reflection analysis. It was found that a weak absorption condition must be satisfied to ensure a linear relationship between absorbance and gas concentration. When the weak absorption condition is not satisfied, a non-linear relation will be observed. A $35 \times 35 \times 35$ cm diffuse integrating cavity was used in the experiment. Different oxygen concentrations were measured by detecting the P9 absorption line at 763.8 nm, based on tunable diode laser absorption spectroscopy. The relationship between the absorption signals and oxygen concentration was linear at low oxygen concentrations and became non-linear when oxygen concentrations were higher than 21%. The absorbance value of this transition point was 0.17, which was considered as the weak absorption condition for this system. This work studied the theoretical reason for the non-linear phenomenon and provided an experimental method to determine the transition point when using a diffuse integrating cavity as a gas cell.

Keywords: diffuse integrating cavity; TDLAS; gas detection; non-linearity

1. Introduction

The measurement of gas concentration using absorption spectroscopy is a technique with great environmental adaptability and commercial potential. To improve the detection sensitivity, a long optical path length is necessary for trace gas measurement because the signal to noise ratio (SNR) improves with increasing optical path length in absorber [1]. Many methods have been developed to extend the optical path length, such as multi-pass cells [2,3], cavity ring-down spectroscopy (CRDS) [4], gas in media absorption spectroscopy (GASMAS) and applications of diffuse integrating cavities [5–7]. Among these methods, integrating cavities have proven advantageous in good stability, effortless laser beam alignment, few interference fringes, and low cost [8,9]. Besides, a diffuse integrating cavity is an attractive choice for spectral techniques with incoherent light sources, such as Fourier transform infrared (FTIR) spectroscopy [10].

Many achievements have been obtained on the light transmission law of diffuse integrating cavities [11,12]. The Beer-Lambert law is usually used to describe the output light radiation from a diffuse cavity, where the effective optical path length (EOPL) is usually introduced to describe the equivalent path length of a photon in the cavity. From previous publications, the EOPL formula has been presented as being related to the launch geometry condition, L_0, reflectivity, ρ, of the inner surface, the single pass average path length, L_{ave}, and the port fraction, f, of the cavity [13,14]. Thus, the EOPL should be independent of the detected gas concentration. The experimental methods used to measure the EOPL and its related parameters, such as ρ, L_{ave}, and f have been established and implemented [15,16]. However, detected absorption signals usually show a nonlinear relation with gas concentration at high gas concentration values [17,18] in applications. To explain such non-linear phenomenon, the EOPL was considered as a variation with gas concentration in some previous research [19,20], in which an absorption term (including gas concentration information) was introduced to describe the varied EOPL in a modified formula. Nevertheless, a varying EOPL makes practical gas concentration measurements complicated. Hence, the EOPL should be independent of the detected gas. That is, a restriction might exist while using the Beer-Lambert law to describe the output radiance from a diffuse cavity, which has not been considered in previous research. If the condition is not satisfied, a non-linear phenomenon will be observed. Thus, two concerns should be addressed preferentially. Firstly, what is the restrictive condition? Secondly, how is the condition value determined from an experiment?

In this study, the relation between absorbance and concentration in gas detection was investigated when a diffuse integrating cavity was used as a gas cell. A beam reflection analysis was implemented to analyze the light transmission process for an arbitrary-shaped diffuse cavity. An approximate condition was proposed and used in the derivation to ensure that the output radiation as a form of the Beer-Lambert law. When this condition was not satisfied, at high gas concentrations, a non-linear phenomenon was observed. Different oxygen concentrations were detected by scanning the P9 absorption line at 763.8 nm based on tunable diode laser absorption spectroscopy (TDLAS). A 35 × 35 × 35 cm diffuse integrating cavity was designed and used as the gas cell. A relationship between absorbance and gas concentration was obtained. The absorbance value at the non-linear transition was determined by experiments.

2. Theory

To fundamentally understand the reason for the non-linear relationship between the absorption signal and gas concentration, the reflection process for a light beam passing through a diffuse integrating cavity was analyzed theoretically. In general, an ideal arbitrary integrating cavity is defined as a cavity in which any two points on the inner wall can be connected to a segment inside the cavity. The inner wall of an integrating cavity is covered with a highly diffuse reflective coating, which can be considered as an ideal Lambertian radiation source. Thus, an incident beam could be diffused continuously. A simplified model of the beam transmission process is shown in Figure 1. The input light radiation flux is denoted by Φ_0, and the first light path, L_0, is the distance between the entrance aperture and the first reflection spot. Neglecting the radiation loss through the entrance and detection windows, the radiant flux after the first reflection can be expressed by Equation (1), according to the Beer-Lambert law:

$$\Phi_1 = \rho \Phi_0 \exp(-\alpha L_0). \tag{1}$$

Here, ρ is the reflectivity of the inner surface, α is the absorption coefficient, which is equal to the product of the particle number density N and the absorption cross section σ.

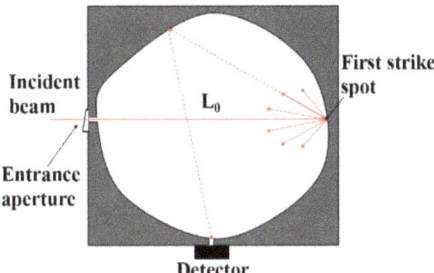

Figure 1. Simplified model of the radiation transmission process in an arbitrary integrating cavity.

A single-pass average path length of L_{ave} was introduced to describe the distance between successive light reflections [21]. If the port fraction of the cavity is f, then light travels a distance L_{ave} and a portion exits from the output aperture. Light exiting the cavity is received by a detector:

$$\Phi_{1out} = f\rho\Phi_0 \exp(-\alpha(L_0 + L_{ave})). \tag{2}$$

The remaining radiation flux is

$$\Phi_1' = (1-f)\rho\Phi_0 \exp(-\alpha(L_0 + L_{ave})). \tag{3}$$

This light radiation is still transmitted in the integrating cavity and part of this light outputs again after traveling L_{ave}. The rest is reflected by the internal wall again. This process repeats continuously. Table 1 shows the output radiation after each reflection from an arbitrary integrating cavity.

Table 1. Output light radiation after each reflection from an arbitrary diffuse integrating cavity.

Reflection Multiplier	Output Radiation from the Aperture
1	$f\rho\Phi_0 \exp(-\alpha(L_0 + L_{ave}))$
2	$f(1-f)\rho^2\Phi_0 \exp(-\alpha(L_0 + 2L_{ave}))$
3	$f(1-f)^2\rho^3\Phi_0 \exp(-\alpha(L_0 + 3L_{ave}))$
...	...
n	$f(1-f)^{n-1}\rho^n\Phi_0 \exp(-\alpha(L_0 + nL_{ave}))$

The total light radiation, Φ_{out}, received by the detector is proportional to the sum of the output radiation:

$$\Phi_{out} = \sum_1^n \Phi_{nout} = f\Phi_0 \exp(-\alpha(L_0 + L_{ave})) \frac{1}{1/\rho - (1-f)\exp(-\alpha L_{ave})}. \tag{4}$$

The absorbance, A, can be expressed as:

$$A = -\ln\left(\frac{\Phi_{out}}{\Phi_0}\right) = -\ln\left[f\exp(-\alpha(L_0 + L_{ave}))\frac{1}{1/\rho - (1-f)\exp(-\alpha L_{ave})}\right]. \tag{5}$$

We found that the absorbance, A, is a complicated expression here. However, a linear relationship between absorbance and gas concentration was expected in gas detection for easy calibration. Thus, some approximations should be introduced to express Φ_{out} in the form of the Beer-Lambert law:

$$\alpha L_{ave} \ll 1, \quad \alpha L_0 \ll 1 \tag{6}$$

$$\alpha\left(L_0 + \frac{1}{1-\rho(1-f)}L_{ave}\right) \ll 1. \tag{7}$$

It was found that Equation (7) is a stricter condition compared to Equation (6). Once Equation (7) is satisfied, Equation (6) must be followed. Thus, the following approximations were obtained in mathematics:

$$\exp(-\alpha L_{ave}) \approx 1 - \alpha L_{ave}, \quad \exp(-\alpha L_0) \approx 1 - \alpha L_0. \tag{8}$$

$$1 - \alpha\left(L_0 + \frac{1}{1-\rho(1-f)}L_{ave}\right) \approx \exp\left(-\alpha\left(L_0 + \frac{1}{1-\rho(1-f)}L_{ave}\right)\right). \tag{9}$$

Thus, Equation (4) can be expressed as:

$$\begin{aligned}\Phi_{out} &= f\Phi_0[1-\alpha(L_0+L_{ave})]\frac{1/\rho-(1-f)-(1-f)\alpha L_{ave}}{[1/\rho-(1-f)]^2-(1-f)^2\alpha^2 L_{ave}^2}\\ &\approx \frac{f}{1/\rho-(1-f)}\Phi_0\left[1-\alpha\left(L_0+\frac{1}{1-\rho(1-f)}L_{ave}\right)\right]\\ &= \Phi'_0\exp\left[-\alpha\left(L_0+\frac{1}{1-\rho(1-f)}L_{ave}\right)\right]. \end{aligned} \tag{10}$$

Here,

$$\Phi'_0 = \frac{f}{1/\rho-(1-f)}\Phi_0. \tag{11}$$

Detailed derivations for Equations (4)–(10) are presented in "Appendix A". These indicate that Equation (10) is in the same form as the Beer-Lambert law. Comparing Equation (10) with the Beer-Lambert law, the expression for the EOPL for an arbitrary diffuse cavity can be obtained as:

$$EOPL = L_0 + \frac{1}{1-\rho(1-f)}L_{ave}. \tag{12}$$

Obviously, Equation (12) is similar to the previously reported EOPL expression [13,14]. The only difference is the coefficient ρ in the numerator, which was replaced by 1 in the previous results. This slight difference was usually smaller than measurement errors and could be ignored.

According to Equation (12), the approximate condition of Equation (7) can be transformed into

$$\alpha \cdot EOPL \ll 1. \tag{13}$$

Here, we define Equation (13) as the weak absorption condition of the diffuse integrating cavity. Thus, the absorbance can be expressed as Equation (14), according to Equation (10).

$$A = \alpha\left(L_0 + \frac{1}{1-\rho(1-f)}L_{ave}\right) = \sigma \cdot N \cdot EOPL. \tag{14}$$

Here, the particle number density, N, denotes the concentration of the test gas and σ is the absorption cross-section. In other words, the absorbance can be expressed as Equation (14) under the weak absorption condition of Equation (13), where the absorbance depends linearly on the gas concentration. On the contrary, if the weak absorption condition was not satisfied, the absorbance will present non-linearly on the gas concentration.

3. Experimental Section

An experiment was designed and implemented to investigate the weak absorption condition of a diffuse integrating cavity. Oxygen was chosen as the sample gas due to its non-toxicity and easy accessibility. To achieve enough absorption for observing the non-linearity, the absorbance of oxygen was estimated using the HITRAN-2016 database [22]. A 35 × 35 × 35 cm diffuse integrating cavity was fabricated for this experiment. The diameters of the input and output light apertures were both 2 mm. The inner surface of the cavity was coated with a 0.4 mm thickness of Avian-D paint (Avian Technologies LLC, New London, NH, USA) and the painted surface was functioned as an ideal Lambertian diffuse scatter. The reflectivity of the coating was 98.3–98.4% at 750–800 nm, according to the supplier [23]. To reduce the port fraction of the cavity, the input light aperture and the gas inlet/outlet holes were combined with a T-junction. The light apertures were glued with a wedge-shaped glass lens to eliminate interference effects. The structure of a cubic integrating cavity is shown in Figure 2a, the in/out light aperture is labeled and an enlarged view of the T-junction is shown. The cavity could be sealed after screwing and gluing the head cover.

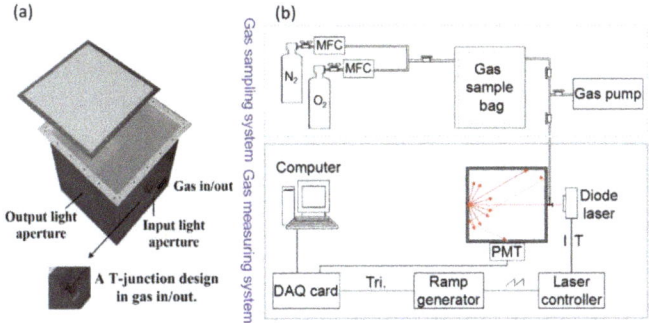

Figure 2. (a) Structure of a cubic diffuse integrating cavity; (b) schematic of the gas-sampling system and TDLAS (tunable diode laser absorption spectroscopy) for measuring the absorption of oxygen with a cubic cavity as the gas cell.

The experimental setup, including a gas-sampling system and a gas-measuring system, is shown in Figure 2b. Different oxygen concentrations were achieved by mixing different proportions of oxygen and nitrogen with mass flow controllers (MFC D07-19C, Sevenstar, Beijing, China). Based on TDLAS, a vertical-cavity surface-emitting tunable diode laser (Laser Components: Single Mode VCSEL 763 nm TO46), with a free running output power of 0.3 mW, was used as the light source. The laser temperature was controlled at 22.47 °C (Temperature Controller: TED 200C, Thorlabs, Newton, MA, USA), and the laser injection current was 1.29 mA (VCSEL Laser Diode Controller: LDC 200, Thorlabs, Newton, MA, USA). The injection current was modulated using a sawtooth wave (10 Hz, 200 mV), corresponding to a current varying from 1.16 to 1.40 mA. The output light from the integrating cavity was detected with a photomultiplier tube (PMTH-S1-1P28, Hamamatsu, Japan). Data were recorded using a data acquisition card (DAQ card: National Instrument Co., NI PCI-6133, Austin, TX, USA). The sawtooth wave was simultaneously put into the DAQ card as the trigger signal. The ambient temperature was controlled at 25 °C for the experiments. The center frequency of the laser was used to scan over the oxygen-absorption P9 line in the A-band at 763.84 nm by comparing the measured laser center wavelength with the calculated oxygen absorption line based on the HITRAN data. Each oxygen concentration was measured 15 times to obtain the detection error. Each signal was averaged 500 times per minute.

4. Results and Discussion

Figure 3 demonstrates the signal of 70% oxygen concentration in the 35 × 35 × 35 cm diffuse integrating cavity. The red curve in Figure 3a denotes the detected signal, Φ_{out}. The black baseline is the incident light radiation, Φ_0, which was obtained by linear fitting based on the data points besides the absorption signal in the red curve. Here, the data points in the range of 250–350 and 700–800 in Figure 3a are used to linearly fit the baseline. Thus, the absorbance can be calculated as shown in Figure 3b.

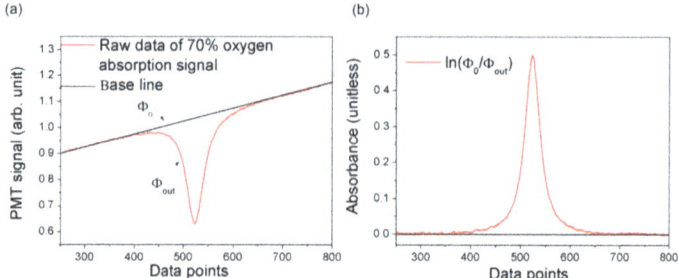

Figure 3. (a) Direct scan PMT (photomultiplier tube) signal for 70% oxygen concentration in the 35 × 35 × 35 cm diffuse integrating cavity; (b) calculated result of the absorbance from the PMT signal.

Here, the peak value of the absorbance was defined as the optical parameter (OP), i.e., the maximum value of $\ln(\Phi_0/\Phi_{out})$. According to Equation (14), the OP is proportional to the optical path length, once the oxygen concentration is constant. Thus, because the oxygen concentration is stable in air, the EOPL of the diffuse cavity was calibrated by comparing the oxygen absorption signal in the cavity and in the air [14–16]. Based on this method, each optical path length was measured 15 times, and the error bars are shown in Figure 4. The EOPL of the 35 × 35 × 35 cm diffuse integrating cavity was 590 ± 9 cm, i.e., the average value was 590 cm and the detection error was 9 cm. The detection limit of this system was ~0.6% for oxygen measurement. Detailed continuous detection results and Allen Variance are shown in Appendix B.

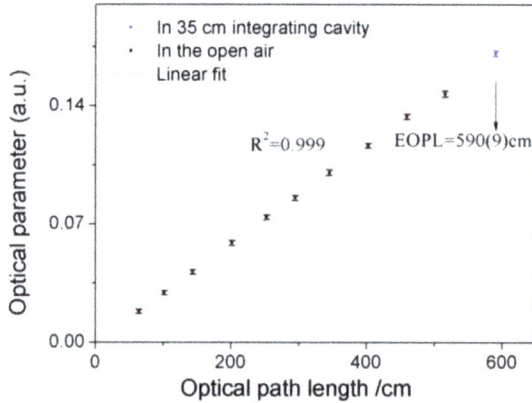

Figure 4. The EOPL (effective optical path length) calibration curve for the cubic integrating cavity between the OP (optical parameter) and optical path lengths in the air.

Then, different oxygen concentrations were measured with the cubic diffuse cavity. The relationship between the OP and oxygen concentration is shown in Figure 5a.

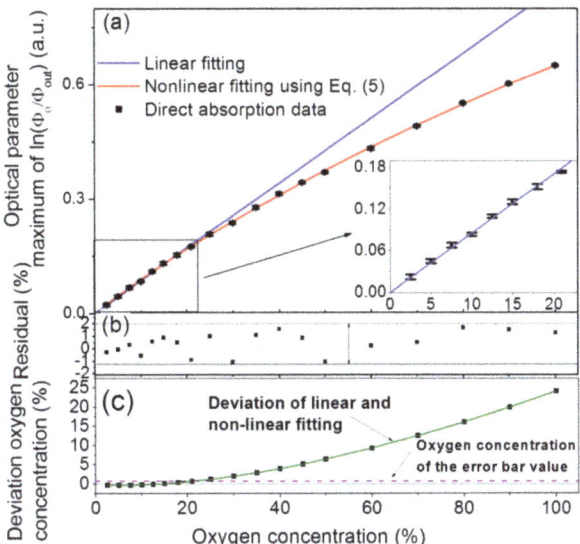

Figure 5. (a) Relationship between the OP and oxygen concentration; the inset shows an enlarged view of the low concentration range. (b) Residual error between the experimental data and the non-linear fitting result. (c) Deviation between the linear and non-linear fitting results.

The black points show the OP values calculated using the detected spectra. It was found that the OP values depended linearly on the concentrations in the low concentration range and were non-linear when the concentration was relatively high. The blue line shows a linear fitting result using the OP data in the 0–18% oxygen concentration range. Obviously, the data gradually diverge from the blue line with an increasing oxygen concentration. The non-linear fitting result (the red curve) was obtained for all absorption data using Equation (5) with the fitted $L_0 = 39 \pm 5$ cm, $\rho = 0.98 \pm 1$, $L_{ave} = 22 \pm 2$ cm, and $f = 0.004 \pm 1$. As shown in Figure 5b, the residual error between the non-linear fitting result and the experimental data was lower than 2%. Thus, the Budget error of the gas sampling system was lower than 2%, including errors of gas flow-rate mixing and gas cylinders. The error of the detection system is analyzed in Appendix B. It indicates that the theoretical predictions by the beam reflection analysis agreed with the experiments. In fact, the theoretical value of L_0 and L_{ave} were 35 and 23.3 cm respectively for this cavity [21,24]. The fitting results agree with the theoretical values within the measured error ranges. The fitting reflectivity of the inner coating was slightly lower than the value provided by the supplier, which was caused by imperfections in the coating (pollution during drying and a humidity difference).

To determine the non-linear transition point, the deviation between the linear (blue curve) and nonlinear (red curve) fitting results was calculated and is shown as the green curve in Figure 5c. The deviation increased with increasing oxygen concentration. We considered that the deviation was resolved when it was larger than the system detection sensitivity. As the signal of each oxygen concentration was measured 15 times, the detection error was obtained as the error bars shown in Figure 5a. The average delta absorbance was 0.0045 for different oxygen concentrations. According to the linear relationship, this detection error denoted a detection sensitivity of 0.6%, which is shown as the dotted line in Figure 5c. Obviously, the deviation was smaller than the error bar value when the oxygen concentration was lower than 21%. The deviation was larger than the error bar value and could be resolved when the oxygen concentration was higher than 21%. Hence, we considered that the transition point was an oxygen concentration of 21%. The absorbance value at the transition point was 0.17. Thus, a linear relationship between the absorbance and gas concentration could be ensured when the absorbance was lower than 0.17 for this system. This result was universal for different gases in such a system. For other detection systems, the transition point is related to the detection error of the system. We have provided an experimental method to determine the transition point of linear and non-linear measurements for a diffused integrating cavity.

5. Conclusions

In summary, this study investigated the relationship between absorbance and gas concentration when a diffuse integrating cavity was used as a gas cell. Theoretically, an approximation for weak absorption was proposed. When the weak absorption condition was not satisfied, a non-linear relation was be observed. Experimentally, different oxygen concentrations were measured by detecting the P9 absorption line at 763.84 nm using a cubic diffuse integrating cavity. A linear relationship between the measured absorbance and the gas concentration was observed at low concentrations. The non-linear transition point was confirmed as an oxygen concentration of 21%, based on experimental results. The absorbance value of the transition point was 0.17. This transition point value was only applicable to this system for oxygen detection. Thus, an approach to determine the applicable condition of a diffuse cavity was proposed. For a different system, it was related to the detection sensitivity and was measured using such method. This work studied the theoretical reason for the non-linear phenomenon and provided an experimental method to determine the transition point when using a diffuse integrating cavity as a gas cell.

Author Contributions: X.Z. performed experiments, analyzed data and wrote the paper; J.Y. and L.W. revised the paper; Z.Z. designed experiments and revised the paper.

Funding: This research was funded by the National Key Research and Development Program of China (2016YFF0102803).

Conflicts of Interest: The authors declare no conflicts of interest.

Appendix A. Detailed Derivation

The total light radiation, Φ_{out}, received by the detector is proportional to the sum of the output radiation:

$$\begin{aligned}\Phi_{out} &= \sum_{n=1}^{n}\Phi_{nout} = f\rho\Phi_0 \exp(-\alpha(L_0+L_{ave}))[1+(1-f)\rho\exp(-\alpha L_{ave})+(1-f)^2\rho^2\exp(-2\alpha L_{ave})\\&+(1-f)^2\rho^2\exp(-2\alpha L_{ave})+(1-f)^3\rho^3\exp(-3\alpha L_{ave})+\cdots+(1-f)^n\rho^n\exp(-n\alpha L_{ave})]\\&= f\rho\Phi_0\exp(-\alpha(L_0+L_{ave}))\frac{1-\rho^n(1-f)^n\exp(-n\alpha L_{ave})}{1-\rho(1-f)\exp(-\alpha L_{ave})}\\&= f\Phi_0\exp(-\alpha(L_0+L_{ave}))\frac{1}{1/\rho-(1-f)\exp(-\alpha L_{ave})}.\end{aligned} \quad (A1)$$

Here, $\rho(1-f)\exp(-\alpha L_{ave})$ is less than 1 and $\rho^n(1-f)^n\exp(-n\alpha L_{ave}) \sim 0$. Thus, we ignored it.

Then, αL_{ave} and αL_0 are usually less than 1 in gas detection (Equation (6) in the main text). The following approximation was obtained mathematically as Equation (A2) (same as Equation (8) in the main text):

$$\exp(-\alpha L_{ave}) \approx 1-\alpha L_{ave},\ \exp(-\alpha L_0) \approx 1-\alpha L_0. \quad (A2)$$

Taking Equation (A2) into (A1), the expression can be written as follow:

$$\begin{aligned}\Phi_{out} &= f\Phi_0[1-\alpha(L_0+L_{ave})]\frac{1}{1/\rho-(1-f)(1-\alpha L_{ave})}\\&= f\Phi_0[1-\alpha(L_0+L_{ave})]\frac{1/\rho-(1-f)-\alpha L_{ave}(1-f)}{[1/\rho-(1-f)]^2-\alpha^2 L_{ave}^2(1-f)^2}.\end{aligned} \quad (A3)$$

As αL_{ave} is less than 1, $\alpha^2 L_{ave}^2(1-f)^2$ was ignored. Then Equation (A3) can be converted into Equation (A4).

$$\begin{aligned}\Phi_{out} &\approx f\Phi_0[1-\alpha(L_0+L_{ave})]\frac{1/\rho-(1-f)-\alpha L_{ave}(1-f)}{[1/\rho-(1-f)]^2}\\ &= \frac{f\Phi_0}{1/\rho-(1-f)}[1-\alpha(L_0+L_{ave})]\left[1-\frac{\alpha L_{ave}(1-f)}{1/\rho-(1-f)}\right]\\ &\approx \frac{f\Phi_0}{1/\rho-(1-f)}\left[1-\alpha(L_0+L_{ave})-\frac{\alpha L_{ave}(1-f)}{1/\rho-(1-f)}\right] \quad (A4)\\ &= \frac{f\Phi_0}{1/\rho-(1-f)}\left[1-\alpha L_0-\alpha L_{ave}\frac{1/\rho}{1/\rho-(1-f)}\right]\\ &= \Phi'_0\left[1-\alpha\left(L_0+\frac{1}{1-\rho(1-f)}L_{ave}\right)\right].\end{aligned}$$

Then, we defined a condition. If Equation (A5) (same as Equation (7) in the main text) is satisfied,

$$\alpha\left(L_0+\frac{1}{1-\rho(1-f)}L_{ave}\right) \ll 1. \quad (A5)$$

Thus,

$$1-\alpha\left(L_0+\frac{1}{1-\rho(1-f)}L_{ave}\right) \approx \exp\left(-\alpha\left(L_0+\frac{1}{1-\rho(1-f)}L_{ave}\right)\right). \quad (A6)$$

Hence, Φ_{out} can be expressed as follows, which is same as the Equation (10) in the main text.

$$\Phi_{out} = \Phi'_0 \exp\left[-\alpha\left(L_0+\frac{1}{1-\rho(1-f)}L_{ave}\right)\right]. \quad (A7)$$

Thus, the output radiation can be expressed in the same form as the Beer-Lambert law.

Appendix B. Detection Error of the System

For each oxygen concentration, the signal was detected 15 times. Figure A1 shows the continuous detection results of the 10, 21, 40 and 60 oxygen concentrations. The detected OP fluctuation ranges for different oxygen concentrations are shown in Table A1, which are plotted as the error bars in Figure 5a. The Allen Variances of the OP for each concentration were calculated and are shown in Table A1. Thus, the detection limit of oxygen was about 0.6% for this system.

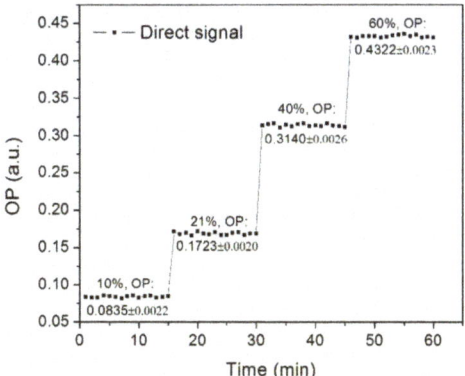

Figure A1. Fluctuations of the measured results under different oxygen concentrations.

Table A1. Fluctuation ranges of the detected OP and Allen Variances for different oxygen concentrations.

Oxygen Concentration (%)	Fluctuated Ranges of OP (ΔOP)	Allen Variances of OP	Detection Limit (ΔC %)
5	0.0045	1.06×10^{-5}	0.53
10	0.0044	1.10×10^{-5}	0.52
15	0.0056	4.23×10^{-5}	0.65
21	0.0040	2.84×10^{-5}	0.47
25	0.0059	5.90×10^{-5}	0.69
30	0.0054	4.3×10^{-5}	0.63
40	0.0052	2.51×10^{-5}	0.61
60	0.0046	1.09×10^{-5}	0.54
80	0.0034	3.2×10^{-5}	0.40
100	0.0032	3.7×10^{-5}	0.38

References

1. Hodgkinson, J.; Tatam, R.P. Optical gas sensing: A review. *Meas. Sci. Technol.* **2013**, *24*, 012004. [CrossRef]
2. Du, Z.H.; Gao, H.; Cao, X.H. Direct high-precision measurement of the effective optical path length of multi-pass cell with optical frequency domain reflectometer. *Opt. Express* **2016**, *24*, 417–426. [CrossRef] [PubMed]
3. Krzempek, K.; Hudzikowski, A.; Gluszek, A.; Dudzik, G.; Abramski, K.; Wysocki, G.; Nikodem, M. Multi-pass cell-assisted photoacoustic/photothermal spectroscopy of gases using quantum cascade laser excitation and heterodyne interferometric signal detection. *Appl. Phys. B* **2018**, *124*, 74–80. [CrossRef]
4. Cone, M.T.; Mason, J.D.; Figueroa, E.; Hokr, B.H.; Bixler, J.N.; Castellanos, C.C.; Noojin, G.D.; Wigle, J.C.; Rockwell, B.A.; Yakovlev, V.V.; et al. Measuring the absorption coefficient of biological materials using integrating cavity ring-down spectroscopy. *Optica* **2015**, *2*, 162–168. [CrossRef]
5. Svensson, T.; Adolfsson, E.; Lewander, M.; Xu, C.T.; Svanberg, S. Disordered, Strongly Scattering Porous Materials as Miniature Multipass Gas Cells. *Phys. Rev. Lett.* **2017**, *107*, 143901. [CrossRef] [PubMed]
6. Davis, M.; Hodgkinson, J.; Francis, D.; Tatam, R.P. Sensitive detection of methane at 3.3 μm using an integrating sphere and interband cascade laser. In Proceedings of the SPIE, Optical Sensing and Detection IV, Brussels, Belgium, 29 April 2016; Volume 9899, p. 98990M. [CrossRef]
7. Ducanchez, A.; Bendoula, R.; Roger, J.M. An investigation into the effects of pressure on gas detection using an integrating sphere as multipass gas absorption cell: Analysis and discussion. *J. Near Infrared Spec.* **2016**, *24*, 405–412. [CrossRef]
8. Lassen, M.; Balslev-Clausen, D.; Brusch, A.; Petersen, J.C. A versatile integrating sphere based photoacoustic sensor for trace gas monitoring. *Opt. Express* **2014**, *22*, 11660–11669. [CrossRef] [PubMed]
9. Zhou, X.; Yu, J.; Wang, L.; Gao, Q.; Zhang, Z.G. Sensitive detection of oxygen using a diffused integrating cavity as a gas absorption cell. *Sens. Actuators B* **2017**, *241*, 1076–1081. [CrossRef]
10. Ogawara, Y.; Bruneau, A.; Kimura, T. Determination of ppb-level CO, CO_2, CH_4, and H_2O in high-purity gases using matrix isolation FT-IR with an integrating sphere. *Anal. Chem.* **1994**, *66*, 4354–4358. [CrossRef]
11. Tranchart, S.; Bachir, I.H.; Destonbes, J.L. Sensitive trace gas detection with near-infrared laser diodes and an integrating sphere. *Appl. Opt.* **1996**, *35*, 7070–7074. [CrossRef] [PubMed]
12. Kirk, J.T.O. Modeling the performance of an integrating-cavity absorption meter: Theory and calculations for a spherical cavity. *Appl. Opt.* **1995**, *34*, 4397–4408. [CrossRef] [PubMed]
13. Labsphere Inc. A Guide to Integrating Sphere Theory and Applications. 1994. Available online: http://www.labsphere.com (accessed on 2 August 2018).
14. Yu, J.; Zheng, F.; Gao, Q.; Li, Y.J.; Zhang, Y.G.; Zhang, Z.G.; Wu, S.H. Effective optical path length investigation for cubic diffuse cavity as gas absorption cell. *Appl. Phys. B* **2014**, *116*, 135–140. [CrossRef]
15. Bergin, S.; Hodgkinson, J.; Francis, D.; Tatam, R.P. Integrating cavity based gas cells: A multibeam compensation scheme for pathlength variation. *Opt. Express* **2016**, *24*, 13647–13664. [CrossRef] [PubMed]
16. Gao, Q.; Zhang, Y.G.; Yu, J.; Zhang, Z.G.; Wu, S.H.; Guo, W. Integrating sphere effective optical path length calibration by gas absorption spectroscopy. *Appl. Phys. B* **2014**, *114*, 341–346. [CrossRef]
17. Hodgkinson, J.; Masiyano, D.; Tatam, R.P. Using integrating spheres as absorption cells: Path-length distribution and application of Beer's law. *Appl. Opt.* **2009**, *48*, 5748–5758. [CrossRef] [PubMed]

18. Hodgkinson, J.; Masiyano, D.; Tatam, R.P. Using integrating spheres with wavelength modulation spectroscopy: Effect of pathlength distribution on 2nd harmonic signals. *Appl. Phys. B* **2013**, *110*, 223–231. [CrossRef]
19. Hawe, E.; Chambers, P.; Fitzpatrick, C.; Lewis, E. CO_2 monitoring and detection using an integrating sphere as a multipass absorption cell. *Meas. Sci. Technol.* **2007**, *18*, 3187–3194. [CrossRef]
20. Hawe, E.; Fitzpatrick, C.; Chambers, P.; Dooly, G.; Lewis, E. Hazardous gas detection using an integrating sphere as a multipass gas absorption cell. *Sens. Actuators A* **2008**, *141*, 414–421. [CrossRef]
21. Fry, E.S.; Musser, J.; Kattawar, G.W.; Zhai, P.W. Integrating cavities: Temporal response. *Appl. Opt.* **2006**, *45*, 9053–9065. [CrossRef] [PubMed]
22. Gordon, I.E.; Rothman, L.S.; Hill, C.; Kochanov, R.V.; Tan, Y.; Bernath, P.F.; Birk, M.; Boudon, V.; Campargue, A.; Chance, K.V.; et al. The HITRAN 2016 molecular spectroscopic database. *J. Quant. Spectrosc. Radiat. Transf.* **2017**, *203*, 3–69. [CrossRef]
23. Avian Technologies LLC. Available online: http://aviantechnologies.com/catalog/avian-d-white-reflectance-coating/ (accessed on 2 August 2018).
24. Yu, J.; Zhang, Y.G.; Gao, Q.; Hu, G.; Zhang, Z.G.; Wu, S.H. Diffuse reflectivity measurement using cubic cavity. *Opt. Lett.* **2014**, *39*, 1941–1944. [CrossRef] [PubMed]

© 2018 by the authors. Licensee MDPI, Basel, Switzerland. This article is an open access article distributed under the terms and conditions of the Creative Commons Attribution (CC BY) license (http://creativecommons.org/licenses/by/4.0/).

Article

High-Efficiency Coupling Method of the Gradient-Index Fiber Probe and Hollow-Core Photonic Crystal Fiber

Chi Wang [1,2], Yue Zhang [1], Jianmei Sun [1], Jinhui Li [2], Xinqun Luan [2,*] and Anand Asundi [3]

[1] Department of Precision Mechanical Engineering, Shanghai University, Shanghai 200444, China; wangchi@shu.edu.cn (C.W.); zyueshu@163.com (Y.Z.); jianmeisun@shu.edu.cn (J.S.)
[2] Science and Technology on Near-surface Detection Laboratory, Wuxi 214035, China; li_jinhui@126.com
[3] School of Mechanical and Aerospace Engineering, Nanyang Technological University, Singapore 639798, Singapore; anand.asundi@pmail.ntu.edu.sg
* Correspondence: stnsdl@126.com; Tel.: +86-137-6473-9726

Received: 22 April 2019; Accepted: 15 May 2019; Published: 20 May 2019

Featured Application: Supplying a high-efficiency coupling method of the gradient-index fiber probe and hollow-core photonic crystal fiber for gas detection.

Abstract: A high-efficiency coupling method using the gradient-index (GRIN) fiber probe and hollow-core photonic crystal fiber (HC-PCF) is proposed to improve the response time and the sensitivity of gas sensors. A coupling efficiency model of the GRIN fiber probe coupled with HC-PCF is analyzed. An optimization method is proposed to guide the design of the probe and five samples of the GRIN fiber probe with different performances are designed, fabricated, and measured. Next, a coupling efficiency experimental system is established. The coupling efficiencies of the probes and single-mode fiber (SMF) are measured and compared. The experimental results corrected by image processing show that the GRIN fiber probe can achieve a coupling efficiency of 80.22% at distances up to 180 μm, which is obviously superior to the value of 33.45% of SMF at the same distance. Moreover, with the increase of the coupling distance, the coupling efficiency of the probe is still higher than that of SMF.

Keywords: hollow-core photonic crystal fiber; GRIN fiber probe; coupling efficiency; gas sensing

1. Introduction

A hollow-core photonic crystal fiber (HC-PCF) has periodical microstructures of air holes surrounding a hollow core where light is confined [1]. Because the light is trapped in HC-PCF by a photonic bandgap in the cladding that is made of spaced air holes instead of internal reflection, it is possible to guide light in a gas-filled core [2]. In addition, its long interaction length can realize resonance and near-resonance light-light and light-matter interactions [3]. Due to its excellent characteristics in gas sensing, it has been extensively studied for gas detection. To realize an all-fiber gas sensor, single-mode fiber (SMF) and HC-PCF are usually utilized for coupling, which has the advantages of a compact structure, easy miniaturization, light weight, anti-interference, and long working range [4–6]. However, the assembly of SMF and HC-PCF has a low coupling efficiency, which will reduce the output light intensity and thus reduce the sensitivity of the detection system according to Lambert-Beer law [7].

Many efforts have been made to improve the coupling efficiency. In one proposed method, the structure or fiber mode of HC-PCF [8,9] is specifically selected, or the micropore collapse effect [10] is reduced, but with limited improvement of the coupling efficiency. In another proposal, the SMF is wedged into HC-PCF by special cutting, to ensure an ultra-high coupling efficiency [11]. However, the structure is too compact and prevents the gas from entering the HC-PCF, resulting in a poor response time of the sensor. In view of this, the coupling efficiency can only be improved by shortening the distance between HC-PCF and SMF. In order to obtain a higher coupling efficiency, HC-PCF should be as close as possible to SMF in consideration of the rapid divergence of SMF-emitted beams. By controlling the distance between SMF and HC-PCF, it is shown that the coupling efficiency is highest when the coupling length is 7.618 µm [12]. However, due to this short distance, the gas diffuses into the HC-PCF very slowly [13], thereby reducing the response time of the sensor. Additionally, the coupling gap is difficult to control for such a short distance.

In view of this, a coupling method that can achieve a high coupling efficiency at a long coupling distance is proposed. By adding a GRIN fiber probe to the SMF, the coupling efficiency can be substantially improved over the desired coupling distance. Compared to the all-fiber structure for gas detection using a GRIN lens with a large volume [14], a GRIN fiber probe is an all-fiber optical lens composed of a single-mode fiber, no-core fiber (NCF), and GRIN fiber. It has an ultra-small structure size and good focusing performance [15]. It has been widely used in the field of optical coherent tomography (OCT) [16–18] and interference sensing measurement [19,20]. The light emitted from the GRIN fiber probe has the characteristics of focusing first and then diverging, which can obtain a smaller beam waist spot at a longer distance, thus overcoming the shortcoming of the short working distance of SMF. This excellent characteristic can be used in the study of HC-PCF gas sensors: a smaller beam waist size helps to improve the coupling efficiency of the probe and HC-PCF, while a longer working distance can maintain a long gap between the two, which is conducive to the entry of gas into HC-PCF, improving the response time of the sensor.

2. Coupling Model of the GRIN Fiber Probe and HC-PCF

The GRIN fiber probe is an all-fiber optical lens which is fused by a single-mode fiber, no-core fiber, and GRIN fiber in turn. The coupling model of the HC-PCF and GRIN fiber probe is shown in Figure 1.

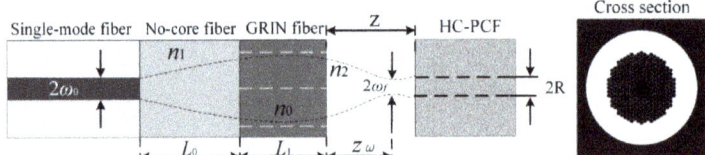

Figure 1. Coupling model of the GRIN fiber probe and HC-PCF.

Symbols are set as follows: λ is the light of the wavelength and ω_0 is the beam radius of the Gaussian beam, the refraction index of the NCF is n_0, the length of the NCF is L_0, the GRIN fiber lens has the refractive index in the center n_1 and the gradient constant g with length L, the refractive index of the transmission medium in the application environment is n_2, and z is the coupling distance or gap width between the GRIN fiber probe and HC-PCF.

If light has no energy loss in the optical fibers, the working distance z_ω and focusing spot size $2\omega_f$ of the GRIN fiber probe are expressed as [21]

$$z_\omega = \frac{S_1 \cos(2gL) + S_2 \sin(2gL)}{S_0 - S_3 \cos(2gL) - S_4 \sin(2gL)} \tag{1}$$

where $S_0 = n_1^2 g^2 + n_1^2 g^2 L_0^2 a^2 + n_0^2 a^2$, $S_1 = -2 n_0 n_2 L_0 a^2$, $S_2 = n_1 n_2 g + n_1 n_2 g L_0^2 a^2 - n_2 n_0^2 a^2/(n_1 g)$, $S_3 = n_1^2 g^2 + n_1^2 g^2 L_0^2 a^2 - n_0^2 a^2$, $S_4 = 2 n_0 n_1 g L_0 a^2$, $a = \lambda/(n_0 \pi \omega_0^2)$;

$$2\omega_f = 2\omega_0 \sqrt{P_0 \cos^2(gL) + P_1 \sin^2(gL) - P_2 \sin(2gL)} \quad (2)$$

where $P_0 = 1 + a^2(L_0 + n_0 z_\omega/n_2)^2$, $P_1 = z_\omega^2 n_1^2 g^2/n_2^2 + a^2(n_0/(n_1 g) - n_1 z_\omega L_0 g/n_2)^2$, $P_2 = z_\omega n_1 g/n_2 - a^2(L_0 + n_0 z_\omega/n_2)(n_0/(n_1 g) - n_1 z_\omega L_0 g/n_2)$.

$\omega(z)$ is set as the radius of the beam from the GRIN fiber probe at distance z and can be given as

$$\omega(z) = \omega_f \sqrt{1 + [\frac{\lambda(z-z_\omega)}{\pi \omega_f^2}]^2} \quad (3)$$

The power of emitted light from the GRIN fiber probe is set to p_0. To avoid a large number of complex calculations for coupling efficiency using an overlap integral [22], a simplified model that only takes distributions of the input beam into consideration is adopted. The emitted light from the GRIN fiber can be approximated as a Gaussian beam, so the power received at the receiving end can be expressed as

$$p_t(z) = \int_0^{\frac{d}{2}} \frac{2p_0}{\pi \omega^2(z)} \exp(-\frac{2r^2}{\omega^2(z)}) 2\pi r dr \quad (4)$$

where d is the mode field diameter (MFD) of HC-PCF. From Equations (3) and (4), the coupling efficiency of the GRIN fiber probe coupled with HC-PCF can be expressed as

$$\eta = \frac{p_t}{p_0} = 1 - \exp(-\frac{d^2/2}{\omega_f^2 + [\frac{\lambda(z-z_\omega)}{\pi \omega_f}]^2}) \quad (5)$$

As a comparison, the spot size of the Gaussian beam after it is emitted from SMF [12] is

$$\omega(z) = \omega_0 \sqrt{1 + [\frac{\lambda z}{\pi \omega_0^2}]^2} \quad (6)$$

and the SMF coupling efficiency is expressed as

$$\eta' = \frac{p_t}{p_0} = 1 - \exp(-\frac{d^2/2}{\omega_0^2 + [\frac{\lambda z}{\pi \omega_0}]^2}) \quad (7)$$

Comparing Equations (5) and (7), the coupling efficiency of the GRIN fiber probe coupled with HC-PCF in Equation (5) achieves the maximum value when the coupling distance z is equal to the working distance z_ω of the probe, while the coupling efficiency of SMF coupled with HC-PCF in Equation (7) achieves the maximum value when z is equal to zero; that is, a GRIN fiber probe can focus an emitted spot at the working distance of the probe when the maximum coupling efficiency is achieved, while SMF achieves the minimum spot and maximum coupling efficiency at zero distance. As a result, the GRIN fiber probe achieves a higher coupling efficiency at a longer distance, which demonstrates the superiority of the GRIN fiber probe over SMF when coupled with HC-PCF.

When the coupling distance z between the GRIN fiber probe and HC-PCF is equal to z_ω, Equation (5) shows that the maximum coupling efficiency is only related to the waist radius ω_f. If γ is set as the ratio of the waist diameter of the GRIN fiber probe to the MFD of HC-PCF $2\omega_f/d$, the maximum efficiency can be expressed as

$$\eta = 1 - \exp[-2(\frac{1}{\gamma})^2] \quad (8)$$

The coupling efficiency η becomes a function of the ratio γ, and it is necessary to analyze the relationship between the two. If the waist diameter is much smaller than MFD of HC-PCF, higher order modes with higher transmission losses are excited and the simplified model may not suitable. The minimum γ is set to 0.9 and changed from 0.9 to 2.5. The corresponding coupling efficiency is shown in Figure 2.

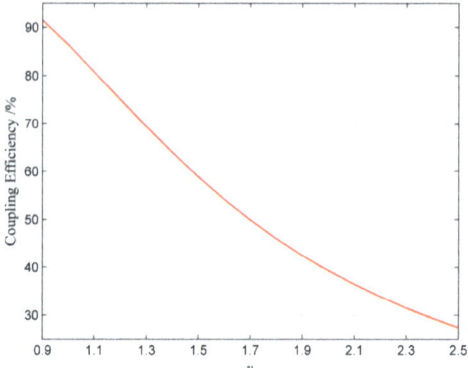

Figure 2. Theoretical coupling efficiency η versus γ.

It can be concluded from Figure 2 that the coupling efficiency is 27.4% when the ratio γ is 2.5, and the coupling efficiency will gradually increase as the ratio decreases. When the ratio is 1, the coupling efficiency reaches 86.5%, and the coupling efficiency reaches 91.5% when the ratio is 0.9. The results show that reducing the waist radius helps to improve the coupling efficiency.

3. The Relationship between Waist Radius and Working Distance of the Probe

Figure 2 shows that the smaller the waist radius ω_f of the GRIN fiber probe is, the higher the coupling efficiency is. However, Equation (8) is based on the premise that the coupling distance between the probe and HC-PCF is equal to the working distance z_ω of the GRIN fiber probe. Because of this, it is necessary to analyze the relationship between the waist radius and working distance. According to Equation (1) and Equation (2) and previous studies [18–20], the nonlinear relationship between the waist radius ω_f and working distance z_ω is complex. Both of them are related to the structural parameters of the GRIN fiber probe, such as the length L_0 of the no-core fiber and the length L of the GRIN fiber.

With the length of GRIN fiber increasing, the working distance and waist radius of the probe change periodically, while both of them become larger with the increase of the length of the no-core fiber. On one hand, to obtain a longer working distance and smaller waist radius, the length of the no-core fiber should be weighed. Reference [21] pointed out that the length of the no-core fiber should not be too long, otherwise the light beam may exceed the inner diameter of the GRIN fiber, which makes the beam enter GRIN fiber cladding and reduces the spot quality. The maximum length L_{0max} of the no-core fiber can be calculated when the parameters of the fibers and beam are known. On the other hand, due to the periodicity pitch characteristics of the GRIN fiber, analysis of the first cycle is enough, so the length of the GRIN fiber can be limited to the first cycle L_{max}. However, previous works have not studied how to design the length of fibers to ensure the longest working distance of the probe at a specified waist radius. In view of this, an optimization solution was proposed to guide the design of the probe.

For a probe with a specified waist radius ω_f, more than one set of no-core fibers and GRIN fibers (L_0, L) is satisfactory because of the periodic property of the function. The set that can generate the largest working distance needs to be identified. The problem above can be summarized as finding the optimal solution $(L_0, L)_{opt}$ from all sets $(L_0, L)_m$ that satisfy a specified waist radius in Equation (2), where the optimal solution can acquire the longest working distance Max(z_ω) in Equation (1). The problem can be converted to an optimization problem:

$$\begin{cases} z_\omega = \frac{S_1 \cos(2gL) + S_2 \sin(2gL)}{S_0 - S_3 \cos(2gL) - S_4 \sin(2gL)} \\ s.t.\ \omega_f = \omega_0 \sqrt{P_0 \cos^2(gL) + P_1 \sin^2(gL) - P_2 \sin(2gL)} \\ 0 < L < L_{max} \\ 0 < L_0 \leq L_{0max} \\ \text{Max}(z_\omega) \end{cases} \quad (9)$$

4. Fabrication and Measurement of GRIN Fiber Probe Samples

To study the coupling efficiency of a GRIN fiber probe coupled with HC-PCF further, and to demonstrate the relationship between the waist radius and working distance of a GRIN fiber probe, GRIN fiber probe samples that meet requirements should be fabricated first. Following the fabricated steps of an ultra-small GRIN fiber probe in the author's research group, samples can be fabricated and measured using the fabricated system and measuring system [21]. The measuring system is the Beam Analyzer produced by Duma Optronics, which has a beam width/position resolution of about 1 μm with a position accuracy of ±15 μm and power resolution of about 0.1 μW with a power measurement accuracy of ±10%.

The parameters of materials are as follows. The center wavelength λ of the light source beam is 1.55 μm, and the SMF core radius ω_0 is 4.5 μm. The no-core fiber has a refractive index n_0 of 1.486. The 50/125GRIN fiber core has a core diameter of 50 μm, outer diameter of 125 μm, refractive index n_1 at the axis of 1.497, and gradient constant g of 5.587 mm^{-1}. The selected no-core fiber and GRIN fiber are both produced by Prime Optical Fiber Corporation (POFC). The refractive index n_2 of air is set as 1. The max length of L_{max} and L_{0max} can be calculated as 0.57 mm and 0.29 mm, respectively, according to Section 3 under these parameters. NKT Photonics' HC19-1550-01 hollow core photonic crystal fiber with a mode field diameter of 13 μm is used. Five probes with different waist radii ω_f were designed and fabricated.

Theoretical and experimental parameters of the fabricated probes are shown in Table 1. The data of No. 1~No. 5 in Table 1 is the theoretical value. The corresponding length of the no-core fiber L_{0opt} and length of the GRIN fiber L_{opt} can be calculated for each γ according to Equation (9). The specified lengths of no-core fibers and GRIN fibers were cut as calculated above and the actual length of fibers was recorded in No. 6~No. 10. After that, GRIN fiber probe samples were fabricated with these cut fibers added to SMF using the fabricated system and the beam of these samples was analyzed by the measuring system. Then, the measured working distance z_ω and the probe waist radius ω_f were recorded in Table 1 from No. 6~No. 10.

Table 1. Parameters of the GRIN fiber probe with a specified waist radius.

Types	No.	γ	$(L_0)_{opt}$ (mm)	$(L)_{opt}$ (mm)	ω_f (μm)	Max(z_ω) (mm)
Theoretical results	1	0.62	0.290	0.213	4.00	0.165
	2	0.92	0.290	0.171	6.00	0.278
	3	1.23	0.290	0.151	8.00	0.376
	4	1.54	0.290	0.139	10.00	0.463
	5	1.85	0.290	0.131	12.00	0.538
Experimental results	6	0.95	0.290	0.213	6.15	0.188
	7	1.35	0.290	0.171	8.80	0.270
	8	1.56	0.290	0.151	10.14	0.320
	9	1.82	0.290	0.139	11.80	0.400
	10	2.33	0.290	0.131	15.16	0.530

The data in Table 1 shows that the actual waist radius and working distance of the probe are slightly different from the theoretical calculation. These errors may come from the error of the measuring device, the limitation of the fabrication device, and human operation error.

5. Coupling Efficiency Experiment and Discussion

To measure the coupling efficiency of the GRIN fiber probe and HC-PCF, an experimental system was built to test the probes above. Both the GRIN fiber probe and HC-PCF are fixed in a V-type groove of micro-displacement platform. The coupling distance between two fibers can be adjusted by a micro-displacement platform. An amplified spontaneous emission (ASE) laser is used as the light source for the GRIN fiber probe and Beam Analyzer is used as a detector to measure the power of emitted light from HC-PCF. The schematic of the experimental system is shown in Figure 3.

ASE-C-11-G of HOYATEK Company was adopted as the ASE light source, which has a central wavelength ranging from 1527 nm to 1565 nm. We selected HC19-1550-01 of NKT Photonics Company as HC-PCF. This fiber has a core radius of 10 μm, outer cladding diameter of 115 μm, and attenuation <0.03 dB/m. We cut 5 cm of this fiber for the experiment to reduce the effect of attenuation. XYZW76H-25-0.25 and MGA2 micro-displacement platforms of Shanghai Lianyi Company were used to adjusted the coupling distance between the two fibers, which can achieve a resolution of 1 μm displacement within 25 mm in the direction of the measurement axis. The experimental system setup is shown in Figure 4a and the detail of the coupling part marked with a red frame is shown in Figure 4b.

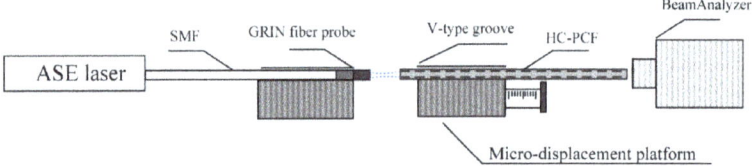

Figure 3. Schematic of the experimental system.

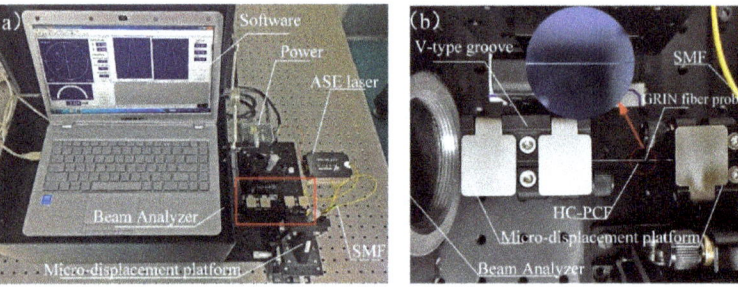

Figure 4. Coupling efficiency testing system: (a) Experimental system setup; (b) detail of coupling part.

As shown in Figure 4a, the GRIN fiber probe connected to the ASE laser was mounted in the V-groove of the micro-displacement platform, and the emitted power p_0 was detected by the Beam Analyzer. After that, the GRIN fiber probe and HC-PCF were installed according to the assembled method in Figure 4b. The micro-displacement platform was adjusted to align two fibers until the emitted power received by the Beam Analyzer was at the maximum.

The micro-displacement platform fixed with HC-PCF was moved from the position where the distance between the GRIN fiber probe and HC-PCF was zero to 1 mm at intervals of 10 μm. The emitted power p_t at each position was measured five times and the average value was used as the result. As a comparison, the coupling efficiency between SMF and HC-PCF was also measured following the same procedure as above by replacing the GRIN fiber probe with SMF.

In fact, the start position can only make two fibers as close to each other as possible, but the distance cannot be zero. It is difficult to tighten two fibers completely because it is likely to damage the porous structure of HC-PCF. Therefore, coupling distance correction of the result is necessary. We used a 40× magnifying glass to take pictures at the start position and measured the initial distance by image processing. The image processing is shown in Figure 5.

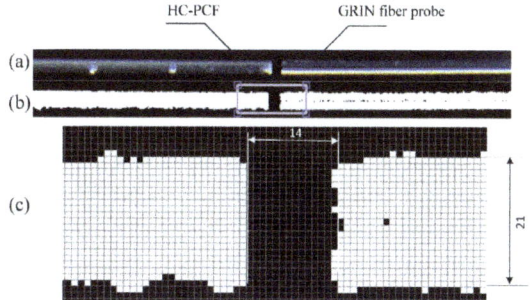

Figure 5. Initial distance measurement between the GRIN fiber probe and HC-PCF: (**a**) original image, (**b**) binarization processing, and (**c**) image in pixels.

Figure 5a is an original image taken by means of a magnifying glass. We extracted the pixel value in a black channel and adopted binarization processing, and the result is showed in Figure 5b. The detail in pixels marked with the frame is shown in Figure 5c. The outer diameter of 125 µm of the GRIN fiber probe that has a clear outline is used as the reference to calculate the initial distance. The initial distance between the two can be calculated as 125*14/21 µm, which is about 80 µm. The initial distance in a comparison experiment that used SMF is processed in the same way.

To avoid accumulative error due to the fabrication errors in Table 1, the actual waist radius and working distance of GRIN fiber probes from No. 6~No. 10 were used as the input for calculation. The theoretical coupling efficiency of the GRIN fiber probe with a ratio γ of 0.95 according to Equation (5) and the experimental results that have been corrected are shown in Figure 6. The theoretical and experimental results of the SMF fiber are plotted in this figure as a comparison.

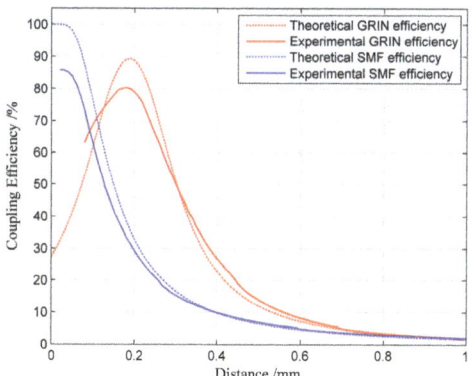

Figure 6. Theoretical and experimental values of coupling efficiency.

It can be seen from Figure 6 that the trend of experimental results is consistent with the theoretical results, but there is little discrepancy between the experimental and theoretical results due to some sources of error. The biggest reason for the discrepancy is that the simplified model only takes distributions of the input beam into consideration, without the distributions of HC-PCF. The second reason is that the MFD of HC-PCF is used as the integral range and the contribution of other areas within the core diameter is neglected. Fresnel reflection, operating error, and straightness error of micro-displacement can also cause deviation.

Figure 6 shows that the coupling efficiency of the probe is lower than that of SMF when the coupling distance is less than 0.1 mm. However, with the increase of coupling distance, the coupling efficiency of SMF decreases gradually, while the coupling efficiency of the probe shows a trend of increasing first and then decreasing. The coupling efficiency of the GRIN fiber probe reaches a maximum value of 80.22% at 180 μm, where the coupling distance is equal to the working distance of the probe, which corresponds to the theoretical calculation in Equation (5). As a comparison, the coupling efficiency of SMF at the same position has decreased to 33.45%. With the increase of coupling distance, the probe can obtain a higher coupling efficiency than SMF and gain a similar value when the distance is longer than 0.8 mm. Considering this, the incident end of HC-PCF can be placed at the range of 0.1 mm to 0.8 mm for this GRIN fiber probe, where the GRIN fiber probe can acquire a higher coupling efficiency than SMF when it is coupled with HC-PCF for gas sensing.

The experimental results show that the coupling model of the GRIN fiber probe can have a much longer working distance than SMF, so a larger gap between two coupled fibers can be achieved. The larger gap may make it easier for gas to enter the HC-PCF, so it takes less time to let the gas fill the HC-PCF before measurement, thereby reducing the system response time. In addition, the model of the GRIN fiber probe can obtain a higher coupling efficiency at the same distance, which can improve the sensitivity of the system. The higher coupling efficiency of the model means the light entering HC-PCF has a higher power, and the power of the output light from HC-PCF passing through the gas is proportional to the power of input light if the gas concentration is constant according to Lambert-Beer law. Therefore, the model of the GRIN fiber probe may display a bigger change in light power than the model of SMF when the concentration changes the same value; that is, the model of the GRIN fiber probe has a higher sensitivity.

The optimization problem used to obtain the maximum working distance for each specified waist radius is proposed in Equation (9). To verify the relationship between the working distance and maximum coupling efficiency, the coupling efficiency experiments for five fabricated fiber probes in Table 1 have been conducted. Theoretical and experimental results are plotted in Figure 7 and measurement errors are shown as error bars.

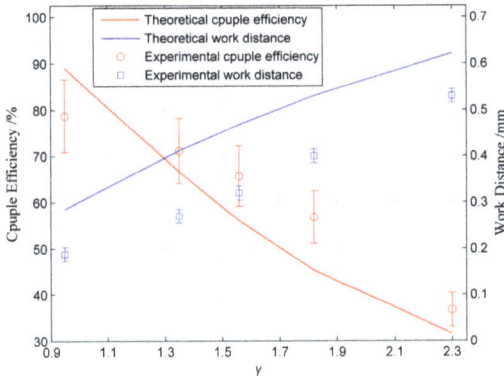

Figure 7. Theoretical and experimental working distance and maximum coupling efficiency versus γ.

The maximum coupling efficiency in Figure 7 is slightly different from theoretical value due to similar errors in Figure 6, and the working distance of the probe is slightly less than the theoretical value because of some errors expounded in Section 4. The results show that the maximum coupling efficiency of the probe decreases and the working distance of the GRIN fiber probe increases with the increase of the coupling distance; that is, the maximum coupling efficiency and working distance are contradictory. Both of them should be balanced according to the application and the optimal choice should be studied in further experiments.

6. Conclusions

A coupling model of the GRIN fiber probe and HC-PCF has been established and a method to calculate the coupling efficiency of the model has been deduced. An optimization problem was proposed as a guideline for the design of the probe to obtain the longest working distance at a specified waist radius. A series of specified probe samples were fabricated and a coupling efficiency experiment was carried out. Image processing was implemented before the experiment to correct the results. In addition, the relationship between the coupling efficiency and working distance was discussed. The theoretical and experimental results show that the maximum coupling efficiency of the probe is obtained when the coupling distance is equal to the working distance of the probe. Although the coupling efficiency of the GRIN fiber probe at a very short distance is lower than that of SMF, the coupling efficiency of the probe is obviously higher than that of SMF with the increase of coupling distance. Therefore, the probe coupled with HC-PCF can obtain a higher coupling efficiency at a longer distance. The longer distance can improve the response time of the system and the higher coupling efficiency can improve the sensitivity of the system. Because of the contradictory relationship between the maximum coupling efficiency and the working distance of the GRIN fiber probe, some theoretical and experimental analyses are necessary in the future to select the optimized GRIN fiber probe for gas sensing.

Author Contributions: C.W. and A.A. conceived and designed the experiment; J.S. performed the experiment; X.L. and J.L. supervised the entire work; C.W. and Y.Z. wrote the paper.

Funding: The authors acknowledge the financial support of the National Natural Science Foundation of China (Grant No. 61773249, No. 41704123), the Natural Science Foundation of Shanghai (Grant No. 16ZR1411700), and the Science and Technology on Near-Surface Detection Laboratory (Grant No. TCGZ2018A007).

Conflicts of Interest: The authors declare no conflict of interest.

References

1. Knight, J.C.; Broeng, J.; Birks, T.A.; Russell, P.S.J. Photonic band gap guidance in optical fibers. *Science* **1998**, *282*, 1476–1478. [CrossRef] [PubMed]
2. Carvalho, J.P.; Magalhães, F.; Ivanov, O.V.; Frazão, O.; Araújo, F.M.; Ferreira, L.A.; Santos, J.L. Evaluation of coupling losses in hollow-core photonic crystal fibres. *Proc. SPIE* **2007**, *6619*, 66191V.
3. Monfared, Y.E. Transient dynamics of stimulated Raman scattering in gas-filled hollow-core photonic crystal fibers. *Adv. Mater. Sci. Eng.* **2018**, *2018*, 1–5. [CrossRef]
4. Zhao, Y.; Jin, W.; Lin, Y.; Yang, F.; Ho, H.L. All-fiber gas sensor with intracavity photothermal spectroscopy. *Opt. Lett.* **2018**, *43*, 1566–1569. [CrossRef] [PubMed]
5. Morel, J.; Marty, P.T.; Feurer, T. All-fiber multi-purpose gas cells and their applications in spectroscopy. *J. Lightwave Technol.* **2010**, *28*, 1236–1240.
6. Wu, Z.F.; Zheng, C.T.; Liu, Z.W.; Yao, D.; Zheng, W.X.; Wang, Y.D.; Wang, F.; Zhang, D.M. Investigation of a slow-light enhanced near-infrared absorption spectroscopic gas sensor, based on hollow-core photonic band-gap fiber. *Sensors* **2018**, *18*, 2192. [CrossRef]
7. Swinehart, D.F. The Beer-Lambert Law. *J. Chem. Educ.* **1962**, *39*, 333. [CrossRef]
8. Sleiffer, V.A.J.M.; Jung, Y.; Baddela, N.K.; Surof, J.; Kuschnerov, M.; Veljanovski, V.; Hayes, J.R.; Wheeler, N.V.; Fokoua, E.R.N.; Wooler, J.P. High capacity mode-division multiplexed optical transmission in a novel 37-cell hollow-core photonic bandgap fiber. *J. Lightwave Technol.* **2014**, *32*, 854–863. [CrossRef]
9. Kiarash Zamani, A.; Digonnet, M.J.F.; Shanhui, F. Optimization of the splice loss between photonic-bandgap fibers and conventional single-mode fibers. *Opt. Lett.* **2010**, *35*, 1938–1940.
10. Zhao, C.L.; Xiao, L.; Demokan, M.S.; Jin, W.; Wang, Y. Fusion splicing photonic crystal fibers and conventional single-mode fibers: Microhole collapse effect. *J. Lightwave Technol.* **2007**, *25*, 3563–3574.

11. Fan, D.; Jin, Z.; Wang, G.; Xu, F.; Lu, Y.; Hu, D.J.J.; Wei, L.; Shum, P.; Zhang, X. Extremely high-efficiency coupling method for hollow-core photonic crystal fiber. *IEEE Photonics J.* **2017**, *9*, 1–8. [CrossRef]
12. Parmar, V.; Bhatnagar, R.; Kapur, P. Optimized butt coupling between single mode fiber and hollow-core photonic crystal fiber. *Opt. Fiber Technol.* **2013**, *19*, 490–494. [CrossRef]
13. Ding, H.; Li, X.; Cui, J.; Yang, L.; Dong, S. An all-fiber gas sensing system using hollow-core photonic bandgap fiber as gas cell. *Instrum. Sci. Technol.* **2011**, *39*, 78–87. [CrossRef]
14. Ma, Y.; Tong, Y.; He, Y.; Jin, X.; Tittel, F.K. Compact and sensitive mid-infrared all-fiber quartz-enhanced photoacoustic spectroscopy sensor for carbon monoxide detection. *Opt. Express* **2019**, *27*, 9302–9312. [CrossRef] [PubMed]
15. Wang, C.; Mao, Y.X.; Fang, C.; Tang, Z.; Yu, Y.J.; Qi, B. Analytical method for designing gradient-index fiber probes. *Opt. Eng.* **2011**, *50*, 62–65.
16. Mao, Y.X.; Chang, S.; Sherif, S.; Flueraru, C. Graded-index fiber lens proposed for ultrasmall probes used in biomedical imaging. *Appl. Opt.* **2007**, *46*, 5887–5894. [CrossRef]
17. Yang, X.; Lorenser, D.; Mclaughlin, R.A.; Kirk, R.W.; Edmond, M.; Simpson, M.C.; Grounds, M.D.; Sampson, D.D. Imaging deep skeletal muscle structure using a high-sensitivity ultrathin side-viewing optical coherence tomography needle probe. *Biomed. Opt. Express* **2014**, *5*, 136–148. [CrossRef]
18. Dirk, L.; Xiaojie, Y.; Sampson, D.D. Ultrathin fiber probes with extended depth of focus for optical coherence tomography. *Opt. Lett.* **2012**, *37*, 1616–1618.
19. Pfeifer, T.; Schmitt, R.; Konig, N.; Mallmann, G.F. Interferometric measurement of injection nozzles using ultra-small fiber-optical probes. *Chin. Opt. Lett.* **2011**, *9*, 37–40.
20. Wang, C.; Xu, L.L.; Zhu, J.; Yuan, Z.W.; Yu, Y.J.; Asundi, A.K. A novel integrated fiber-optic interferometer model and its application in micro-displacement measurement. *Opt. Lasers Eng.* **2016**, *86*, 125–131. [CrossRef]
21. Bi, S.B.; Wang, C.; Zhu, J.; Yuan, Z.W.; Yu, Y.J.; Valyukh, S.; Asundi, A. Influence of no-core fiber on the focusing performance of an ultra-small gradient-index fiber probe. *Opt. Lasers Eng.* **2018**, *107*, 46–53. [CrossRef]
22. Hoo, Y.L.; Jin, W.; Ju, J.; Ho, H.L. Loss analysis of single-mode fiber/photonic-crystal fiber splice. *Microw. Opt. Technol. Lett.* **2004**, *40*, 378–380. [CrossRef]

© 2019 by the authors. Licensee MDPI, Basel, Switzerland. This article is an open access article distributed under the terms and conditions of the Creative Commons Attribution (CC BY) license (http://creativecommons.org/licenses/by/4.0/).

Article

Real-Time Vision through Haze Based on Polarization Imaging

Xinhua Wang [1,2], Jihong Ouyang [1,*], Yi Wei [3], Fei Liu [2,3,*] and Guang Zhang [2]

1. College of Computer Science and Technology, Jilin University, Changchun 130012, China; xinhuajlu@163.com
2. State Key Laboratory of Applied Optics, Changchun Institute of Optics, Fine Mechanics and Physics, Chinese Academy of Sciences, Changchun 130033, China; zhangguang0920@163.com
3. School of Physics and Optoelectronic Engineering, Xidian University, Xi'an 710071, China; wei18334704740@163.com
* Correspondence: ouyj@jlu.edu.cn (J.O.); feiliu@xidian.edu.cn (F.L.)

Received: 19 November 2018; Accepted: 25 December 2018; Published: 3 January 2019

Abstract: Various gases and aerosols in bad weather conditions can cause severe image degradation, which will seriously affect the detection efficiency of optical monitoring stations for high pollutant discharge systems. Thus, penetrating various gases and aerosols to sense and detect the discharge of pollutants plays an important role in the pollutant emission detection system. Against this backdrop, we recommend a real-time optical monitoring system based on the Stokes vectors through analyzing the scattering characteristics and polarization characteristics of both gases and aerosols in the atmosphere. This system is immune to the effects of various gases and aerosols on the target to be detected and achieves the purpose of real-time sensing and detection of high pollutant discharge systems under bad weather conditions. The imaging system is composed of four polarizers with different polarization directions integrated into independent cameras aligned parallel to the optical axis in order to acquire the Stokes vectors from various polarized azimuth images. Our results show that this approach achieves high-contrast and high-definition images in real time without the loss of spatial resolution in comparison with the performance of conventional imaging techniques.

Keywords: real-time observation; optical sensing; stokes vectors; information processing technology

1. Introduction

Environmental pollution has become a matter of great concern in recent years. In order to protect the environment, the inspection of the environmental quality, especially the sensing and detection of the emission of pollutants, is a problem that must be faced [1,2]. In view of this situation, many countries have set up environmental stations to monitor companies with high pollutant emissions [3]. The optical sensing and detection method has become the most popular monitoring method owing to its high resolution, abundant information, and use of simple equipment. However, in bad weather conditions, such as hazy and rainy weather conditions, the atmospheric aerosols and various gases, especially the haze particles, have a serious scattering effect on the light wave [4,5]. The image obtained under these conditions will go through a serious degradation process, which will reduce the efficiency of optical sensing and detecting, as well as the monitoring distance [6,7]. Therefore, without increasing the number of monitoring stations, how to improve the sense and detect distance of the monitoring station under the condition of bad weather and how to remove the influence of gases and aerosols on the monitoring efficiency have become urgent problems to be solved.

This paper proposes a real-time optical sensing and detection system based on the Stokes vectors through analyzing the scattering characteristics and polarization characteristics of gases and aerosols in the atmosphere. This system can sense and detect the emission of pollutants in real time in bad weather and improve the monitoring distance of the environmental monitoring system. In this study, we first analyzed the principles of polarization imaging algorithms and established a physical model for polarization images through gases and aerosols in bad weather based on the Stokes vectors. Next, we developed a real-time optical sensing and detection system with four cameras aligned parallel to the optical axis. By solving the linear polarization parameters of the Stokes matrix, we achieved real-time registration of different polarization images. Further, we achieved visual enhancement of polarization degree images using an image registration algorithm based on the speeded up robust features (SURF) algorithm and a bilinear interpolation based on the contrast-limited adaptive histogram equalization (CLAHE) algorithm. Our results show that the proposed system can achieve real-time high-contrast, high-definition imaging under bad weather conditions without a loss of spatial resolution.

2. Physical Model for Polarization Image

To construct our physical model, we consider the situation in which two degraded orthogonally polarized azimuth images are first acquired by rotating the polarizer installed in front of the detector in bad weather that has various gases and aerosols present. A clear scene is subsequently acquired by effectively splitting the azimuth images based on the differences of polarization properties between the atmospheric light and the target light [8]. The total intensity incident on the detector can be expressed as:

$$I_{Total} = T + A = L_{Object}e^{-\beta z} + A_\infty \left(1 - e^{-\beta z}\right) \tag{1}$$

The total light intensity I_{Total} received by the detector consists of the target light T and the atmospheric light A. Target light T is the target radiation L_{Object} that reaches the detector after being scattered by various gases and aerosols, and it exponentially decays during transmission [9], as shown in the attenuation model in Figure 1a. Atmospheric light A refers to the sunlight received by the detector after the scattering of various gases and aerosols, and it increases with the detection distance, as shown in the atmospheric light model in Figure 1b. The relationship between I_{Total} T, A, and L_{Object} is given by Equation (1), where A_∞ denotes the atmospheric light intensity from infinity, β denotes the atmospheric attenuation coefficient, and z denotes the distance between the target and the detector.

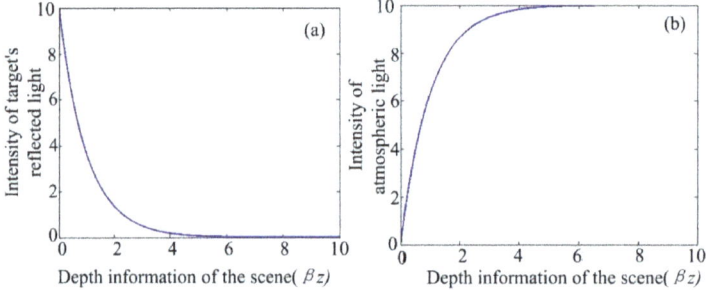

Figure 1. (a) Attenuation model, and (b) atmospheric light model.

During polarization imaging, two orthogonally polarized azimuth images are acquired at the detector by rotating the polarizer. In particular, I_{min} denotes the minimum atmospheric light intensity, whereas its orthogonal counterpart, I_{max}, denotes the maximum atmospheric light intensity [6]. The expressions for I_{min}, I_{max}, and I_{Total} are as follows:

$$\begin{aligned} I_{max} &= I_T/2 + I_B^{max} \\ I_{min} &= I_T/2 + I_B^{min} \end{aligned} \tag{2}$$

$$I_{Total} = I_{max} + I_{min} \tag{3}$$

Upon selecting an area with uniform atmospheric light without targets and setting $A_{max}(x, y)$ and $A_{min}(x, y)$, the polarization of the atmospheric light can be expressed as Equation (4), where $A_{max}(x, y)$ and $A_{min}(x, y)$ denotes the brightest and darkest atmospheric light, respectively, and (x, y) denotes image pixel point coordinate:

$$p = \frac{A_{max}(x,y) - A_{min}(x,y)}{A_{max}(x,y) + A_{min}(x,y)} \qquad (4)$$

Substitution of Equations (2)–(4) into Equation (1) yields the mathematical model for a high-definition scenario, as follows:

$$L_{Object} = \frac{(I_{Total} - (I_{max} - I_{min})/p)}{1 - A/A_\infty} \qquad (5)$$

From Equation (5), we note that the polarization image model uses the vector vibration direction difference between the target and the atmospheric light to filter the effects of the atmospheric light. High-contrast and high-definition images of targets under smoggy conditions are acquired by comparing the differences of the orthogonally linear-polarized azimuth images acquired by rotating the polarizers.

3. Real-Time Optical Sensing and Detection System

The polarization image model primarily relies on the random rotation of the polarizer installed in front of the detector for obtaining the maximum and minimum polarization azimuth images, thus limiting its real-time detection results. In this study, we propose a polarization image acquisition method based on the Stokes vectors. According to the Stokes principle, a real-time optical sensing and detection system is designed, which is capable of acquiring the four Stokes parameters in real time, which are then used to solve for the clear scene based on the polarization image model. In addition, to ensure that all the Stokes vectors are targeting the same field of view, the system uses a calibration platform based on a luminescence theodolite to complete the calibration of the common optical axis of the four-camera array.

3.1. Polarization Image Acquisition Method Based on Stokes Vectors

In 1852, Stokes proposed a detection method that utilized light intensity to describe the polarization characteristics. This method, known as the Stokes vector [10] method, allows for a more intuitive and convenient observation and acquisition of the polarization information of light. The Stokes matrix $S = [S_0, S_1, S_2, S_3]^T$ is commonly used to describe the polarization state of a light wave, where S_0 denotes the total intensity of radiation, S_1 the intensity difference between vertically and horizontally polarized light waves, S_2 the intensity difference between two 45°-polarized light waves, and S_3 the intensity difference between right-handed and left-handed circularly polarized light waves. The expression S is as follows:

$$S = \begin{bmatrix} S_0 \\ S_1 \\ S_2 \\ S_3 \end{bmatrix} = \begin{bmatrix} I_{0°} + I_{90°} \\ I_{0°} - I_{90°} \\ I_{45°} - I_{135°} \\ 2I_{\lambda/4, 45°} - (I_{0°} + I_{90°}) \end{bmatrix} \qquad (6)$$

where $I_{0°}$, $I_{45°}$, $I_{90°}$, and $I_{135°}$ denote the intensities of light corresponding to polarization directions of 0°, 45°, 90°, and 135°; these intensities of light are obtained by installing polarizers that have different light transmission directions (including 0°, 45°, 90°, and 135°) in front of the detector. $I_{\lambda/4, 45°}$ means the right-handed circularly polarized light. This is achieved by installing a polarizer that transmits 45° light in front of the detector, and then installing a wave plate whose fast axis direction is 0° in front of the polarizer.

Under natural conditions, the circular component can be neglected because there is no circular polarization properties in nature. Hence, the stokes vector can be reduced to a vector containing only linear polarization components $S' = [S_0, S_1, S_2]^T$. Therefore, only the intensities of light corresponding to different polarization directions, namely $I_{0°}$, $I_{45°}$, $I_{90°}$, and $I_{135°}$, need to be measured. Subsequently, linear components S_0, S_1, and S_2 of the target-light Stokes matrix can be obtained using Equation (7). The polarization state of the light wave can be expressed as follows:

$$S' = \begin{bmatrix} S_0 \\ S_1 \\ S_2 \end{bmatrix} = \begin{bmatrix} I_{0°} + I_{90°} \\ I_{0°} - I_{90°} \\ I_{45°} - I_{135°} \end{bmatrix} \tag{7}$$

After the polarization information of the light wave is obtained through measurement, it is visualized using polarization degree images P or polarization angle images θ. In general, the degree of polarization (DoP) can be calculated by means of the Stokes vectors, and the degree of polarization of the actual measurement, that is, the degree of linear polarization (DoLP), can be calculated using Equation (8):

$$DoLP = \frac{\sqrt{S_1^2 + S_2^2}}{S_0} \tag{8}$$

Further, I_{max} and I_{min}, that is, the maximum and minimum light intensities, respectively, required for the polarization-based imaging model, can be calculated using Equation (9):

$$\begin{aligned} I_{max} &= \frac{(1+DoLP)S_0}{2} \\ I_{min} &= \frac{(1-DoLP)S_0}{2} \end{aligned} \tag{9}$$

3.2. Design of Real-Time Optical Sensing and Detection System

As per the polarization image model, the real-time optical sensing and detection system with four cameras aligned parallel to the optical axis was designed. Analyzers were mounted and integrated inside each lens at angles of 0°, 45°, 90°, and 135° relative to the polarization direction of the incident light. The structure of our optical system is shown in Figure 2.

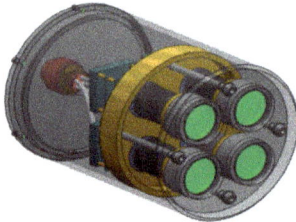

Figure 2. Real-time optical sensing and detection system.

A calibration platform, using a luminous theodolite, was adopted for the calibration of the common optical axis of the four-camera array. During calibration, via controlling the linear motion of a two-dimensional rail, the exit pupil of the theodolite was aligned with the entrance pupils of the cameras separately. Next, with the crosshair lit, the orientation of the optical axis of each camera was adjusted individually to center the crosshair image on the target surface of the camera. High-precision calibration of the common optical axis was thereby completed, and the requirements for subsequent image registration were fulfilled. According to the above analysis, we next constructed a four-axis calibration platform, which was composed of the following elements: a luminous theodolite (1), a horizontal rail (2), a vertical rail (3), a load platform (4), a right angle fixed block (5), and a servo drive (6). Each part is marked on Figure 3.

Figure 3. Four-axis calibration platform.

In addition, the cameras were equipped with external triggers for synchronized exposure (trigger IN) and image capture (trigger OUT) from the four cameras. Transistor–transistor logic (TTL) levels that had both trigger IN and OUT interfaces were employed such that one signal could be used to trigger multiple cameras. That is, trigger OUT was enabled on the first camera, while the remaining three were enabled with trigger IN, such that synchronized imaging using multiple cameras was enabled. Compared with other polarization imaging systems, such as a focal-plane polarization imaging system [11], micropolarizer arrays polarization imaging system [12], and so on [13–16], the proposed system has the advantages of low cost and simple imaging system. This system can obtain the polarization images without losing the light energy.

4. Target Enhancement Algorithms Based on Polarization Information

4.1. Polarization Image Registration Algorithm Based on SURF

The real-time optical sensing and detection system described in this study used four cameras aligned parallel to the optical axis, which can cause misalignment and rotation of pixel units because of different shooting positions and directions of the cameras. Because the solutions of intensity, polarization difference, and polarization degree images are based on pixel units, misalignment of pixel units between polarized azimuth images can result in false edges or blurring.

In our study, by the adoption of the SURF-based real-time image registration algorithm for subpixel-precision registration of the acquired linear polarization degree images, the pixels between the polarization images were aligned via the procedure shown in Figure 4, which constituted three steps: feature detection and extraction, feature vector matching, and spatial transformation model parameter estimation [17,18]. The detailed algorithm procedure is listed below.

Step 1: Use the high-precision four-dimensional calibration platform to pre-calibrate the overlapping imaging areas of the adjacent cameras.

Step 2: Use fast Hessian detectors to extract the feature points of the reference image and the image to be registered within the overlapping area, and generate SURF description vectors.

Step 3: Use the fast approximate nearest neighbors (FANN) algorithm to obtain the initial matching point pair, and sort the Euclidean distance of the feature vectors of the pair.

Step 4: Use the progressive sample consensus (PROSAC) algorithm to perform spatial transformation model parameter estimation, which derives the geometric spatial transformation relationship between the reference image and the image to be registered.

This algorithm is invariant to changes in image size, rotation, and illumination. The registration speed is within milliseconds, and the accuracy is at the level of 0.1 pixel.

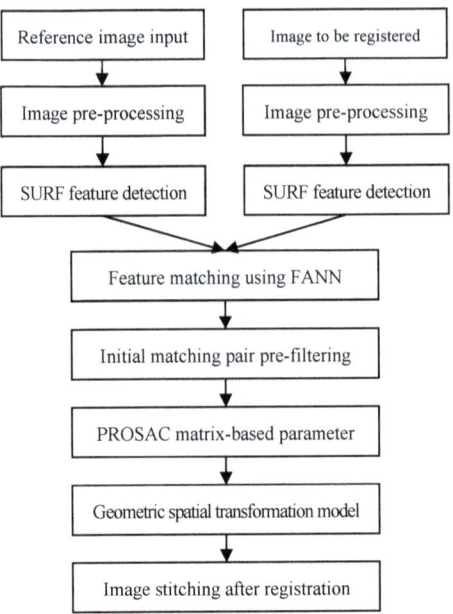

Figure 4. Flowchart of image registration algorithm.

4.2. CLAHE Image Enhancement Algorithm Based on Bilinear Interpolation

While the overall quality and contrast were significantly improved in the polarization reconstructed scenes, local details were not adequately enhanced, and overexposure was noted in the sky portion of the image [19]. To resolve these issues, we used a bilinear interpolation CLAHE algorithm to further enhance the polarization reconstructed images via the following steps:

Step 1: Divide the image into several area blocks to perform individual contrast enhancement based on the block's histogram. The local contrast enhancement is represented by Equation (10):

$$x'_{i,j} = m_{i,j} + k(x_{i,j} - m_{i,j}) \tag{10}$$

where $x_{i,j}$ and $x'_{i,j}$ denote the grayscale values before and after enhancement, respectively, and $m_{i,j} = \frac{1}{m \times n} \sum_{(i,j) \in W} x_{i,j}$ denotes the average grayscale value of the pixels in block W. Parameter k can be represented as in Equation (11):

$$k = k' \left[\left(\frac{\sigma_{i,j}^2}{\sigma_n^2} \right) - 1 \right] \tag{11}$$

where k' denotes the scale factor, σ_n^2 the noise variance of the whole image, and $\sigma_{i,j}^2$ the grayscale variance of block W.

Step 2: Stitch neighboring regions by means of bilinear interpolation to effectively eliminate the artifacts between neighboring regions introduced post local contrast enhancement. Assume that the values of four points $K_{11} = (x_1, y_1)$, $K_{12} = (x_1, y_2)$, $K_{21} = (x_2, y_1)$, and $K_{22} = (x_2, y_2)$ of a function $f(x)$ are known; then, the value of a point $H = (x, y)$ of the $f(x)$ function can be derived by linear interpolation.

First, linear interpolation is performed along the x-direction to obtain the value as represented by Equations (12) and (13):

$$f(H_1) \approx \frac{x_2-x}{x_2-x_1} f(K_{11}) + \frac{x-x_1}{x_2-x_1} f(K_{21})$$
$$H_1 = (x, y_1)$$
(12)

$$f(H_2) \approx \frac{x_2-x}{x_2-x_1} f(K_{12}) + \frac{x-x_1}{x_2-x_1} f(K_{22})$$
$$H_2 = (x, y_2)$$
(13)

Subsequently, the same operation is performed along the y-direction as represented in Equation (14):

$$f(H) \approx \frac{y_2 - x}{y_2 - y_1} f(H_1) + \frac{y - y_1}{y_2 - y_1} f(H_2)$$
(14)

5. Results and Discussion

Testing Environment

All images in this study were acquired by use of a real-time optical sensing and detection system with four cameras aligned parallel to the optical axis. Figure 5 shows the photograph of the imaging system. The specifications of the cameras were as follows: 60° field of view, 8-mm focal length, and 4 million pixels. Different from other optical imaging systems, polarization imaging system only requires the installation of a polarizer in front of ordinary industrial cameras to achieve clear scene imaging. In this study, the type of camera is GS3-U3-41C6C-C, which is produced by Pointgrey based in vancouver, Canada, and the type of polarizer mounted in front of the camera isLPVISC100, this polarizer is produced by Thorlabs in the American state of New Jersey.

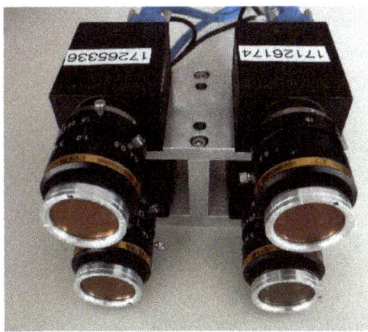

Figure 5. Real-time polarization dehazing imaging system.

MATLAB R2017b(developed by MathWorks in Massachusetts, America) was used to execute the real-time polarization image processing and enhancement algorithm, and the program was run on a computer with a Windows 7 (64-bit) operating system and an Intel Core i7-4790K processor with 32 GB RAM, this processor is produced by Dell based in Texas, America. The time required to process a frame of image on the system was 30 ms.

The raw images obtained from the real-time optical sensing and detection system using four cameras parallel to the optical axis are shown in Figure 6. These images were acquired under hazy weather conditions, and the image quality was seriously degraded, and the presence of atmospheric light, which was produced via haze particles, seriously reduced the image contrast and the detection distance of the system. In our study, we first applied the conventional polarization image enhancement algorithm proposed by Y.Y. Schechner [20], and the result is shown in Figure 6h. The algorithms proposed by Y.Y. Schechner are the most classical of all polarization imaging algorithms [21–23], and their processing effect can represent the processing effect of most algorithms. When compared with the raw images, the images reconstructed by the conventional algorithm showed obvious visual enhancements as the atmospheric light at both the near and far ends of the images are removed, and the contrast was also improved. However, this enhancement algorithm resulted in overexposure in the sky region, thereby limiting the visual perception of the image. Figure 6i shows the reconstructed image obtained by applying the registration and enhancement algorithm proposed in this study. We first note that this algorithm effectively eliminates the effects of atmospheric light on the image and improved the contrast of the image. Further, the proposed algorithm reduced not only the effects of atmospheric light for objects at the near ends, but also for those at the far ends, thus creating a perception of stereovision. In addition, a comparison of Figure 6h,i demonstrates that the proposed algorithm not only avoided overexposure in the sky but also enriched the details of the image.

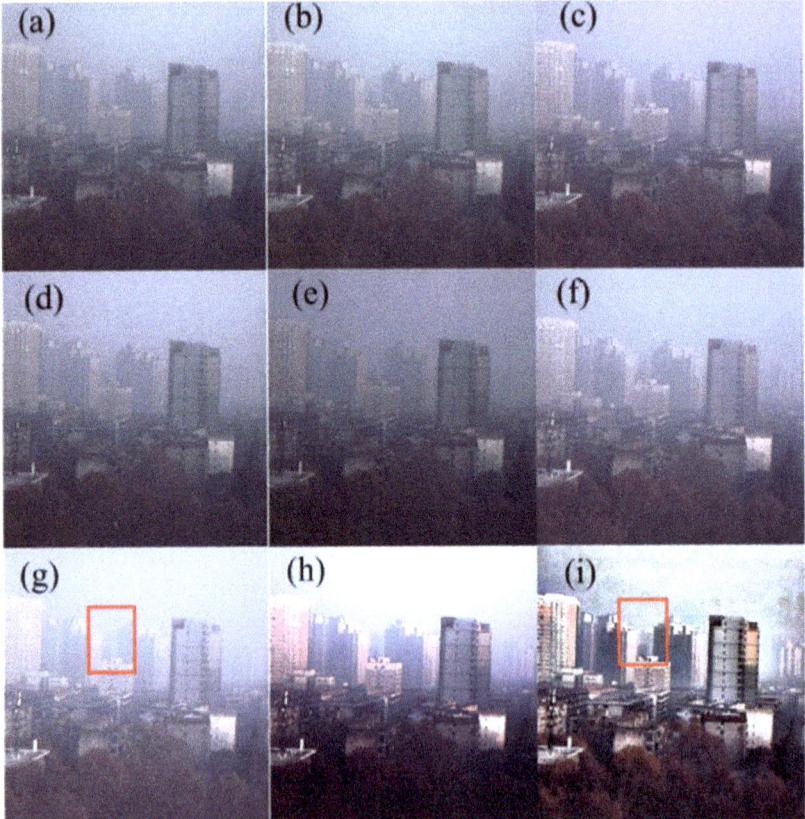

Figure 6. (**a–d**) Polarization images corresponding to polarization angles of 0°, 45°, 90°, and 135°; (**e**) Polarization image with minimum airlight; (**f**) Polarization image with maximum airlight; (**g**) Total intensity image; (**h**) Detection result obtained using the method of Schechner; (**i**) Detection result obtained with proposed method; The zoomed-in view of the region of interest marked with a red rectangle in (**g**) and (**i**) will be shown in Figure 7a,b.

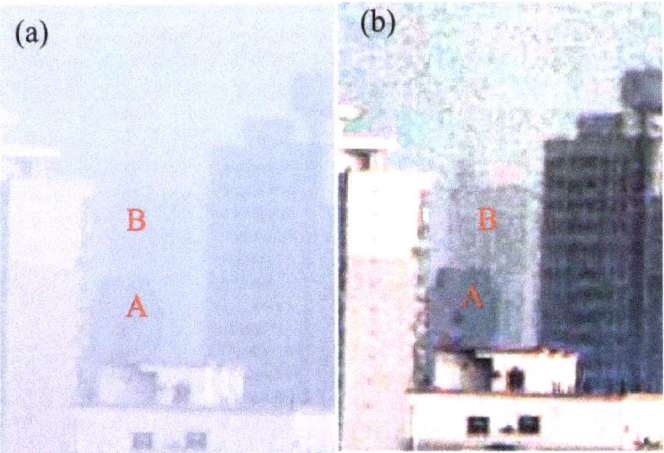

Figure 7. (a,b) The zoomed-in view of the region of interest marked with a red rectangle in Figure 6g,i; A and B are two buildings with different distances from the detector in the scene.

The zoomed-in view of the region of interest marked with a red rectangle in Figure 6g,i is shown in Figure 7a,b. The distance between building A and the detector was 1.6 km, and we can only observe the outline of the building under the influence of haze, which means the spatial resolution of Figure 7a at the detection distance of 1.6 km is 15 m. However, we can see two windows of building A in Figure 7b, which means the spatial resolution of Figure 7b at the detection distance of 1.6 km was 2 m. Furthermore, building B, which is completely invisible in Figure 7a, is highlighted in Figure 7b. These changes suggest that our imaging system can improve spatial resolution in severe weather conditions. However, the limitation of the proposed method is also obvious. There was a large amount of noise in the distant area. This was because when using Equation (1) to enhance the reflected light energy of distant targets, the noise existing there also tended to be amplified in line with the same trend [24].

Figure 8 shows the intensity distribution curve in the pixels of row 415 (counting from top to bottom) across the background and the target from Figure 6g–i, which provides an intuitive demonstration of the difference between the atmospheric light and the target light in the imaging scene, as well as changes in the contrast. When compared with the results obtained with the Schechner algorithm (red curve), the reconstruction result of our algorithm (blue curve) increases the atmospheric light-target light difference to a certain extent and enhances the contrast of the image. In Figure 8, the fluctuations of the blue curve are stronger than those of the red one, particularly at the target location. The fluctuation of the pixel intensity curve after reconstruction using the proposed algorithm increased significantly, which indicates that this algorithm is superior to that of the Schechner algorithm in improving the image contrast.

Figure 9a–c shows the grayscale histograms corresponding to Figure 6g–i, respectively. This rendering of information provides a more intuitive representation of the characteristics of Figure 6g–i. When compared with the raw image that has its histogram concentrated in the right half of the panel, the reconstructed image exhibited a wider and more evenly distributed histogram. This result confirms that the proposed algorithm can effectively reduce the effects of scattering due to aerosols and various gases to increase image contrast and enrich image details. Figure 9d–f illustrates the pixel intensity distributions in the red, green and blue (RGB) channels of Figure 6g–i, respectively. We note that the pixel intensity distribution in Figure 9f was optimized over the distributions of the raw image and the image processed by means of the conventional polarization imaging algorithm. In addition, the proposed algorithm increased the dynamic range and stereovision of the image.

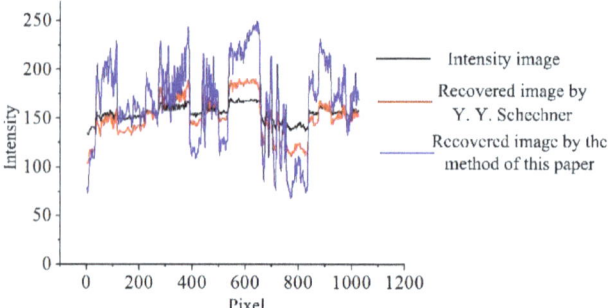

Figure 8. Horizontal line plot at vertical position pixel 415 in Figure 6g–i.

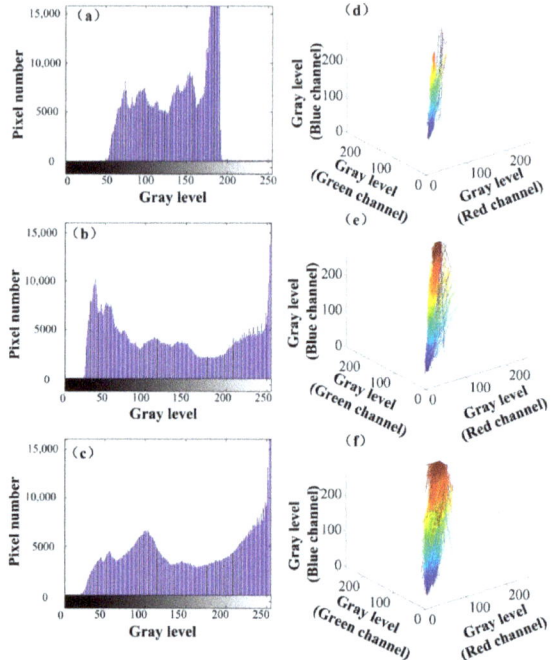

Figure 9. (a–c) Gray histograms corresponding to Figure 6g–i; (d–f) Pixel intensity distributions of channels R, G, and B of Figure 6g–i, respectively.

Figure 10 shows the results of the detection under different weather conditions. Figure 10a is the total intensity image, which was captured under rainy and foggy weather. Figure 10b is the detection result of Figure 10a using the proposed method, which proved the effectiveness of our method in this weather condition. Figure 10c shows the total intensity image of contaminated polluted lake water, which was captured under dense fog weather, where the pollution of the lake water was difficult to be observed because of the influence of haze, the overall contrast of the image was low, and the detection distance was limited. After the detection by our imaging system, as shown in Figure 10d, the influence of aerosols and various gases was removed. This system could directly observe the pollution of the lake, so that we could control the pollution of the lake. In addition, the detection distance of the imaging system was improved, which greatly reduces the cost.

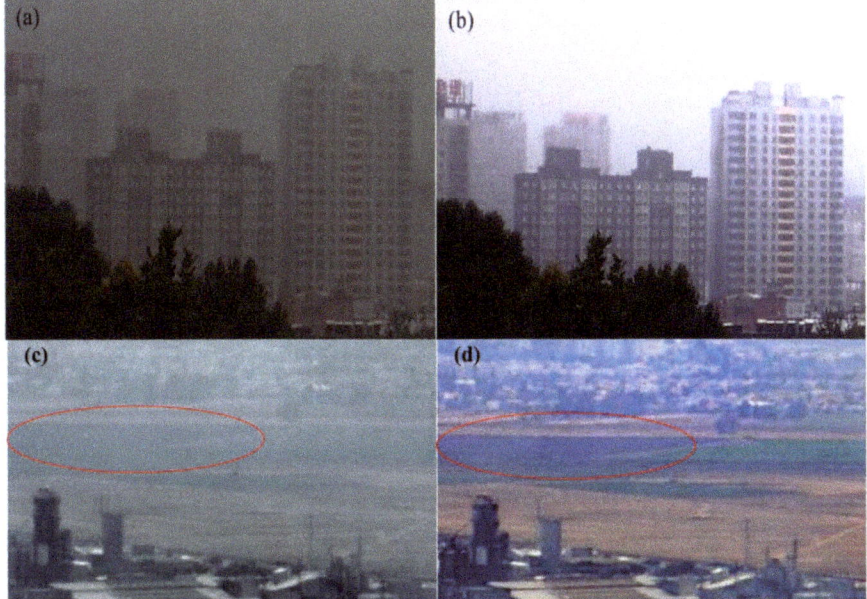

Figure 10. (a) Total intensity image captured under rainy and foggy weather; (b) Detection result of (a) using the proposed method; (c) Total intensity image of contaminated polluted lake water (the region marked with a red circle) captured under dense fog weather; and (d) Detection result of (c) using the proposed method, the region marked with a red circle shows that the pollution level of the lake can be directly observed.

To validate the proposed algorithm, we selected multiple scenes for imaging, as shown in Figure 11. When compared with the scenes processed by the Schechner algorithm [25], we note that the proposed algorithm could effectively avoid overexposure in the blank regions of the sky and considerably enhance the visual effect of the image. In addition, the proposed algorithm achieved far superior image detail enhancement for both near and far objects in the image, consequently providing an improved sense of stereovision.

To provide an objective evaluation of the effect of sensing and detection with various scenes, we next adopted four commonly used image quality assessment metrics to compare algorithms; the corresponding results are listed in Table 1. The mean gradient assesses high-frequency information such as edges and details of the image; a higher gradient indicates an image with clearer edges and more details. The edge strength is the amplitude of the edge point gradient. The image contrast represents the ratio between the bright and dark areas of the general image; an image with a higher contrast indicates more levels between bright/dark gradual changes, and thus, it contains more information. Overall, the quality of the image processed by the proposed algorithm was significantly improved. When compared with the raw image, the processed image generally had its contrast increased by a factor of ≈ 10, mean gradient increased by a factor of 4, and edge strength increased by a factor of ≈ 3. These results demonstrate that the processed images provided improved quality in terms of contrast, details, and definition, which agrees well with the conclusion of the previous subjective assessment.

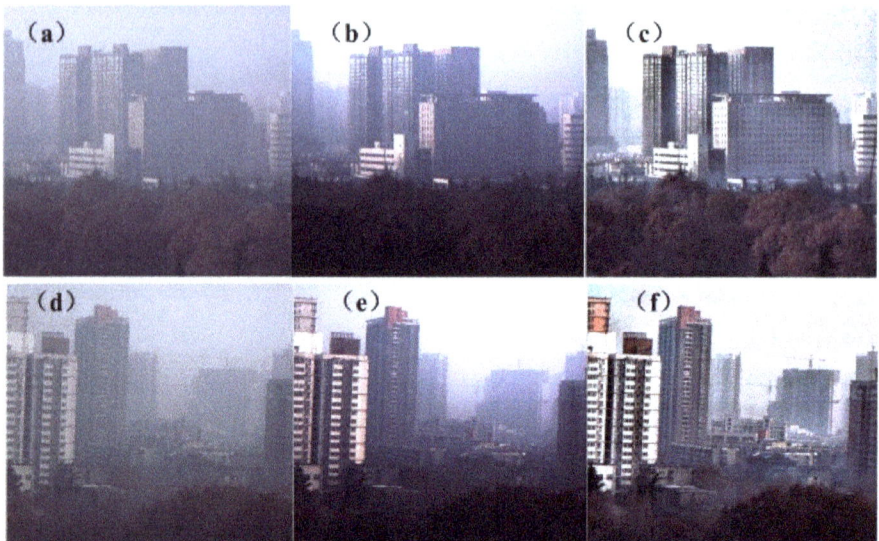

Figure 11. (**a,d**) Total intensity images of different scenes; (**b,e**) Detection results obtained by the method of Schechner; and (**c,f**) Detection results obtained with the proposed method.

Table 1. Objective evaluation of dehazed images.

Image	Metrics	Mean Gradient	Edge Strength	Contrast
Figure 6g	Total light intensity image	0.0049	0.0523	1.3455
	Result by the conventional algorithm	0.0084	0.0897	5.0434
	Result by the proposed algorithm	0.0233	0.2443	17.5323
Figure 8a	Total light intensity image	0.0066	0.07	1.9452
	Result by the conventional algorithm	0.0095	0.101	5.3448
	Result by the proposed algorithm	0.0245	0.2575	22.656
Figure 8d	Total light intensity image	0.0048	0.0505	0.9295
	Result by the conventional algorithm	0.0055	0.0589	1.6168
	Result by the proposed algorithm	0.0155	0.165	9.1692

6. Conclusions

In this study, we proposed a real-time polarization imaging algorithm and investigated its performance, both theoretically and experimentally. We designed and constructed a real-time optical sensing and detection system based on the Stokes vectors underpinned by the principles of polarization imaging, wherein optical analyzers at different angles relative to the polarization direction of the incident light were integrated into four independent cameras individually. Linear polarization components were calculated by use of the Stokes vectors, followed by real-time image registration of images with different polarization components based on the SURF algorithm and subsequent visualization of the polarization images. Finally, we adopted an improved image enhancement algorithm using the CLAHE-based bilinear interpolation to generate real-time high-contrast and high-definition images. Our experimental results further reinforce the conclusion that the proposed method can acquire high-contrast and high-definition images in real time without loss of spatial resolution, which improves the detection range of environmental monitoring stations in hazy weather conditions

Author Contributions: This research article contains five authors, including X.W., J.O., Y.W., F.L. and G.Z. X.W. and F.L. jointly designed the overall architecture and related algorithms. G.Z. and Y.W. conceived and designed the experiments. J.O. coordinated the overall plan and direction of the experiments and related skills. X.W. wrote this paper.

Funding: This research was funded by Jilin Province Science and Technology Development Plan Project (approval number: 20160209006GX, 20170309001GX); Self-fund of State Key Laboratory of Applied Optics (approval number: 2016Y6133FQ164); and Open-fund of State Key Laboratory of Applied Optics (approval number: CS16017050001).

Acknowledgments: The authors would like to thank Pingli Han from XiDian university for her useful writing advises. The authors would also like to thank Xuan Li, Kui Yang and Xin Li for the help about the experiment.

Conflicts of Interest: The authors declare no conflict of interest.

References

1. Xie, C.; Nishizawa, T.; Sugimoto, N.; Matsui, I.; Wang, Z. Characteristics of aerosol optical properties in pollution and Asian dust episodes over Beijing, China. *Appl. Opt.* **2008**, *47*, 4945–4951. [CrossRef]
2. Edner, H.; Ragnarson, P.; Spännare, S.; Svanberg, S. Differential optical absorption spectroscopy (DOAS) system for urban atmospheric pollution monitoring. *Appl. Opt.* **1993**, *32*, 327–333. [CrossRef] [PubMed]
3. Xu, F.; Lv, Z.; Lou, X.; Zhang, Y.; Zhang, Z. Nitrogen dioxide monitoring using a blue LED. *Appl. Opt.* **2008**, *47*, 5337–5340. [CrossRef] [PubMed]
4. Guo, F.; Cai, Z.X.; Xie, B.; Tang, J. Review and prospect of image dehazing techniques. *J. Comput. Appl.* **2010**, *30*, 2417–2421. [CrossRef]
5. Huang, B.; Liu, T.; Han, J.; Hu, H. Polarimetric target detection under uneven illumination. *Opt. Express* **2015**, *23*, 23603–23612. [CrossRef]
6. Chen, C.W.; Fang, S.; Huo, X.; Xia, X.S. Image dehazing using polarization effects of objects and airlight. *Opt. Express* **2014**, *22*, 19523–19537. [CrossRef]
7. Liu, F.; Cao, L.; Shao, X.; Han, P.; Bin, X. Polarimetric dehazing utilizing spatial frequency segregation of images. *Appl. Opt.* **2015**, *54*, 8116–8122. [CrossRef]
8. Han, P.; Liu, F.; Yang, K.; Ma, J.; Li, J.; Shao, X. Active underwater descattering and image recovery. *Appl. Opt.* **2017**, *56*, 6631–6638. [CrossRef] [PubMed]
9. Liu, F.; Shao, X.; Gao, Y.; Xiangli, B.; Han, P.; Li, G. Polarization characteristics of objects in long-wave infrared range. *JOSA A* **2016**, *33*, 237–243. [CrossRef]
10. Sadjadi, F.A. Invariants of polarization transformations. *Appl. Opt.* **2007**, *46*, 2914–2921. [CrossRef]
11. Zhang, M.; Wu, X.; Cui, N.; Engheta, N.; van der Spiegel, J. Bioinspired focal-plane polarization image sensor design: From application to implementation. *Proc. IEEE* **2014**, *102*, 1435–1449. [CrossRef]
12. Zhao, X.; Boussaid, F.; Bermak, A.; Chigrinov, V.G. High-resolution thin "guest-host" micropolarizer arrays for visible imaging polarimetry. *Opt. Express* **2011**, *19*, 5565–5573. [CrossRef] [PubMed]
13. Sarkar, M.; Bello, D.S.S.; van Hoof, C.; Theuwissen, A.J. Biologically inspired CMOS image sensor for fast motion and polarization detection. *IEEE Sens. J.* **2013**, *13*, 1065–1073. [CrossRef]
14. Garcia, M.; Edmiston, C.; Marinov, R.; Vail, A.; Gruev, V. Bio-inspired color-polarization imager for real-time in situ imaging. *Optica* **2017**, *4*, 1263–1271. [CrossRef]
15. Maruyama, Y.; Terada, T.; Yamazaki, T.; Uesaka, Y.; Nakamura, M.; Matoba, Y.; Komori, K.; Ohba, Y.; Arakawa, S.; Hirasawa, Y. 3.2-MP Back-Illuminated Polarization Image Sensor with Four-Directional Air-Gap Wire Grid and 2.5-µm Pixels. *IEEE Trans. Electron Devices* **2018**, *65*, 2544–2551. [CrossRef]
16. Garcia, M.; Davis, T.; Blair, S.; Cui, N.; Gruev, V. Bioinspired polarization imager with high dynamic range. *Optica* **2018**, *5*, 1240–1246. [CrossRef]
17. Lowe, D.G. Distinctive image features from scale-invariant key-points. *Int. J. Comput. Vis.* **2004**, *60*, 91–110. [CrossRef]
18. Muja, M.; Lowe, D.G. Scalable nearest neighbor algorithms for high dimensional data. *IEEE Trans. Pattern Anal. Mach. Intell.* **2014**, *36*, 2227–2240. [CrossRef]
19. Liang, J.; Ren, L.; Qu, E.; Hu, B.; Wang, Y. Method for enhancing visibility of hazy images based on polarimetric imaging. *Photonics Res.* **2014**, *2*, 38–44. [CrossRef]
20. Shwartz, S.; Namer, E.; Schechner, Y.Y. Blind haze separation. In Proceedings of the 2006 IEEE Computer Society Conference on Computer Vision and Pattern Recognition, New York, NY, USA, 17–22 June 2006; Volume 2, pp. 1984–1991. [CrossRef]

21. Nishino, K.; Kratz, L.; Lombardi, S. Bayesian defogging. *Int. J. Comput. Vis.* **2012**, *98*, 263–278. [CrossRef]
22. Ancuti, C.O.; Ancuti, C. Single image dehazing by multi-scale fusion. *IEEE Trans. Image Process.* **2013**, *22*, 3271–3482. [CrossRef]
23. Pust, N.J.; Shaw, J.A. Wavelength dependence of the degree of polarization in cloud-free skies: Simulations of real environments. *Opt. Express* **2012**, *20*, 15559–15568. [CrossRef] [PubMed]
24. Schechner, Y.Y.; Averbuch, Y. Regularized image recovery in scattering media. *IEEE Trans. Pattern Anal. Mach. Intell.* **2007**, *29*, 1655–1660. [CrossRef] [PubMed]
25. Schechner, Y.Y.; Narasimhan, S.G.; Nayar, S.K. Polarization-based vision through haze. *Appl. Opt.* **2003**, *42*, 511–525. [CrossRef] [PubMed]

© 2019 by the authors. Licensee MDPI, Basel, Switzerland. This article is an open access article distributed under the terms and conditions of the Creative Commons Attribution (CC BY) license (http://creativecommons.org/licenses/by/4.0/).

MDPI
St. Alban-Anlage 66
4052 Basel
Switzerland
Tel. +41 61 683 77 34
Fax +41 61 302 89 18
www.mdpi.com

Applied Sciences Editorial Office
E-mail: applsci@mdpi.com
www.mdpi.com/journal/applsci

www.ingramcontent.com/pod-product-compliance
Lightning Source LLC
LaVergne TN
LVHW071940080526
838202LV00064B/6641